Time-of-Flight and Structured Light Depth Cameras

Pietro Zanuttigh • Giulio Marin • Carlo Dal Mutto
Fabio Dominio • Ludovico Minto
Guido Maria Cortelazzo

Time-of-Flight and Structured Light Depth Cameras

Technology and Applications

Pietro Zanuttigh
Department of Information Engineering
University of Padova
Padova, Italy

Giulio Marin
Department of Information Engineering
University of Padova
Padova, Italy

Carlo Dal Mutto
Aquifi Inc.
Palo Alto, CA, USA

Fabio Dominio
Department of Information Engineering
University of Padova
Padova, Italy

Ludovico Minto
Department of Information Engineering
University of Padova
Padova, Italy

Guido Maria Cortelazzo
3D Everywhere s.r.l.
Padova, Italy

ISBN 978-3-319-30971-2 ISBN 978-3-319-30973-6 (eBook)
DOI 10.1007/978-3-319-30973-6

Library of Congress Control Number: 2016935940

Printed on acid-free paper

This Springer imprint is published by Springer Nature
The registered company is Springer International Publishing AG Switzerland

"Cras ingens iterabimus aequor" (Horace, Odes, VII)
In memory of Alberto Apostolico (1948–2015)
unique scholar and person

Preface

This book originates from three-dimensional data processing research in the Multimedia Technology and Telecommunications Laboratory (LTTM) at the Department of Information Engineering of the University of Padova. The LTTM laboratory has a long history of research activity on consumer depth cameras, starting with Time-of-Flight (ToF) depth cameras in 2008 and continuing since, with a particular focus on recent structured light and ToF depth cameras like the two versions of Microsoft Kinect™. In the past years, the students and researchers at the LTTM laboratory have extensively explored many topics on 3D data acquisition, processing, and visualization, all fields of large interest for the computer vision and the computer graphics communities, as well as for the telecommunications community active on multimedia.

In contrast to a previous book by some of the authors, published as Springer Briefs in Electrical and Computer Engineering targeted to specialists, this book has been written for a wider audience, including students and practitioners interested in current consumer depth cameras and the data they provide. This book focuses on the system rather than the device and circuit aspects of the acquisition equipment. Processing methods required by the 3D nature of the data are presented within general frameworks purposely as independent as possible from the technological characteristics of the measurement instruments used to capture the data. The results are typically presented by practical exemplifications with real data to give the reader a clear and concrete idea about the actual processing possibilities.

This book is organized into three parts, the first devoted to the working principles of ToF and structured light depth cameras, the second to the extraction of accurate 3D information from depth camera data through proper calibration and data fusion techniques, and the third to the use of 3D data in some challenging computer vision applications.

This book comes from the contribution of a great number of people besides the authors. First, almost every student who worked at the LTTM laboratory in the past years gave some contribution to the know-how at the basis of this book and must be acknowledged. Among them, in particular, Alvise Memo must be thanked for his help with the acquisitions from a number of different depth cameras and for his review of this book. Many other students and researchers have contributed, and we would like to thank also Mauro Donadeo, Marco Fraccaro, Giampaolo Pagnutti, Luca Palmieri, Mauro Piazza, and Elena Zennaro. We consider a major contribution to this book the proofreading by Ilene Rafii which improved not only the quality of the English language but also the readability of this book in many parts. The authors would like to acknowledge 3DEverywhere which in 2008 purchased the first ToF camera with which the research about depth sensors at the LTTM laboratory began. Among the 3DEverywhere people, a special thank goes to Enrico Cappelletto, Davide Cerato, and Andrea Bernardi. We would also like to thank Gerard Dahlman and Tierry Oggier for the great collaboration we received from Mesa Imaging, Arrigo Benedetti (now with Microsoft, formerly with Canesta), and Abbas Rafii (with Aquifi) who helped the activity of the LTTM laboratory in various ways.

This book also benefited from the discussions and the supportive attitude of many colleagues, among which we would like to recall David Stoppa and Fabio Remondino (with FBK), Roberto Manduchi (with U.C.S.C.), Stefano Mattoccia (with the University of Bologna), Marco Andreetto (with Google), and Tim Droz and Mitch Reifel (with SoftKinetic).

Padova, Italy
January 2016

Pietro Zanuttigh
Giulio Marin
Carlo Dal Mutto
Fabio Dominio
Ludovico Minto
Guido Maria Cortelazzo

Contents

Chapter 1
Introduction

The acquisition of the geometric description of dynamic scenes has traditionally been a challenging task which required state of the art technology and instrumentation only accessible by research labs or major companies until professional-grade and consumer-grade depth cameras arrived in the market. Both professional-grade and consumer-grade depth cameras mainly belong to two technological families, one based on the *active triangulation* working principle and the other based on the *Time-of-Flight* working principle. The cameras belonging to the active triangulation family are usually called *structured light* depth cameras, while the cameras belonging to the second family are usually called *matricial Time-of-Flight* depth cameras, or simply ToF depth cameras, as in the remainder of this book.

Structured light depth cameras are the most diffused depth cameras in the market. Among them, the most notable example is the Primesense camera used in the first generation of Microsoft Kinect™. ToF depth cameras have historically been considered professional-grade (e.g., Mesa Imaging SwissRanger), however, recently they have also appeared as consumer-grade products, such as the first and second generation of Microsoft Kinect™, from now on called Kinect™ v1 and v2.

In several technical communities, especially those of computer vision, artificial intelligence, and robotics, a large interest has risen for these devices, along with the following questions: "What is a ToF camera?", "How does the Kinect™ work?", "Are there ways to improve the low resolution and high noise characteristics of ToF cameras data?", "How far can I go with the depth data provided by a 100–150 dollar consumer-grade depth camera with respect to those provided by a few thousand dollars professional-grade ToF camera?". This book tries to address these and other similar questions from a data user's point of view, as opposed to a technology developer's perspective.

This first part of this book describes the technology behind structured light and ToF cameras. The second part focuses on how to best exploit the data produced

P. Zanuttigh et al., *Time-of-Flight and Structured Light Depth Cameras*,
DOI 10.1007/978-3-319-30973-6_1

by structured light and ToF cameras, i.e., on the processing methods best suited to depth information. The third part reviews a number of applications where depth data provide significant contributions.

This book leverages on the depth nature of the data to present approaches that are as device-independent as possible. Therefore, we refer as often as possible to *depth cameras* and make the distinction between structured light and ToF cameras only when necessary. We focus on the depth nature of the data, rather than on the devices themselves, to establish a common framework suitable for current data from both structured light and ToF cameras, as well as data from new devices of these families that will reach the market in the next few years. Although structured light and ToF cameras are functionally equivalent depth cameras, i.e, providers of depth data, there are fundamental technological differences between them which cannot be ignored. These differences strongly impact noise, artifacts and production costs.

The synopsis of distance measurement methods in Fig. 1.1, derived from [17], offers a good framework to introduce these differences. For the purposes of this book, the reflective optical methods of Fig. 1.1 are typically classified into *passive* and *active*. Passive range sensing refers to 3D distance measurement by way of radiation (typically, but not necessarily, in the visible spectrum) already present in the scene. Stereo-vision systems are a classical example of this family of methods. Active sensing refers instead to 3D distance measurement obtained by projecting some form of radiation in the scene, as done for instance by structured light and ToF depth cameras.

The operation of structured light and ToF depth cameras involves a number of different concepts about imaging systems, ToF sensors and computer vision. These

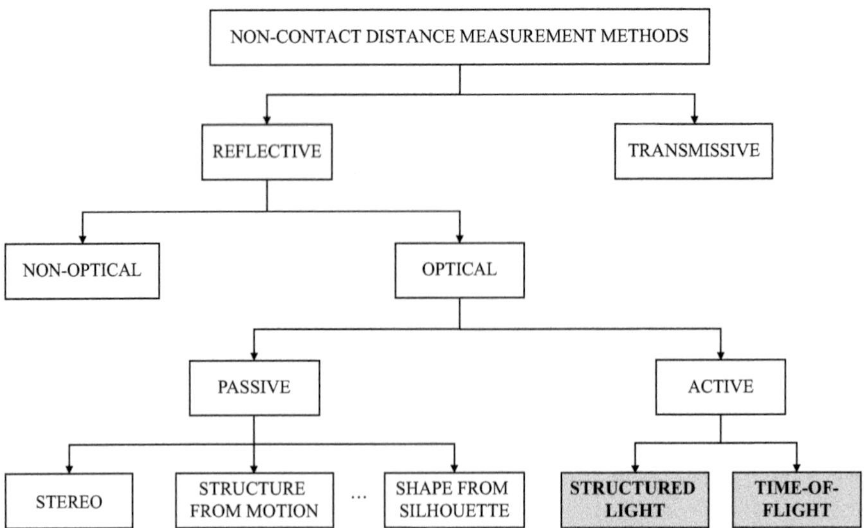

Fig. 1.1 Taxonomy of distance measurement methods (derived from [17])

concepts are recalled in the next two sections of this chapter to equip the reader with the notions needed for the remainder of the book; the next two sections can be skipped by readers already acquainted with structured light and ToF systems operation.

The depth or distance measurements taken by the systems of Fig. 1.1 can typically be represented by depth maps, i.e., data with each spatial coordinate (u, v) associated with the corresponding depth value z, and the depth maps can be combined into full all-around 3D models [14] as will be seen in Chap. 7. Data made by a depth map together with the corresponding color image are also referred to as RGB-D data.

1.1 Basics of Imaging Systems

1.1.1 Pin-Hole Camera Model

Let us consider a 3D reference system with axes x, y and z, called *Camera Coordinates System (CCS)*, with origin at O, called *center of projection*, and a plane parallel to the (x, y)-plane intersecting the z-axis at negative z-coordinate f, called *sensor* or *image plane* S as shown in Fig. 1.2. The axes' orientations follow the so called right-hand convention. Consider also a 2D reference system

$$\begin{aligned} u &= x + c_x \\ v &= y + c_y \end{aligned} \tag{1.1}$$

associated with the sensor, called *S-2D reference system*, oriented as shown in Fig. 1.2a. The intersection c of the z-axis with the sensor plane has coordinates $\mathbf{c} = [c_x, c_y]^T$. The set of sensor points p, called *pixels*, of coordinates $\mathbf{p} = [u, v]^T$ obtained from the intersection of the rays connecting the center of projection O with all the 3D scene points P with coordinates $\mathbf{P} = [x, y, z]^T$, is the scene footprint on the sensor S.

The relationship between P and p, called *central* or *perspective projection*, can be shown by triangle similarity (see Fig. 1.2b, c) to be

$$\begin{cases} u - c_x = f\dfrac{x}{z} \\ v - c_y = f\dfrac{y}{z} \end{cases} \tag{1.2}$$

where the distance $|f|$ between the sensor plane and the center of projection O is typically called *focal length*. In the adopted notation, f is the negative coordinate of the location of the sensor plane with respect to the z-axis. The reader should be aware that other books adopt a different notation, where f denotes the focal length, hence it is a positive number and the z coordinate of the sensor plane is denoted as $-f$.

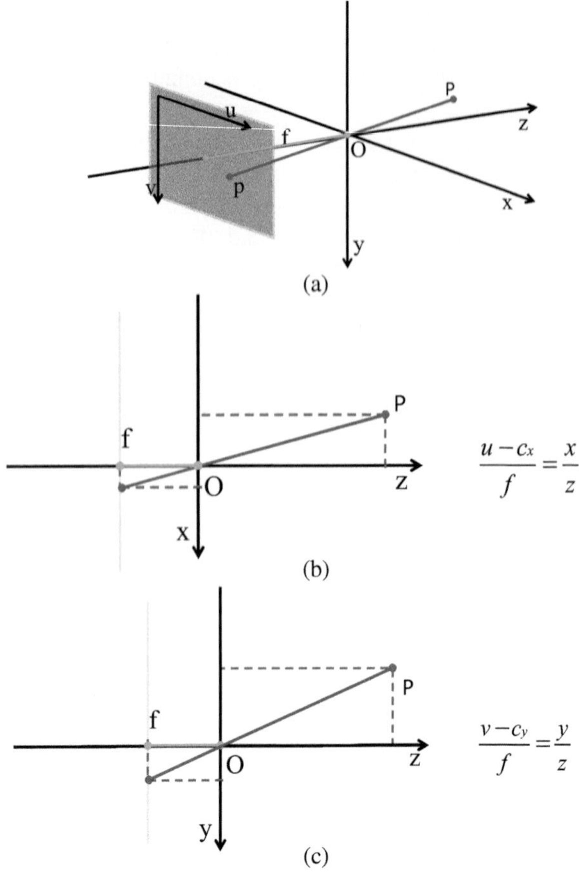

Fig. 1.2 Perspective projection: (**a**) scene point P projected to sensor pixel p; (**b**) horizontal section of (**a**); (**c**) vertical section of (**a**)

The perspective projection (1.2) is a good description of the geometric relationship between the coordinates of the scene points and the corresponding location in an image obtained by a pin-hole imaging device with the pin-hole positioned at center of projection O. Such a system allows a single light ray to go through the pin-hole at O. For a number of reasons, in imaging systems it is more practical to use optics, i.e., suitable sets of lenses, instead of pin-holes. Quite remarkably, the ideal model of an optical system, called *thin-lens model*, maintains the relationship (1.2) between the coordinates of P and of p if the lens' optical center (or *nodal point*) is in O and the lens' optical axis, i.e., the line orthogonally intersecting the lens at its nodal point, is orthogonal to the sensor. If a thin lens replaces a pin-hole in Fig. 1.2c, the optical axis coincides with the z-axis of the CCS.

1.1.2 Camera Geometry and Projection Matrix

Projective geometry associates to each 2D point p with Cartesian coordinates $\mathbf{p} = [u, v]^T$ of a plane a 3D representation called *2D homogeneous coordinates* $\tilde{\mathbf{p}} = [hu, hv, h]^T$, where h is any real constant. The usage of $h = 1$ is rather common and $[u, v, 1]^T$ is often called the *extended vector* of p [57].

The coordinates $\mathbf{p} = [u, v]^T$ can be obtained by dividing $\tilde{\mathbf{p}} = [hu, hv, h]^T$ by its third coordinate h. Vector $\tilde{\mathbf{p}}$ can be interpreted as the 3D ray connecting the sensor point p with the center of projection O.

In a similar way each 3D point P with Cartesian coordinates $\mathbf{P} = [x, y, z]^T$ can be represented in 3D homogeneous coordinates by a $4D$ vector $\tilde{\mathbf{p}} = [hx, hy, hz, h]^T$ where h is any real constant. Vector $[x, y, z, 1]^T$ is often called the *extended vector* of P.

The coordinates $\mathbf{P} = [x, y, z]^T$ can be obtained by dividing $\tilde{\mathbf{P}} = [hx, hy, hz, h]^T$ by its fourth coordinate h. An introduction to projective geometry suitable to computer vision applications can be found in [32].

The homogeneous coordinates representation of p allows one to rewrite the non-linear relationship (1.2) in a convenient matricial form:

$$z \begin{bmatrix} u \\ v \\ 1 \end{bmatrix} = \begin{bmatrix} f & 0 & c_x \\ 0 & f & c_y \\ 0 & 0 & 1 \end{bmatrix} \begin{bmatrix} x \\ y \\ z \end{bmatrix}. \tag{1.3}$$

Note that the left side of (1.3) represents p in 2D homogeneous coordinates but the right side of (1.3) represents P in 3D Cartesian coordinates. It is straightforward to add a column with all 0 entries at the right of the matrix in order to represent P in homogeneous coordinates as well. This latter representation is more common than (1.3), which nevertheless is often adopted for its simplicity [57].

Digital sensor devices are typically planar matrices of rectangular sensor cells hosting photoelectric conversion systems based on CMOS or CCD technology in the case of digital cameras or video cameras, or single ToF receivers in the case of ToF cameras, as explained in Sect. 1.4. Customarily, they are modeled as a rectangular lattice Λ_S with horizontal and vertical step-size k_u and k_v respectively, as shown in Fig. 1.3a.

Given the finite sensor size, only a rectangular window of Λ_S made by N_C columns and N_R rows is of interest for imaging purposes.

In order to deal with normalized lattices with origin at $(0, 0)$ and unitary pixel coordinates $\mathbf{u_S} \in [0, \ldots, N_C - 1]$ and $\mathbf{v_S} \in [0, \ldots, N_R - 1]$ in both the u and v direction, relationship (1.3) is replaced by

$$z \begin{bmatrix} u \\ v \\ 1 \end{bmatrix} = \mathbf{K} \begin{bmatrix} x \\ y \\ z \end{bmatrix} \tag{1.4}$$

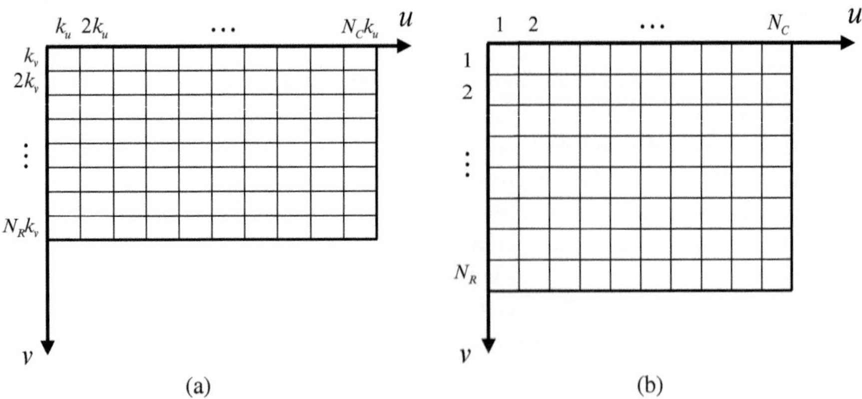

Fig. 1.3 2D sensor coordinates: (**a**) rectangular window of a non-normalized orthogonal lattice; (**b**) rectangular window of a normalized orthogonal lattice

where $\mathbf{K}$ is the intrinsic parameters matrix defined as

$$\mathbf{K} = \begin{bmatrix} f_x & \alpha & c_x \\ 0 & f_y & c_y \\ 0 & 0 & 1 \end{bmatrix} \approx \begin{bmatrix} f_x & 0 & c_x \\ 0 & f_y & c_y \\ 0 & 0 & 1 \end{bmatrix} \tag{1.5}$$

with $f_x = fk_u$ the x-axis focal length of the optics, $f_y = fk_v$ the y-axis focal length of the optics, c_x and c_y the (u, v) coordinates of the intersection of the optical axis with the sensor plane. All these quantities are expressed in [pixel], i.e., since f is in [mm], k_u and k_v are assumed to be [pixel]/[mm]. Notice also that an additional parameter α (*axis skew*) is sometimes used to account for the fact that the two axes in the sensor lattice are not perfectly perpendicular, however since it is typically negligible we will not consider it in the rest of the book and approximate $\mathbf{K}$ by the r.h.s. of (1.5). The symbol $\approx$ within (1.5) denotes approximation.

In many practical situations it is convenient to represent the 3D scene points not with respect to the CCS, but with respect to a different easily accessible reference system conventionally called *World Coordinate System (WCS)*, in which a scene point denoted as P has coordinates $\mathbf{P}_W = [x_W, y_W, z_W]^T$. The relationship between the representation of a scene point with respect to the CCS, denoted as $\mathbf{P}$, and its representation with respect to the WCS, denoted as $\mathbf{P}_W$, is

$$\mathbf{P} = \mathbf{R}\mathbf{P}_W + \mathbf{t} = \begin{bmatrix} \mathbf{r}_1^T \\ \mathbf{r}_2^T \\ \mathbf{r}_3^T \end{bmatrix} \mathbf{P}_W + \begin{bmatrix} t_1 \\ t_2 \\ t_3 \end{bmatrix} \tag{1.6}$$

where $\mathbf{R}$ and $\mathbf{t}$ are a suitable rotation matrix and translation vector, respectively. For future usage let us introduce an explicit notation for the rows $\mathbf{r}_i^T$, $i = 1, 2, 3$ of $\mathbf{R}$ and the components t_i, $i = 1, 2, 3$ of $\mathbf{t}$. By representing $\mathbf{P}_W$ at the right side in homogeneous coordinates $\tilde{\mathbf{P}}_W = [hx_W, hy_W, hz_W, h]^T$ and choosing $h = 1$, the relationship (1.6) can be rewritten as

$$\mathbf{P} = [\mathbf{R} \mid \mathbf{t}]\tilde{\mathbf{P}}_W. \tag{1.7}$$

In this case, the relationship between a scene point represented in homogeneous coordinates with respect to the WCS and its corresponding pixel in homogeneous coordinates, from (1.4), becomes

$$\tilde{\mathbf{p}} \cong \begin{bmatrix} u \\ v \\ 1 \end{bmatrix} \cong \frac{1}{z}\mathbf{K}\mathbf{P} \cong \frac{1}{z}\mathbf{K}[\mathbf{R} \mid \mathbf{t}]\tilde{\mathbf{P}}_W \cong \frac{1}{z}\mathbf{M}\tilde{\mathbf{P}}_W \cong \frac{1}{z}\mathbf{M}\begin{bmatrix} x_W \\ y_W \\ z_W \\ 1 \end{bmatrix} \tag{1.8}$$

where the 3×4 matrix

$$\mathbf{M} = \mathbf{K}[\mathbf{R} \mid \mathbf{t}] = \begin{bmatrix} \mathbf{m}_1^T \\ \mathbf{m}_2^T \\ \mathbf{m}_3^T \end{bmatrix} \tag{1.9}$$

is called *projection matrix*. Projection matrix $\mathbf{M}$ depends on the intrinsic parameters matrix $\mathbf{K}$ and on the extrinsic parameters $\mathbf{R}$ and $\mathbf{t}$ of the imaging system. A projection matrix $\mathbf{M}$ is said to be in normalized form if its bottom row is exactly $\mathbf{m}_3^T = [\mathbf{r}_3^T \mid t_3]$. It is straightforward to see that if $\mathbf{M}$ is in normalized form, (1.8) holds with the equality sign: in this case $z = \mathbf{r}_3^T\tilde{\mathbf{P}}_W + t_3$ assumes the value of the depth of P_W with respect to the camera reference system. By denoting with $\mathbf{m}_i^T$, $i = 1, 2, 3$ the rows of $\mathbf{M}$, the image coordinates (u, v) of point P from (1.8) can be written as

$$\begin{cases} u = \dfrac{\mathbf{m}_1^T\tilde{\mathbf{P}}_W}{\mathbf{m}_3^T\tilde{\mathbf{P}}_W} = \dfrac{\tilde{\mathbf{P}}_W^T\mathbf{m}_1}{\tilde{\mathbf{P}}_W^T\mathbf{m}_3} \\ v = \dfrac{\mathbf{m}_2^T\tilde{\mathbf{P}}_W}{\mathbf{m}_3^T\tilde{\mathbf{P}}_W} = \dfrac{\tilde{\mathbf{P}}_W^T\mathbf{m}_2}{\tilde{\mathbf{P}}_W^T\mathbf{m}_3} \end{cases}. \tag{1.10}$$

The symbol $\cong$ within (1.8) denotes that in general, the equality holds up to a multiplicative constant since it involves homogeneous coordinates. In this sense $\mathbf{M}$ is also defined up to a multiplicative constant since it has 12 parameters but just 11 degrees of freedom: 5 from $\mathbf{K}$ (4 excluding the skew parameter), 3 from $\mathbf{R}$ and 3 from $\mathbf{t}$.

From a set of J known 2D-3D correspondence values $(p^j, P^j), j = 1, \ldots, J$ from (1.10) one may derive a set of $2J$ homogeneous linear equations

$$\begin{cases} \mathbf{m}_3^T\tilde{\mathbf{P}}_W^j u^j - \mathbf{m}_1^T\tilde{\mathbf{P}}_W^j = 0 \\ \mathbf{m}_3^T\tilde{\mathbf{P}}_W^j v^j - \mathbf{m}_2^T\tilde{\mathbf{P}}_W^j = 0 \end{cases} \quad j = 1, \ldots, J \tag{1.11}$$

from which $\mathbf{M}$ can be computed. In principle, $J = 6$ correspondences suffice since $\mathbf{M}$ has 12 entries; in practice, one should use $J \gg 6$ in order to effectively deal with noise and non-idealities. However, this method, typically called *Direct Linear Transform* (DLT), only minimizes a target with algebraic significance, and is not invariant with respect to Euclidean transformations. Therefore the result of the DLT is typically used as starting point for a nonlinear minimization either in L_2 or L_∞ directly addressing Eqs. (1.10), for example

$$\min_{\mathbf{K},\mathbf{R},\mathbf{t}} \sum_{j=1}^{J} |p^j - f(\mathbf{K},\mathbf{R},\mathbf{t},P^j)|^2 \tag{1.12}$$

where $f(\mathbf{K},\mathbf{R},\mathbf{t},P^j)$ is a function that given $\mathbf{K}$, $\mathbf{R}$ and $\mathbf{t}$, projects P in the image plane, as in (1.8). More details on the estimation of $\mathbf{K}$, $\mathbf{R}$ and $\mathbf{t}$ will be provided in Chap. 4.

1.1.3 Lens Distortions

As a consequence of distortions and aberrations of real optics, the coordinates $\hat{\mathbf{p}} = (\hat{u}, \hat{v})$ of the pixel actually associated with scene point P with coordinates $\mathbf{P} = [x, y, z]^T$ in the CCS system do not satisfy relationship (1.4). The correct pixel coordinates (u, v) of (1.4) can be obtained from the distorted coordinates $(\hat{u}, \hat{v})$ actually measured by the imaging system, by inverting suitable distortion models, such as

$$\mathbf{p}_T = \Psi^{-1}(\hat{\mathbf{p}}_T) \tag{1.13}$$

where $\Psi(\cdot)$ denotes the distortion transformation.

Anti-distortion model (1.14), also called the *Heikkila model* [33], has become popular since it adequately corrects the distortions of most imaging systems and effective methods exist for computing its parameters:

$$\begin{bmatrix} u \\ v \end{bmatrix} = \Psi^{-1}(\hat{\mathbf{p}}_T) = \begin{bmatrix} \hat{u}(1 + k_1 r^2 + K_2 r^4 + k_3 r^6) + 2d_1\hat{u}\hat{v} + d_2(r^2 + 2\hat{u}^2) \\ \hat{v}(1 + k_1 r^2 + k_2 r^4 + k_3 r^6) + d_1(r^2 + 2\hat{v}^2) + 2d_2\hat{u}\hat{v} \end{bmatrix} \tag{1.14}$$

where $r = \sqrt{(\hat{u} - c_x)^2 + (\hat{v} - c_y)^2}$, parameters k_i with $i = 1, 2, 3$ are constants accounting for radial distortion and d_i with $i = 1, 2$ account for tangential distortion. A number of other more complex models, e.g. [18], are also available.

Distortion parameters

$$\mathbf{d} = [k_1, k_2, k_3, d_1, d_2] \tag{1.15}$$

are intrinsic camera parameters to be considered together with $[f, k_u, k_v, c_x, c_y]$. Equation (1.12) can be modified to also account for distortion, in this case, the projection function f becomes $f(\mathbf{K}, \mathbf{R}, \mathbf{t}, \mathbf{d}, P^j)$.

The estimation of intrinsic and extrinsic parameters of an imaging system by suitable methods such as [16] and [6] is called *geometric calibration* and is discussed in Chap. 4.

1.2 Stereo Vision Systems

This section and the previous one summarize basic computer vision concepts necessary for understanding the rest of this book and can be skipped by readers familiar with computer vision. Readers interested in a more extensive presentation of these topics are referred to computer vision textbooks such as [15, 20, 22, 24, 26, 27, 32, 45, 48, 55, 57, 61].

1.2.1 Two-view Stereo Systems

A stereo vision, or *stereo*, system is made by two standard cameras partially framing the same scene, namely the left camera L, also called *reference camera*, and the right camera R, also called *target camera*. Each camera is assumed to be calibrated, with calibration matrices $\mathbf{K}_L$ and $\mathbf{K}_R$ for the L and R cameras respectively. As previously seen, each has its own 3D CCS and 2D reference systems, as shown in Fig. 1.4. Namely, the L camera has CCS with coordinates (x_L, y_L, z_L), also called *L-3D reference system*, and a 2D reference system with coordinates (u_L, v_L). The R camera has CCS with coordinates (x_R, y_R, z_R), also called *R-3D reference system*,

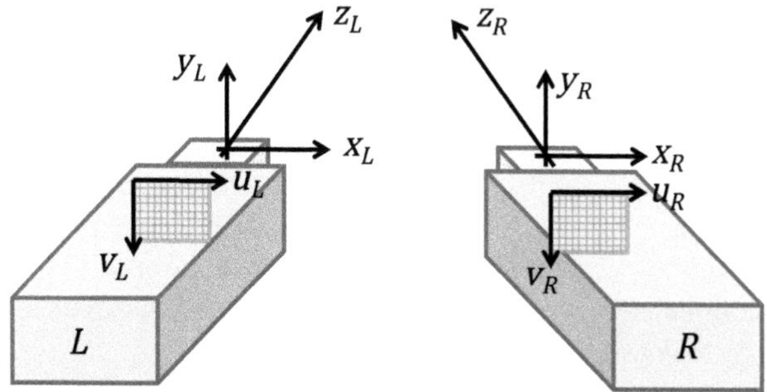

Fig. 1.4 Stereo vision system coordinates and reference systems

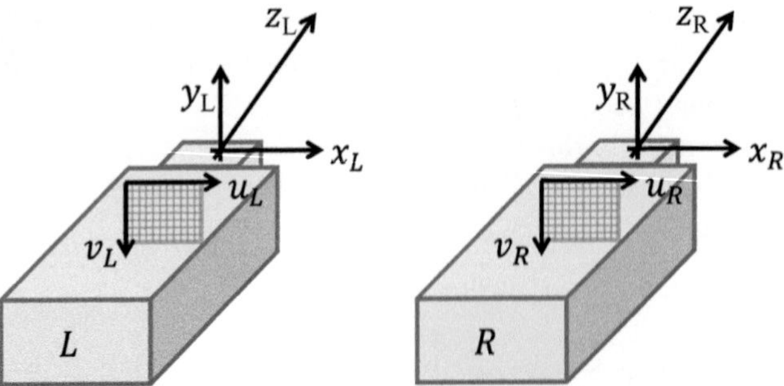

Fig. 1.5 Rectified stereo system

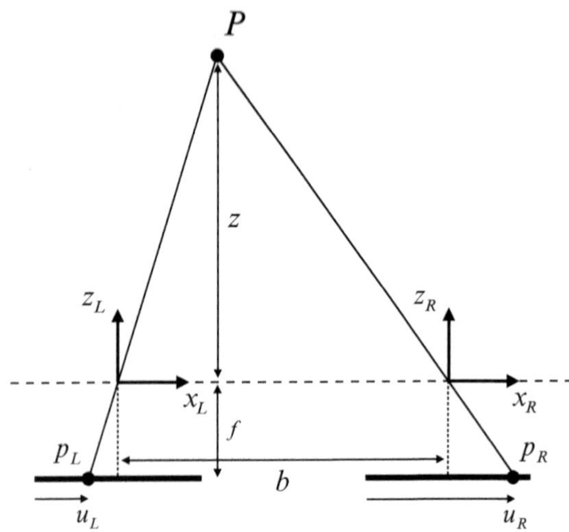

Fig. 1.6 Triangulation with a rectified stereo system

and a 2D reference system with coordinates (u_R, v_R). The two cameras may be different, but in this book they are assumed to be identical, with $\mathbf{K} = \mathbf{K}_L = \mathbf{K}_R$, unless explicitly stated. A common convention is to consider the L-3D reference system as the reference system of the stereo vision system and to denote it as *S-3D reference system.*

Let us momentarily consider the case of a calibrated and rectified stereo vision system, i.e., a stereo vision system made by two identical standard cameras with coplanar and aligned imaging sensors and parallel optical axes as shown in Fig. 1.5. In rectified stereo vision systems points p_L and p_R have the same vertical coordinates. By denoting

$$d = u_L - u_R \tag{1.16}$$

the difference between their horizontal coordinates, called *disparity*, a 3D point P with coordinates $\mathbf{P}_L = [x_L, y_L, z_L]^T$ with respect to the S-3D reference system, is projected to the pixels p_L and p_R of the L and R cameras with coordinates

$$\mathbf{p}_L = \begin{bmatrix} u_L \\ v_L \end{bmatrix} \qquad \mathbf{p}_R = \begin{bmatrix} u_R = u_L - d \\ v_R = v_L \end{bmatrix} \tag{1.17}$$

respectively. Furthermore, let $\mathbf{P}_R = [x_R, y_R, z_R]^T$ denote the coordinate of P with respect to the R-3D reference system and let $(\mathbf{R}, \mathbf{t})$ denote the rigid transformation mapping the R-3D reference system to the L-3D reference system, which is also the S-3D reference system, i.e.,

$$\mathbf{P}_R = \mathbf{R}\mathbf{P}_L + \mathbf{t}. \tag{1.18}$$

By introducing normalized image coordinates

$$\begin{aligned}
\tilde{\mathbf{q}}_L &= \begin{bmatrix} u'_L \\ v'_L \\ 1 \end{bmatrix} = \mathbf{K}^{-1}\tilde{\mathbf{p}}_L = \begin{bmatrix} \frac{1}{f} & 0 & -\frac{c_x}{f} \\ 0 & \frac{1}{f} & -\frac{c_y}{f} \\ 0 & 0 & 1 \end{bmatrix} \begin{bmatrix} u_L \\ v_L \\ 1 \end{bmatrix} = \begin{bmatrix} \frac{u_L - c_x}{f} \\ \frac{v_L - c_y}{f} \\ 1 \end{bmatrix} \\
\tilde{\mathbf{q}}_R &= \begin{bmatrix} u'_R \\ v'_R \\ 1 \end{bmatrix} = \mathbf{K}^{-1}\tilde{\mathbf{p}}_R = \begin{bmatrix} \frac{1}{f} & 0 & -\frac{c_x}{f} \\ 0 & \frac{1}{f} & -\frac{c_y}{f} \\ 0 & 0 & 1 \end{bmatrix} \begin{bmatrix} u_R \\ v_R \\ 1 \end{bmatrix} = \begin{bmatrix} \frac{u_R - c_x}{f} \\ \frac{v_R - c_y}{f} \\ 1 \end{bmatrix}
\end{aligned} \tag{1.19}$$

the Cartesian coordinates of P with respect to the L and R 3D reference system can be written as

$$\begin{aligned}
\mathbf{P}_L &= z_L\mathbf{K}^{-1}\tilde{\mathbf{p}}_L = z_L\tilde{\mathbf{q}}_L \\
\mathbf{P}_R &= z_R\mathbf{K}^{-1}\tilde{\mathbf{p}}_R = z_R\tilde{\mathbf{q}}_R
\end{aligned} \tag{1.20}$$

and (1.18) can be rewritten as

$$z_R\tilde{\mathbf{q}}_R - z_L\mathbf{R}\tilde{\mathbf{q}}_L = \mathbf{t} \tag{1.21}$$

or

$$\begin{cases} z_R u'_R - z_L \mathbf{r}_1^T \tilde{\mathbf{q}}_L = t_1 \\ z_R v'_R - z_L \mathbf{r}_2^T \tilde{\mathbf{q}}_L = t_2 \\ z_R - z_L \mathbf{r}_3^T \tilde{\mathbf{q}}_L = t_3 \, . \end{cases} \tag{1.22}$$

By substituting in the first equation of (1.22) $z_R = t_3 + z_L \mathbf{r}_3^T \tilde{\mathbf{q}}_L$, derived from the third equation, one obtains

$$z_L = \frac{t_1 - u'_R t_3}{u'_R \mathbf{r}_3^T \tilde{\mathbf{q}}_L - \mathbf{r}_1^T \tilde{\mathbf{q}}_L} \, . \tag{1.23}$$

Equation (1.23) shows that the depth, i.e., the z coordinate, of $3D$ point P with respect to the L-3D reference system denoted z_L can be obtained upon knowledge of the left image coordinate $\tilde{\mathbf{p}}_L$ and of the right image coordinate $\tilde{\mathbf{p}}_R$ of point P, assuming the stereo system calibration parameters are known. These parameters are the external calibration parameters $(\mathbf{R}, \mathbf{t})$, relating the position of the right camera to the left camera, and the internal calibration parameters $\mathbf{K}$, concerning both cameras of the rectified stereo system. The procedure computing the stereo system calibration parameters will be seen in Chap. 4. Such procedure delivers as output the left and right camera projection matrices, which within the assumed conventions respectively result in

$$\mathbf{M}_L = \mathbf{K}_L \left[\mathbf{I} \mid \mathbf{0} \right] \qquad \mathbf{M}_R = \mathbf{K}_R \left[\mathbf{R} \mid \mathbf{t} \right] . \tag{1.24}$$

The methods indicated in Sect. 1.2.1.3 can be used to solve the so-called correspondence problem, i.e., the automatic determination of image points $\tilde{p}_L$ and $\tilde{p}_R$, called conjugate points. *Triangulation* or *computational stereopsis* is the process by which one may compute the 3D coordinates $\mathbf{P}_L = [x_L, y_L, z_L]^T$ of a scene point P from (1.20), from the knowledge of conjugate points $\tilde{\mathbf{p}}_L$ and $\tilde{\mathbf{p}}_R$, obtained by solving the correspondence problem, i.e., as

$$\mathbf{P}_L = \begin{bmatrix} x_L \\ y_L \\ z_L \end{bmatrix} = \mathbf{K}_L^{-1} \begin{bmatrix} u_L \\ v_L \\ 1 \end{bmatrix} z \tag{1.25}$$

where $\mathbf{K}_L^{-1}$ is the inverse of the intrinsic parameters matrix (1.5) of camera L (or R) of the stereo system.

In the case of a rectified system, where $\mathbf{K} = \mathbf{K}_L = \mathbf{K}_R$, as shown in Fig. 1.6, the parameters entering $\mathbf{M}_R$ in (1.24) are $\mathbf{R} = \mathbf{I}$ and $\mathbf{t} = [-b, 0, 0]^T$ and it can be readily seen that from $\mathbf{r}_3^T \tilde{\mathbf{q}}_L = 1$ and $\mathbf{r}_1^T \tilde{\mathbf{q}}_L = u'_L$, expression (1.23) becomes

$$z_L = \frac{-b}{u'_R - u'_L} = -\frac{bf}{u_R - u_L} = \frac{bf}{d} \tag{1.26}$$

where d is the disparity defined in (1.16). Equation (1.26), which shows that disparity is inversely proportional to the depth value z of P, can be directly obtained from the similarities of the triangles inscribed within the triangle with vertices at P, p_L and p_R of Fig. 1.6. Indeed, from the established CCS conventions, one can write for the L camera

$$\frac{u_L - c_x}{x_L} = \frac{f}{z_L} \tag{1.27}$$

and for the R camera

$$\frac{u_R - c_x}{x_L - b} = \frac{f}{z_R} = \frac{f}{z_L} \tag{1.28}$$

since in the case of rectified stereo systems $x_R = x_L - b$ and $z_R = z_L$. By substituting (1.27) in (1.28) one obtains

$$z_L = \frac{u_L - c_x}{u_R - c_x} z_L - \frac{bf}{u_R - c_x} \tag{1.29}$$

which gives (1.26). The above derivation is what justifies the name of *triangulation* for the procedure adopted to infer the 3D coordinates of a scene point P from its conjugate image points p_L and p_R.

The procedure actually used for triangulation or stereopsis can be summarized in very general terms as follows. Since

$$\begin{cases} \tilde{\mathbf{p}}_L \cong \frac{1}{z}\mathbf{M}_L \tilde{\mathbf{P}}_L \\ \tilde{\mathbf{p}}_R \cong \frac{1}{z}\mathbf{M}_R \tilde{\mathbf{P}}_L \end{cases} \tag{1.30}$$

where $\mathbf{M}_L = [\mathbf{m}_{1L}^T,\ \mathbf{m}_{2L}^T,\ \mathbf{m}_{3L}^T]$ and $\mathbf{M}_R = [\mathbf{m}_{1R}^T,\ \mathbf{m}_{2R}^T,\ \mathbf{m}_{3R}^T]$ are the perspective projection matrices of the L and R camera of (1.24) and $\tilde{\mathbf{P}}_L$ represents the coordinates of P with respect to the S-3D reference system, assumed to be the L-3D reference system, by (1.10) expression (1.30) can be rewritten as

$$\begin{bmatrix} \mathbf{m}_{3L}^T u_L - \mathbf{m}_{1L}^T \\ \mathbf{m}_{3L}^T v_L - \mathbf{m}_{2L}^T \\ \mathbf{m}_{3R}^T u_R - \mathbf{m}_{1R}^T \\ \mathbf{m}_{3R}^T v_R - \mathbf{m}_{2R}^T \end{bmatrix} \tilde{\mathbf{P}}_L = \mathbf{0}_{4\times 1} \tag{1.31}$$

which, since $\mathbf{p}_L$, $\mathbf{p}_R$, $\mathbf{M}_L$, and $\mathbf{M}_R$ are assumed known, corresponds to a linear homogeneous system of four equations in the unknown coordinates of P. Clearly

(1.31) gives a non-trivial solution only if the system matrix has rank 3. This condition may not always be verified because of noise. The so-called *linear-eigen* method [31] based on singular value decomposition overcomes such difficulties. As already seen for the estimate of **M** by the DLT method, since the estimate of P returned by (1.31) complies only with an algebraic criterion, it is typical to use it as a starting point for the numerical optimization of (1.31), in terms of

$$\min_{\tilde{\mathbf{P}}_L} \left\{ \left(u_L - \frac{\mathbf{m}_{1L}^T \tilde{\mathbf{P}}_L}{\mathbf{m}_{3L}^T \tilde{\mathbf{P}}_L} \right)^2 + \left(v_L - \frac{\mathbf{m}_{2L}^T \tilde{\mathbf{P}}_L}{\mathbf{m}_{3L}^T \tilde{\mathbf{P}}_L} \right)^2 + \right.$$
$$\left. + \left(u_R - \frac{\mathbf{m}_{1R}^T \tilde{\mathbf{P}}_L}{\mathbf{m}_{3R}^T \tilde{\mathbf{P}}_L} \right)^2 + \left(v_R - \frac{\mathbf{m}_{2R}^T \tilde{\mathbf{P}}_L}{\mathbf{m}_{3R}^T \tilde{\mathbf{P}}_L} \right)^2 \right\}. \quad (1.32)$$

Equation (1.32) can be interpreted as a variation of (1.12), where the reprojection error is jointly minimized in both cameras. Note that the goal of (1.32) is to find the coordinates of P, rather than **M**, as in (1.12).

1.2.1.1 Epipolar Geometry

Figure 1.7 schematically represents the stereo system of Fig. 1.5 and evidences only some elements of special geometric significance, such as the optical centers C_L and C_R and the image planes of the two cameras. It shows that given p_L, its conjugate p_R must lie on the plane defined by p_L, C_L, and C_R, called *epipolar plane*, and similarly for p_R. This geometric constraint implies that given p_L, its conjugate point p_R can only be sought along the intersection of the epipolar plane through p_L, C_L,

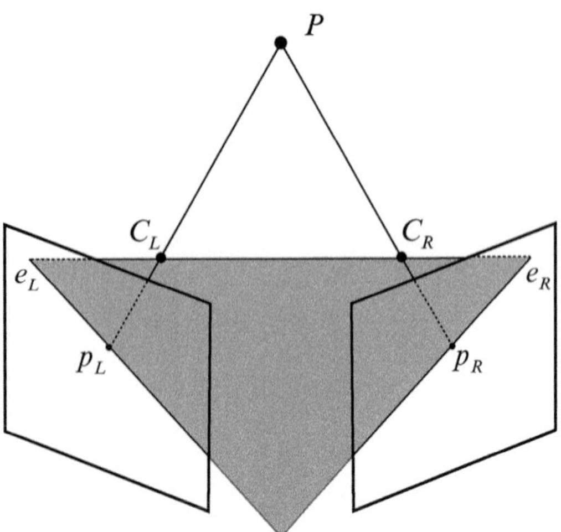

Fig. 1.7 Epipolar geometry

and C_R with the right image plane, which is a line, called the *right epipolar line* of p_L. Similar reasoning applies to p_R. Epipolar geometry, which reduces the search for the conjugate point from planar to linear, is formalized by the Longuet-Higgins equation

$$\tilde{\mathbf{p}}_R^T \mathbf{F} \tilde{\mathbf{p}}_L^T = 0 \tag{1.33}$$

where 3×3 matrix $\mathbf{F}$ is called the *fundamental matrix* [44]. In practical settings due to noise and inaccuracies the equation does not perfectly hold and it can be replaced by the search for the conjugate points that minimize (1.33). The homogeneous equation of the epipolar line of $\mathbf{p}_L$ from (1.33) is $\tilde{\mathbf{p}}_R^T \mathbf{F}$ and similarly $\tilde{\mathbf{p}}_L^T \mathbf{F}$ is the equation of the epipolar line of $\mathbf{p}_R$.

Since the epipolar plane is defined by P, C_L, and C_R, it varies with P. Therefore, there are infinite epipolar planes forming infinite epipolar lines on the left and right image. It is worth noting that since every epipolar plane, i.e., the epipolar plane defined by any P, includes C_L and C_R, all epipolar planes include the baseline connecting C_L and C_R. Furthermore, the baseline intersects the left and right image planes at two points called *left epipole* e_L and *right epipole* e_R. Indeed, e_L and e_R belong to the bundle of all the left and right epipolar lines, since every epipolar plane defined by any P must include rays $p_L\ P$ and $p_R\ P$.

1.2.1.2 Epipolar Rectification

A stereo system is called rectified if it has parallel image planes, as in Fig. 1.6. This configuration is of special interest, since the epipoles become points at infinity; therefore, the epipolar lines, bounded to intersect the epipoles, become parallel lines as shown in Fig. 1.8. Such a geometry further simplifies the search for the conjugate of $\mathbf{p}_L = [u_L, v_L]^T$ in the right image, which epipolar geometry already turns from a 2D search to a 1D search, to a search on the horizontal right image line of equation $y = v_L$.

Figure 1.8 emphasizes that the projection matrices $\mathbf{M}_L$ and $\mathbf{M}_R$ and the left and right images I_L and I_R of the stereo system with vergent cameras differ from the those of the rectified system, respectively denoted as $\mathbf{M}'_L$, $\mathbf{M}'_R$ and I'_L, I'_R. In a rectified stereo system (Fig. 1.6), the left and the right projection matrices are

$$\mathbf{M}_L = \mathbf{K}\left[\mathbf{I} \mid \mathbf{0}\right] \qquad \mathbf{M}_R = \mathbf{K}\left[\mathbf{I} \mid [b, 0, 0]^T\right]. \tag{1.34}$$

There exist methods for computationally rectifying vergent stereo systems, such as the algorithm of [28] which first computes $\mathbf{M}'_L$ and $\mathbf{M}'_R$ from $\mathbf{M}_L$ and $\mathbf{M}_R$ and then rectifies the images, i.e., it computes I'_L and I'_R upon $\mathbf{M}'_L$ and $\mathbf{M}'_R$. Figure 1.9 shows an example of image rectification. In current practice, it is typical to apply computational stereopsis to rectified images, which is equivalent to computationally turning actual stereo systems into rectified stereo systems.

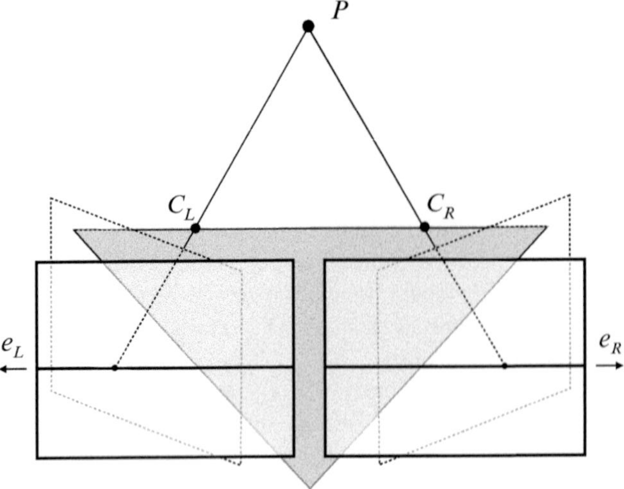

Fig. 1.8 Epipolar rectification

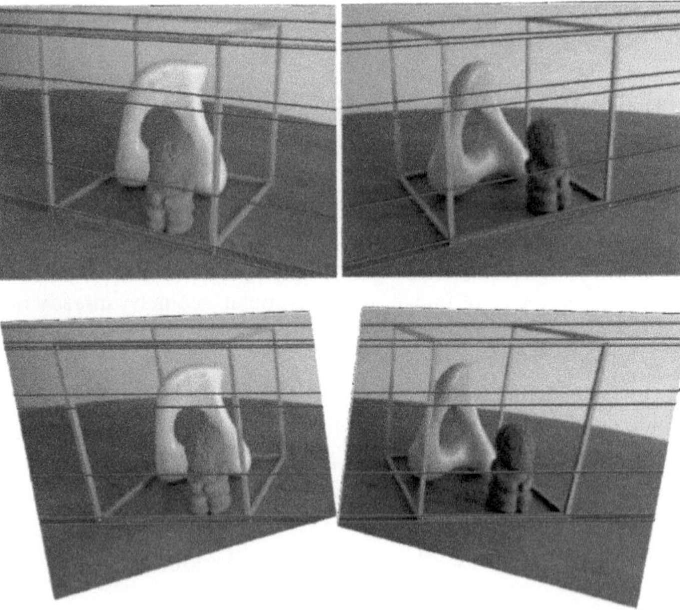

Fig. 1.9 *Top:* pair of images acquired by a vergent stereo system [41]. *Bottom:* rectified images. *Red lines* highlight some epipolar lines

1.2.1.3 The Correspondence Problem

The triangulation procedure assumes the availability of a pair of conjugate points p_L and p_R. This represents a delicate and tricky assumption for the triangulation procedure, first of all because such a pair may not exist due to occlusions. Even if it exists, it may not be straightforward to find it.

Indeed, the *correspondence problem*, i.e. the detection of conjugate points between the stereo image pairs, is one of the major challenges of stereo vision algorithms. The methods proposed for this task can be classified according to various criteria. A first distinction concerns dense and sparse stereo algorithms. The former, representing current trends [51], are methods aimed at finding a conjugate point for every pixel of the left (right) image, of course within the limits imposed by occlusions. The latter are methods which do not attempt to find a conjugate for every pixels.

A second distinction concerns the methods suited for short baseline and wide baseline stereo systems. The former implicitly assume the two images share considerable similarity characteristics hence, in principle, can adopt simpler methods with respect to the latter.

The third distinction concerns *local* and *global* approaches. Local methods consider only local similarity measures between the region surrounding p_L and regions of similar shape around all the candidate conjugate points p_R of the same row. The selected conjugate point is the one which maximizes the similarity measure, a method typically called *Winner Takes All (WTA)* strategy. Conversely, global methods do not consider each couple of points on their own, but instead estimate all of the disparity values at once, exploiting global optimization schemes. Global methods based on Bayesian formulations are currently receiving great attention in dense stereo. Such techniques generally model the scene as a Markov Random Field (MRF), and include within a unique framework clues coming from local comparisons between the two images and scene depth smoothness constraints. Global stereo vision algorithms typically estimate the disparity image by minimizing a cost function made by a *data term* representing the cost of local matches, similar to the computation of local algorithms (e.g., covariance) and a *smoothness term* defining the smoothness level of the disparity image by explicitly or implicitly accounting for discontinuities [57].

Wide baseline stereo methods traditionally rest on salient point detection techniques such as Harris corner detector [29]. *Scale Invariant Feature Transform* (SIFT) [42], which offers a robust salient point detector and an effective descriptor of the detected points, gave a truly major contribution to this field [47] and inspired a number of advances in related areas. An application of wide baseline matching which recently received major attention, as reported below, is 3D reconstruction from a generic collection of images of a scene [54].

It is finally worth recalling that although specific algorithms may have a considerable impact on the solution of the correspondence problem, the ultimate quality of 3D stereo reconstruction inevitably also depends on scene characteristics. This can be readily realized considering the case of a scene without geometric or color features, such as a straight wall of uniform color. The stereo images of such a scene will be uniform, and since no corresponding points can be detected from them, no depth information about the scene can be obtained by triangulation. As we mention in Sect. 1.3, active triangulation used in the so called *structured light systems* offers an effective and reliable way to cope with the correspondence problem issues.

1.2.2 N-view Stereo Systems and Structure from Motion

The situation of a 2-camera system shown in Fig. 1.5 can be generalized to that of an N-camera system. Figure 1.10 schematically shows N cameras and the relative CSS and image plane with a notation slightly different from the one adopted for the 2-camera stereo systems, since the image coordinates of the n-th camera are denoted as $[u_n, v_n]$ and the 3D coordinates with respect to the n-th CCS are denoted as $[x_n, y_n, z_n]$ with $n = 1, \ldots, N$. Extending the conventional notation of the 2-camera case, the CCS of the first and leftmost camera is typically adopted as WCS of the N-view system.

If the image coordinates on each camera $\mathbf{p}_n$, $n = 1, \ldots, N$ of the 3D point P and the projection matrices of each camera

$$\mathbf{M}_n = \begin{bmatrix} \mathbf{m}_{1n}^T \\ \mathbf{m}_{2n}^T \\ \mathbf{m}_{3n}^T \end{bmatrix} \quad n = 1, \ldots, N \tag{1.35}$$

are known, triangulation expression (1.31) generalizes to the homogeneous system of 2N equations in the four unknowns given by the homogeneous coordinates of P

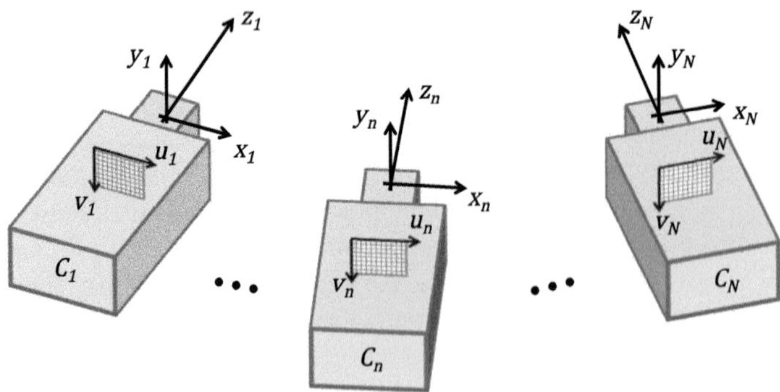

Fig. 1.10 N-camera system

$$\begin{bmatrix} \mathbf{m}_{31}^T u_1 - \mathbf{m}_{11}^T \\ \mathbf{m}_{31}^T v_1 - \mathbf{m}_{21}^T \\ \vdots \\ \mathbf{m}_{3k}^T u_n - \mathbf{m}_{1n}^T \\ \mathbf{m}_{3k}^T v_n - \mathbf{m}_{2n}^T \\ \vdots \\ \mathbf{m}_{3N}^T u_N - \mathbf{m}_{1N}^T \\ \mathbf{m}_{3N}^T v_N - \mathbf{m}_{2N}^T \end{bmatrix} \tilde{\mathbf{P}}_L = \mathbf{0}_{2N\times 1} \tag{1.36}$$

and nonlinear refinement (1.32) becomes

$$\min_{\tilde{\mathbf{P}}_L} \sum_{n=1}^{N} \left\{ \left(u_n - \frac{\mathbf{m}_{1n}^T \tilde{\mathbf{P}}_L}{\mathbf{m}_{3n}^T \tilde{\mathbf{P}}_L} \right)^2 + \left(v_n - \frac{\mathbf{m}_{2n}^T \tilde{\mathbf{P}}_L}{\mathbf{m}_{3n}^T \tilde{\mathbf{P}}_L} \right)^2 \right\}. \tag{1.37}$$

Once again, (1.37), like (1.32), can be seen as an extension of (1.12).

The $N = 3$ case corresponds to a special geometric structure and can be treated by trifocal tensor theory (see Chaps. 15 and 16 of [32] for a systematic treatment of this topic). Trinocular systems enjoy a number of interesting properties which attracted considerable attention. Even though the $N > 3$ case does not have the structure of the $N = 3$ case, it may benefit from specific geometric results enabling effective N-view stereo computational methods (Chap. 18 of [32]). On the practical side it is worth keeping in mind that the number of correspondences on N images typically decreases rapidly as N increases.

The geometry of Fig. 1.10 also corresponds to the situation of a single camera moving to different positions at subsequent times $1, \ldots, N$ in order to capture a static scene. This represents a situation of great interest which can be dealt with by an approach typically called *Structure From Motion* (SFM), for reasons which will become apparent soon. To this end, a quick digression is useful in order to introduce the concept of the essential matrix.

Consider a 2-view stereo system with the two cameras having identical and known intrinsic parameters represented by the calibration matrix $\mathbf{K}$. Denote normalized perspective projection matrices (1.24) defined with respect to normalized image coordinates $\mathbf{q}_L = \mathbf{K}^{-1}\mathbf{p}_L$ and $\mathbf{q}_R = \mathbf{K}^{-1}\mathbf{p}_R$, defined in (1.19), as

$$\mathbf{N}_L = \left[\mathbf{I} \mid \mathbf{0} \right] \qquad \mathbf{N}_R = \left[\mathbf{R} \mid \mathbf{t} \right] \tag{1.38}$$

with $\mathbf{t} = \left[t_x, t_y, t_z \right]^T$. Longuet-Higgins equation (1.33) rewritten with respect to normalized image coordinates becomes

$$\mathbf{q}_R^T \mathbf{K}^{-T} \mathbf{F} \mathbf{K} \mathbf{q}_L = \mathbf{q}_R^T \mathbf{E} \mathbf{q}_L = 0 \tag{1.39}$$

where matrix

$$\mathbf{E} = \mathbf{K}^{-T}\mathbf{F}\mathbf{K} \tag{1.40}$$

is called *essential matrix*. It can be proved that **E** can also be rewritten as

$$\mathbf{E} = \mathbf{T}\mathbf{R} \tag{1.41}$$

where

$$\mathbf{T} = \begin{bmatrix} 0 & -t_z & t_y \\ t_z & 0 & -t_x \\ -t_y & t_x & 0 \end{bmatrix} \tag{1.42}$$

is an antisymmetrical singular matrix obtained from the component of **t** as in (1.38). Hence **E** codifies the extrinsic parameters **R** and **t**, also called *motion* in the literature, as from (1.18), they represent the displacement of the left camera in order to assume the position of the right camera.

Matrix **E** could be computed in a straightforward manner from (1.39) given eight correspondences, however, since it enjoys strong structural properties [35] for which it has only 5 degrees of freedom, one could exploit five correspondences only [40]. Once **E** is known, the parameters **R** and **t** can be extracted from it [35].

The structure from motion (SFM) approach assumes a single moving camera with known intrinsic parameters **K**. It simultaneously computes the 3D scene points, called *structure*, and the extrinsic parameters **R** and **t** describing the camera position at subsequent times. If the image acquired by the camera at time t_n is denoted as I_n, the basic steps of SFM procedures are the following:

1. compute correspondences (p^j_n, p^j_{n+1}) with $j = 1, \ldots, J$ between image I_n and I_{n+1}
2. from (p^j_n, p^j_{n+1}) with $j = 1, \ldots, J$ compute the essential matrix $\mathbf{E}_n$ relative to the 2-view stereo system associated with (I_n, I_{n+1})
3. from $\mathbf{E}_n$ compute extrinsic parameters $\mathbf{R}_n$ and $\mathbf{t}_n$ (motion) characterizing the camera displacement at time t_{n+1} with respect to the camera position at time t_n
4. from $\mathbf{R}_n$ and $\mathbf{t}_n$ instantiate $\mathbf{M}_n$ and $\mathbf{M}_{n+1}$ similar to (1.38)
5. from $\mathbf{M}_n$ and $\mathbf{M}_{n+1}$ triangulate 3D points P^j (structure) from (p^j_n, p^j_{n+1}), $j = 1, \ldots, J$

If SFM is implemented by a video camera, subsequent images are similar since they correspond to positions close in time. The estimate of point p_{n+1} corresponding to p_n in this case is a tracking problem for which a number of short-baseline matching techniques have been proposed [13, 43, 58].

If SFM is implemented by a photo camera, however, the displacement of subsequent frames are usually wider than those of video cameras and the correspondences are obtained by wide-baseline saliency detectors/descriptors methods like SIFT [42] and similar approaches [47].

1.2.3 Calibrated and Uncalibrated 3D Reconstruction

Calibrated 3D reconstruction is the procedure by which one computes $\mathbf{M}_L$ and $\mathbf{M}_R$ from knowledge of the stereo system calibration parameters $\mathbf{K}_L$, $\mathbf{K}_R$, $\mathbf{R}$ and $\mathbf{t}$. Then, by triangulation, the pair of conjugate points (p_L, p_R) are reprojected into their corresponding 3D point P. Calibrated reconstruction delivers an Euclidean scene reconstruction, i.e., a 3D reconstruction preserving distances and angles of the original scene imaged by I_L and I_R.

SFM procedures exploit the properties of the essential matrix $\mathbf{E}$ in order to reconstruct 3D scene geometry from a temporal sequence or just from a collection of images upon knowledge of the intrinsic parameters. A similar result can also be obtained without knowing intrinsic parameters by exploiting the properties of the fundamental matrix $\mathbf{F}$ which embeds all the epipolar geometry information, as it can be computed directly from the stereo system's intrinsic and extrinsic parameters $\mathbf{K}_L$, $\mathbf{K}_R$, $\mathbf{R}$, $\mathbf{t}$. Similar to what has been seen for $\mathbf{E}$, Eq. (1.33) allows to compute $\mathbf{F}$ from a set of correspondences (p_L^j, p_R^j) with $j = 1, \ldots, J$. Now, let us recall that 3×3 matrix $\mathbf{F}$, in spite of having nine entries, has only 7 degrees of freedom, and that there are techniques for computing $\mathbf{F}$ which use just $J = 7$ and $J = 8$ correspondences (p_L^j, p_R^j) [30]. For a review of the methods suitable to compute $\mathbf{F}$, see [63].

Since $\mathbf{F}$ can be used to instantiate a pair of projection matrices $\mathbf{M}_L$ and $\mathbf{M}_R$ (or more than a pair in the case of N-view stereo), and $\mathbf{F}$ can be directly computed from a set of correspondences (p_L^j, p_R^j), $j = 1, \ldots, J$ on a pair of stereo images, 3D reconstruction can be achieved solely based on image information without knowledge of the extrinsic or intrinsic parameters of the stereo system, like in the case of SFM. Such techniques, called *uncalibrated reconstruction* methods, in general cannot preserve the distance and angle information of the imaged scene, because if only the left hand side $\tilde{\mathbf{p}}_L$ (or $\tilde{\mathbf{p}}_R$) of (1.30) is known and $\mathbf{M}_L$ (or $\mathbf{M}_R$) and $\tilde{\mathbf{P}}_L$ satisfy (1.30) on the right side, under such circumstances $\mathbf{M}_L\mathbf{W}^{-1}$ and $\mathbf{W}\tilde{\mathbf{P}}_L$ will also satisfy it for any 4×4 invertible matrix $\mathbf{W}$. Since such a matrix $\mathbf{W}$ represents a 3D projective transformation, the 3D reconstruction provided by uncalibrated methods is called *projective reconstruction* [49] and has limited usefulness. Fortunately, there are special conditions reasonable for practical applications under which the intrinsic parameters can be recovered from the $\mathbf{F}$ matrices and the projective reconstruction can be turned Euclidean up to a scale factor and an arbitrary choice of the WCS. An example of such a condition is given by at least three images taken by systems with the same intrinsic parameters, equivalent to using a single system for taking at least three images. The intrinsic parameters matrix $\mathbf{K}$ can be directly computed exploiting a property of the essential matrix, from the set of fundamental matrices $\mathbf{F}_i$ associated with pairs of images I_n and I_{n+1} by the method of Mendonca and Cipolla [46] or derived as a second step, following a projective reconstruction, by the so called *Euclidean promotion* approach [34]. The former method rests on a numerical optimization and it requires a reasonable starting point for $\mathbf{K}$. The latter assumes that the 3D points are visible from all the images, a condition seldom satisfied in practical cases. Furthermore, noise and wrong matches can seriously compromise the reconstruction quality of these methods.

Sequential structure from motion approaches of the type of the *Bundler* method [53] received considerable attention as they proved to be very effective in practice. Such methods can provide either Euclidean reconstruction, if the intrinsic camera parameters are known (in this case they compute the essential matrices), or projective reconstruction, if the intrinsic camera parameters are not available (in this case they compute the fundamental matrices). The various elements characterizing such methods are mentioned next.

First, structure and motion are incrementally computed by adding one image at a time to the set of images under analysis. The initial set is made by selecting a pair of images according to a balance between the number of candidate matches and image separation (to avoid bad conditioning due to excessive proximity). The next image to include in the image analysis set is determined based on the maximum overlay of the new image with respect to images already analyzed.

A salient point detector is applied to each image in the set and candidate corresponding points are determined on the basis of their descriptors. The matrix **E** (or **F**) is computed by robust estimation algorithms, such as RANSAC [25]. At each iteration, a matrix **E** (or **F**) is computed on a random set of correspondences. The consensus is given by the number of correspondences satisfying epipolar constraints. The matrix associated with the highest consensus is selected, thus eliminating the effect of the outliers and retaining only robust correspondences on each image pair. Such correspondences are propagated to the other images giving so called "tracks", which typically must comply with various constraints.

Each matrix **E** (or **F**) is used to instantiate a pair of projection matrices from which one computes the 3D points associated with the validated correspondences.

As a new image is added to the set of images under analysis, 3D structure information is updated and extended in two ways: (1) the 3D position of the points computed from the previous correspondences is recomputed with the image points of the new image (extra data improve the accuracy of triangulation); (2) new 3D points are triangulated from the correspondences provided by the new image.

Bundle adjustment [37, 59] is used for global tuning of the estimated **E** (or **F**) matrices in order to handle inevitable error propagation as the number of images increases.

1.3 Basics of Structured Light Depth Cameras

As previously noted, the reliability of the correspondences remains a critical step of computational stereopsis. Structured light depth camera systems address this issue and provide effective solutions, as will be seen in this section and in greater detail in Chap. 2.

In triangulation or computational stereopsis procedures, the main conceptual and structural point at the basis of triangulation is of geometric nature and is shown in

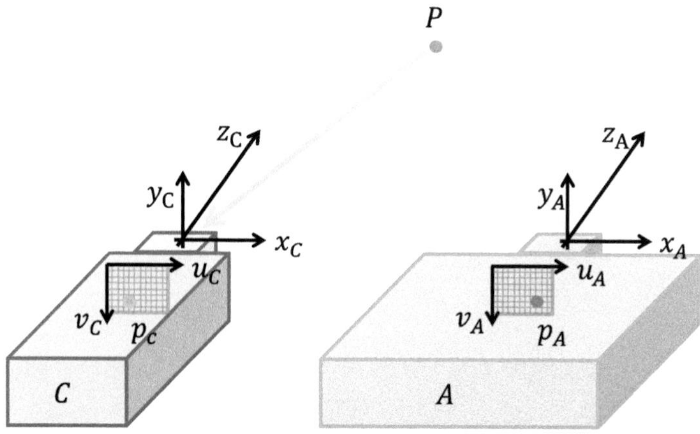

Fig. 1.11 Active triangulation by a system made of a camera C and a light projector A

the triangle arrangement between rays Pp_L, Pp_R and p_Lp_R in Fig. 1.5. The fact that p_L and p_R in standard (passive) stereo systems are a result of the light reflected by P towards the two cameras is a totally secondary element.

Let us also recall that since from a perspective geometry standpoint [32], image points are equivalent to rays exiting a center of projection, any device capable of projecting rays between its center of projection and the scene points is functionally equivalent to a standard camera, modeled as a passive pin-hole system in Sect. 1.1. Therefore, *light projectors* or *illuminators*, devices in which each pixel p_A illuminates a scene point P by its specific light value thus creating a spatial pattern, can be modeled as active pin-hole systems where light rays connecting the center of projection and the scene point P through pixel p_A (as shown in Fig. 1.11) are emitted, rather than received as in standard cameras. Triangulation or computational stereopsis also remains applicable if one of the two cameras of a stereo system, as in Fig. 1.6, is replaced by a projector, granted by triangle arrangement Pp_C, Pp_A and p_Cp_A of Fig. 1.11. The active, rather than passive, nature of ray Pp_A does not affect the reasoning behind the demonstration of triangulation. Such an arrangement made by a camera C and a projector A as shown in Fig. 1.11, is called a *structured light system*. More formally, let us associate to camera C a CCS system with coordinates (x_C, y_C, z_C) also called *C-3D reference system* and a 2D reference system with coordinates (u_C, v_C). Similarly let us associate to projector A a CCS with coordinates (x_A, y_A, z_A) also called *A-3D reference system* and a 2D reference system with coordinates (u_A, v_A) as shown in Fig. 1.11. As in the 2-view stereo case, assume the C-3D reference system also serves as reference system of the structured light system, denoted as S-3D system.

Figure 1.12 exemplifies the operation of a structured light system: projector A illuminates the scene with a given pattern and camera C acquires the image of the scene with the superimposed pattern. In particular, in Fig. 1.12 a light pattern pixel p_A with coordinates $\mathbf{p_A} = [u_A, v_A]^T$ in the A-2D reference system, is projected to 3D

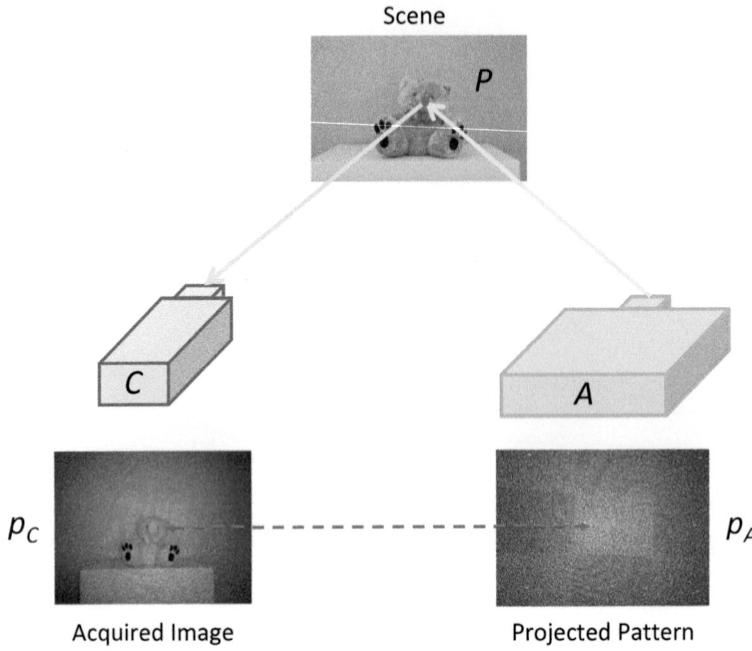

Fig. 1.12 Structured light systems operation: pixel p_A is coded in the pattern. The pattern is projected to the scene and acquired by C. The 3D point associated with p_A is P and the conjugate point of p_A is p_C. A correspondence estimation algorithm estimates the conjugate points upon which triangulation can be instantiated

scene point P with coordinates $\mathbf{P}_C = [x_C, y_C, z_C]^T$ in the C-3D (and S-3D) reference system. Figure 1.12 clearly shows that if P is not occluded, it projects the light radiant power received by the projector to the pixel p_C of camera C establishing triangle $p_C P p_A$ which, as previously observed, recreates the triangulation premises of Fig. 1.6.

From the Cartesian coordinates of P with respect to the C-3D and A-3D reference systems and relative normalized coordinates

$$\begin{aligned} \mathbf{P}_C &= z_C \mathbf{K}^{-1} \tilde{\mathbf{p}}_C = z_C \tilde{\mathbf{q}}_C \\ \mathbf{P}_A &= z_A \mathbf{K}^{-1} \tilde{\mathbf{p}}_A = z_A \tilde{\mathbf{q}}_A \end{aligned} \tag{1.43}$$

relationship

$$\mathbf{P}_A = \mathbf{R}\mathbf{P}_C + \mathbf{t} \tag{1.44}$$

can be rewritten as

$$z_A \tilde{\mathbf{q}}_A - z_C \mathbf{R} \tilde{\mathbf{q}}_C = \mathbf{t} \tag{1.45}$$

or

$$\begin{cases} z_A u'_A - z_C \mathbf{r}_1^T \tilde{\mathbf{q}}_C = t_1 \\ z_A v'_A - z_C \mathbf{r}_2^T \tilde{\mathbf{q}}_C = t_2 \\ z_A - z_C \mathbf{r}_3^T \tilde{\mathbf{q}}_C = t_3. \end{cases} \tag{1.46}$$

Different from the 2-view stereo case, note that in the common case of a projected pattern with vertical stripes, the second equation of (1.46) cannot be used since v'_A is unknown. Similarly, in the case of horizontal stripes, one cannot use the first equation of (1.46) since u'_A is not known.

Assuming a pattern with known u'_A, by substituting in the first equation of (1.46) $z_A = t_3 + z_C \mathbf{r}_3^T \mathbf{q}_C$ derived from the third equation of (1.46), one obtains

$$z_C = \frac{t_1 - t_3 u_A}{u_A \mathbf{r}_3^T \tilde{\mathbf{q}}_C - \mathbf{r}_1^T \tilde{\mathbf{q}}_C} \tag{1.47}$$

in a manner very similar to the derivation of (1.23). Indeed, (1.47) and (1.23) are identical apart from the adopted notational convention. The left camera 3D coordinate system is assumed to be the 3D system reference system S-3D, both in stereo and structured light systems, except that in stereo systems, the coordinates with respect to the S-3D reference system are denoted by subscript L (referring to left camera L) and in structured light systems these are denoted by subscript C (again referring to the camera this time denoted by C).

Structured light systems have the same structural geometry of standard passive stereo systems, thus calibration and rectification procedures [60] can also be applied to them in order to simplify the depth estimation process. In the case of rectification, pixel p_A with coordinates $\mathbf{p}_A = [u_A, v_A]^T$ of the projected pattern casts a ray that intersects the acquired scene at a certain 3D location $\mathbf{P}_C = [x_C, y_C, z_C]^T$. If both the projective distortion of A and C are compensated, p_C has coordinates

$$\mathbf{p}_C = \begin{bmatrix} u_C = u_A + d \\ v_C = v_A \end{bmatrix} \tag{1.48}$$

as in (1.17) with disparity value

$$d = u_C - u_A \tag{1.49}$$

defined exactly as in (1.16) apart from the different notation adopted for the coordinate system. Furthermore, since the parameters entering the projection matrix of the illuminator M_A are $R = I$ and $\mathbf{t} = [b, 0, 0]^T$, from $\mathbf{r}_3^T \tilde{\mathbf{q}}_C = 1$ and $\mathbf{r}_1^T \tilde{\mathbf{q}}_C = u'_C$, (1.47) becomes

$$z_C = \frac{b}{u'_A - u'_C} = \frac{bf}{u_A - u_C} = \frac{b|f|}{d} \tag{1.50}$$

with disparity d defined by (1.49), exactly as in (1.26).

Since we have established that all triangulation expressions derived for a 2-camera stereo system also apply to structured light systems made by an illuminator and a single camera, let us now consider the advantages of the latter with respect to the former. As previously noted, in passive stereo systems made by a pair of cameras, the possibility of identifying conjugate points depends completely on the visual characteristics of the scene. In particular, in the case of a featureless scene, like a flat wall of uniform color, a stereo system could not establish any point correspondence between the image pair and could not give any depth information about the scene. On the contrary, in the case of a structured light system the light pattern pixel p_A of the projector "colors" the scene point P to which it projects with its radiant power. Assuming a straight wall without occlusions, the pixel p_C of the camera C where P is projected receives from P the "color" of p_A and becomes recognizable among its neighboring pixels. This enables the possibility of establishing a correspondence between conjugate points p_A and p_C. Structured light systems can therefore also provide depth information in scenes without geometry and color features where standard stereo systems fail to give any depth data.

The characteristics of the projected patterns are fundamental for the solution of the correspondence problem and for the overall system performance. The illuminators mainly belong to two families, namely, *static* illuminators, which project a static pattern/texture into the scene, and *dynamic* illuminators, which project a pattern/texture that varies in time.

The projection of sequences of different light patterns was typical of early structured light systems and confined their usage to still scenes. These types of systems were and continue to be the common choice for 3D modeling of still scenes. For the 3D modeling methods used in this field see [14, 19, 39]. In general, active techniques are slower and more expensive than passive methods but much more accurate and robust. In order to measure distances of dynamic scenes, i.e., scenes with moving objects, recent structured light methods focused on reducing the number of projected patterns to a few units or single pattern [50, 64, 65].

In the literature the definition of structured light systems and of light-coded systems is not unequivocal. For clarity's sake, in this book we refer to structured light systems as systems with an illuminator projecting a pattern characterized by some structure, as opposed to an illuminator that emits the same amount of light in all directions (e.g., a diffuse light emitter). This structure can be either in the form of a continuous texture (e.g., Intel RealSense R200 [3]), in the form of a striped pattern (e.g., Intel RealSense F200 [3]) or in the form of a pattern characterized by dots (e.g., Primesense cameras). A detailed description of patterns will be provided in Chap. 2.

It is also clear that a system with two cameras $C1$ and $C2$ and a projector A, as shown in Fig. 1.13, is a variation of a structured light system by which the coordinates of point P in principle can be obtained by any of the types of

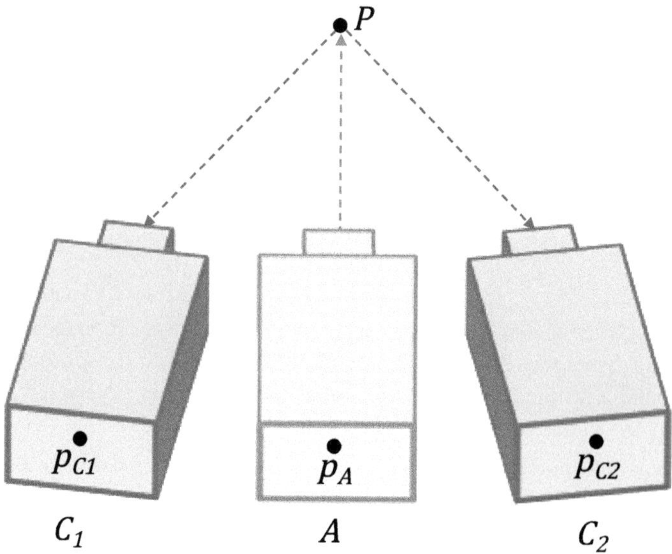

Fig. 1.13 Structured light system with two cameras and a projector

triangulation seen so far, or by a combination of them. Indeed, P can be computed by triangulation upon knowledge of either conjugate points p_{C1} and p_{C2}, points p_{C1} and p_A, or points p_{C2} and p_A. Given this, the projector could simply color the scene by a still "salt and pepper" pattern meant to strengthen or create features or use more effective methods reported in Chap. 2.

Let us anticipate, though, that one can demonstrate the complete functional equivalence between the various structured light systems configurations, namely the single camera, the two cameras and the so called space-time stereo systems [23, 36]. This topic, useful for conceptual purposes and for a unified treatment of the various structured light depth camera products, will be further discussed in Chap. 2.

1.4 Basics of ToF Depth Cameras

1.4.1 ToF Operation Principle

ToF and LIDAR (Light Detection And Ranging) devices operate on the basis of the RADAR (Radio Detection And Ranging) principle, which rests on the fact that the electro-magnetic radiation travels in air at light speed $c \approx 3 \times 10^8 [\mathrm{m/s}]$. Hence, the distance $\rho\,[\mathrm{m}]$ covered in time $\tau\,[\mathrm{s}]$ by an optical radiation is $\rho = c\tau$. Figure 1.14 shows the typical ToF measurement scheme: the radiation $s_E(t)$ emitted at time 0 by the ToF transmitter (or illuminator) TX on the left travels straight towards the scene

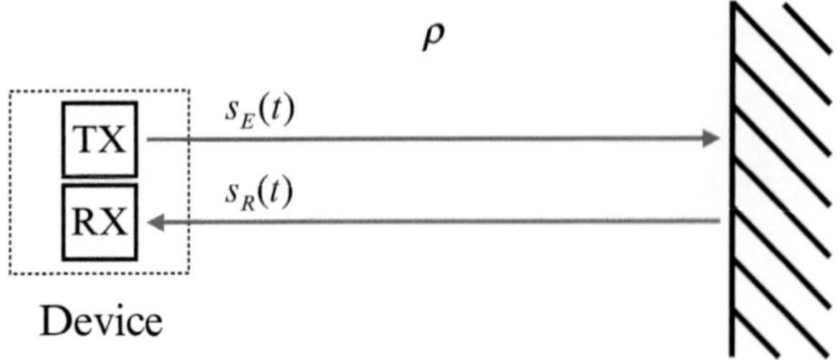

Fig. 1.14 Scheme of principle of ToF measurement

for a distance ρ. It is then echoed or back-reflected by a point P on the scene surface and travels a distance ρ. At time τ it reaches the ToF receiver (or sensor) RX, ideally co-positioned with the transmitter, as signal $s_R(t)$. Since at time τ the path length covered by the radiation is 2ρ, the relationship between ρ and τ is

$$\rho = \frac{c\tau}{2} \tag{1.51}$$

which is the basic expression of a ToF camera's distance measurement.

In stereo or structured light systems, occlusions are inevitable due to the presence of two cameras, or a camera and a projector, in different positions. Additionally, the distance between the camera positions (i.e. the baseline) improves the distance measurement accuracy. This is an intrinsic difference with respect to ToF, in which measurements are essentially occlusion-free, because the ToF measurement scheme assumes the transmitter and receiver are collinear and ideally co-positioned. In common practice such a requirement is enforced by placing them as close together as possible.

In spite of the conceptual simplicity of relationship (1.51), its implementation presents tremendous technological challenges because it involves the speed of light. For example, since

$$c = 3 \times 10^8 \, \frac{[m]}{[s]} = 2 \times 150 \, \frac{[m]}{[ps]} = 2 \times 0.15 \, \frac{[\text{mm}]}{[\text{ps}]} \tag{1.52}$$

it takes 6.67 [ns] to cover a 1 [m] path and distance measurements of nominal resolution of 1 [mm] need time measurement mechanisms with accuracy superior to $6.67 \div 7$ [ps], while a nominal resolution of 1 [cm] needs accuracy superior to 70 [ps]. The accurate measurement of round-trip time τ is the fundamental challenge in ToF systems and can be solved by two approaches: direct methods, addressing either the measurement of time τ by pulsed light or of phase φ in case of continuous wave operation (see Fig. 1.15), and indirect methods deriving τ (or φ as an intermediate step) from time-gated measurements of signal $s_R(t)$ at the receiver. The former methods are considered next while the latter methods will be discussed in detail in Chap. 3.

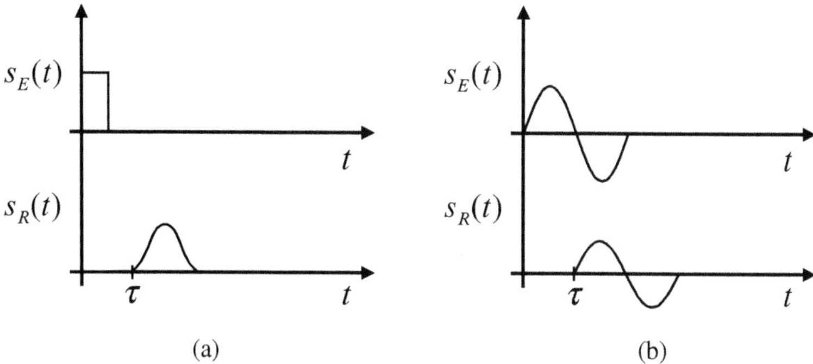

Fig. 1.15 Schematic operation of direct ToF measurement: (**a**) pulse modulation; (**b**) CW (sinusoidal) modulation

1.4.2 Direct ToF Measurement Methods

1.4.2.1 Direct Pulse Modulation

Figure 1.15a shows a square pulse $s_E(t)$ transmitted at time 0 and the received signal $s_R(t)$ arriving at time τ. Pulsed light has several advantages since it allows for a high signal to noise ratio due to the characteristics of pulse signals, which concentrate high energy values in a short time. High signal to noise ratio gives good resilience with respect to background illumination, enabling long distance measurements, reducing average optical power (with eye safety and power consumption advantages) and eliminating the need for high sensibility detectors at the receiver front-end.

The receiver back-end has a measurement device called a stop-watch, typically implemented by a counter with a start coinciding with the transmission of $s_E(t)$ and a stop coinciding with at the detection of $s_R(t)$. The accurate detection of $s_R(t)$ represents the most challenging task of the receiver. Indeed, the actual shape and values of $s_R(t)$ are hardly predictable since they do not only depend on the target distance and material but on the atmospheric attenuation of the signal path, especially for long distances. Therefore, simple fixed thresholding on the receiver optical power or straightforward correlation are not adequate, also because pulses with sharp rise and fall times at the transmitter side are not easy to implement. In addition, pulse signaling requires large bandwidth and high dynamic range at the receiver sensors.

Light-emitting diodes (LEDs) and laser diodes are commonly used for the pulse generation of ToF systems with repetition rates on the order of tens of KHz. Current commercial products for architectural and topographic surveys [1, 4, 8, 10] covering ranges from tens of meters to kilometers use direct ToF measurements based on pulse modulation.

1.4.2.2 Direct CW Modulation

Direct continuous wave (CW) modulation is not as common as direct pulse modulation. However, because CW modulation is commonly used also with indirect ToF measurements as we will discuss in Chap. 3, we introduce its basic principles here for sinusoidal signals.

Figure 1.15b shows a sinusoid $s_E(t) = A_E \sin(2\pi f_c t + \varphi_c)$ transmitted at $t = 0$ and the received attenuated sinusoidal $s_R(t)$ which arrives at time $t = \tau$

$$\begin{aligned} s_R(t) &= A_R \sin\left[2\pi f_c\left(t + \tau\right) + \varphi_c\right] \\ &= A_R \sin\left[2\pi f_c t + \varphi_c + \Delta\varphi\right] \\ &= \frac{A_R}{A_E} s_E(t + \tau) \end{aligned} \tag{1.53}$$

where

$$\Delta\varphi = 2\pi f_c \tau \tag{1.54}$$

is the phase difference between $s_E(t)$ and $s_R(t)$.

The round-trip transmitter to target point distance ρ from (1.54) and (1.51) can be directly computed from the phase difference as

$$\rho = c\frac{\Delta\varphi}{4\pi f_c}. \tag{1.55}$$

Many methods can be employed to estimate the phase difference between sinusoids as this is a classical telecommunications and control issue (see, for example, the phase-locked loop literature). It is important to note that the measurement of $\Delta\varphi$ only requires knowledge of f_c but not φ_c, hence the transmitter can operate with incoherent sinusoids.

An important characteristic of direct ToF systems which differs from stereo and structured light systems is that the measurement accuracy is distance independent, only depending on the accuracy of the time or phase measurement devices. Such an important characteristic will be retained also by indirect ToF measurement systems as discussed in Chap. 3.

1.4.3 Surface Measurement by Single Point and Matricial ToF Systems

ToF systems made by a single transmitter and receiver, as schematically shown in Fig. 1.14, are typically used in range-finders for point-wise measurements. Such systems can also be mounted on time sequential 2D (or 1D) scanning mechanisms for measuring distance on surfaces or along lines. It is typical to move a single ToF transmitter and receiver system along a vertical linear support placed on a rotating platform, with motion both in vertical and horizontal directions, as in the case of the scanning systems used for topographic or architectural surveys (e.g., [1, 4, 8, 10]). Since any time sequential scanning mechanism requires time in order to scan different scene points, such systems are intrinsically unsuited to acquire dynamic scenes, i.e., scenes with moving objects. Nevertheless, fast ToF scanning systems have been studied and some products, targeting automotive applications, reached the market [9]. 3D scanning mechanisms are bulky and unsuited to vibrations, as well as expensive, since they require high precision mechanics and electro-optics.

The progress of microelectronics fostered a different surface measurement solution, based on the concept of scanner-less ToF systems [38] and appeared in the literature under various names such as Range IMaging (RIM) cameras, ToF depth cameras, or ToF cameras, which are the denominations used in this book. Unlike systems which acquire scene geometry one point at a time by a point-wise ToF sensor mounted on time sequential scanning mechanisms, ToF cameras estimate the scene geometry in a single shot by a matrix of $N_R \times N_C$ in-pixel ToF sensors where all the pixels independently but simultaneously measure the distance of the scene point in front of them. ToF cameras deliver depth maps, i.e., measurement matrices with entries giving the distance between the matrix sensor and the imaged scene points, at video rates.

Even though a ToF camera may be conceptually interpreted as a matricial organization of many single devices, each made by an emitter and a co-positioned receiver as discussed previously, in practice implementations based on a simple juxtaposition of a multitude of single-point measurement devices are not feasible. Currently, it is not possible to integrate $N_R \times N_C$ emitters and $N_R \times N_C$ receivers in a single chip, especially for high values of N_R and N_C as required by imaging applications. However, each receiver does not require a specific co-positioned emitter; instead, a single emitter may provide an irradiation that is reflected back by the scene and collected by a multitude of receivers close to each other. Once the receivers are separated from emitters, the former can be integrated in a $N_R \times N_C$ matrix of lock-in pixels commonly called *ToF camera sensor* or more simply, *ToF sensor*.

The implementation of lock-in pixels and their integration into matricial configurations are fundamental issues of current ToF systems research and industrial development. Different technological choices lead to different ToF camera types. An in depth review of state of the art ToF technology can be found in [56].

1.4.4 ToF Depth Camera Components

ToF depth cameras lend themselves to a countless variety of different solutions, however, all the current implementations share the same structure shown in Fig. 1.16 made by the following basic components:

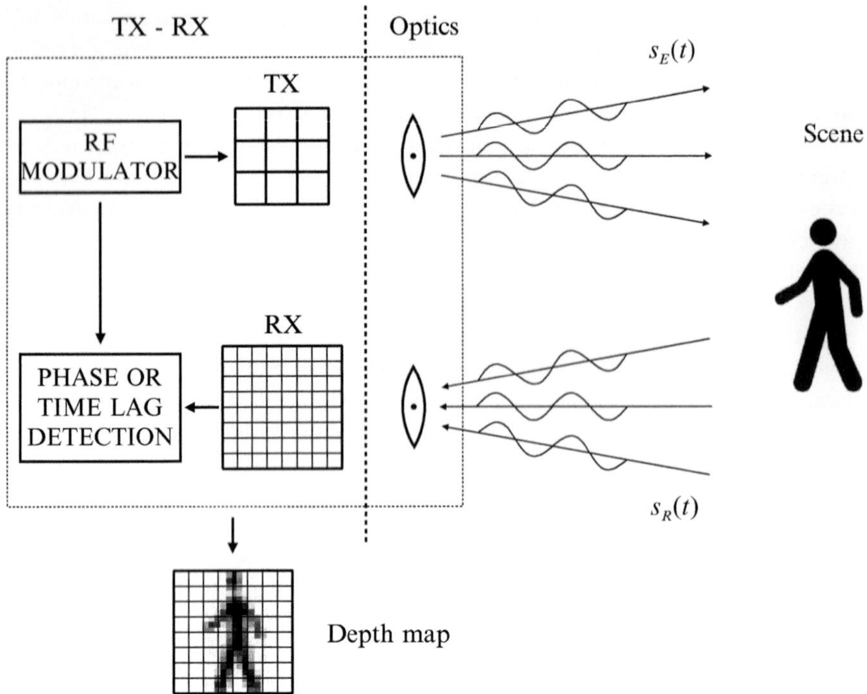

Fig. 1.16 Basic ToF depth camera structure

(a) The transmitter is typically made by an array of laser emitters or LEDs with an optics allowing illumination of the whole scene by the same signal $s_E(t)$.
(b) The receiver can be interpreted as a matrix of single ToF receivers coupled with an imaging system framing a scene frustum. Each pixel receives the signal $s_R(t)$ echoed by a surface point P.
(c) The emitted light is modulated by frequencies of some tens of MHz (in general in the high HF or low VHF bands), e.g., 16, 20, 30, 60, 120 MHz and each pixel of the receiver sensor matrix typically estimates the phase difference $\Delta\varphi$ between $s_E(t)$ and $s_R(t)$. Such quantities are computed by time-gated measurements (indirect measurement approach)

A number of clarifications and comments about the above listed ToF cameras features are presented in the following sections.

1.4.4.1 Modulation Methods for ToF Depth Cameras

The choice of modulation is the best starting point in order to understand the characteristics of current ToF depth cameras. Indeed, the modulation operation determines the basic transmitter and receiver functions and structure. Although in principle many modulation types suit ToF depth cameras, in practice, all current commercial ToF depth camera products [2, 5, 7] adopt only one type of CW modulation, namely homodyne amplitude modulation with either a sinusoidal or square wave modulating signal $m_E(t)$. This is because current microelectronic technology solutions for homodyne AM are more mature than others for commercial applications. The advantages of AM modulation, besides its effective implementability by current CMOS solutions, are that it uses a single modulation frequency f_m and does not require a large bandwidth. A major disadvantage is that it offers little defense against multipath and other propagation artifacts (see Chap. 5 of [56]) as will be seen in Chap. 3.

Other modulation types than AM could be usefully employed in ToF depth cameras and their implementation is being actively investigated [56]. Other candidate modulation types include pulse modulation and pseudo-noise modulation. The former, as already mentioned, is the preferred choice for single transmitter and receiver ToF systems. Although in principle it would be equally suitable for matrix ToF sensors, in practice its application is limited by the difficulties associated with implementing effective stop-watch at pixel level within matrix arrangements. Current research approaches this issue in various ways (see Chaps. 2 and 3 of [56]). Pseudonoise modulation would be very effective against multipath, as other applications such as indoor radio localization [21] indicate.

CW modulation itself offers alternatives to homodyne AM, such as heterodyne AM or frequency modulation (FM) with chirp signals. Such properties, although reported in ToF measurement literature, are still problematic for matrix ToF sensor electronics.

The remainder of this section considers the basic characteristics of ToF depth camera transmitters and receivers assuming the underlying modulation is Continuous Wave Amplitude Modulation (CWAM).

1.4.4.2 ToF Depth Camera Transmitter Basics

Lasers and LEDs are the typical choice for the light sources at the transmitter since they are inexpensive and can be easily modulated by signals within the high HF or low VHF bands up to some hundreds of MHz. The LED emissions typically used are in the near infrared (NIR) range, with wavelength around $\lambda_c = 850$ [nm], corresponding to

$$f_c = \frac{c}{\lambda_c} = 3 \times 10^8 \,\frac{[\mathrm{m}]}{[\mathrm{s}]}\,\frac{1}{850 \times 10^{-9}\,[\mathrm{m}]} \approx 352\ [\mathrm{THz}]. \tag{1.56}$$

The transmitter illuminates the scene by an optical 2D wavefront signal which, for simplicity, can be modeled as

$$s_E(t) = m_E(t)\ \cos(2\pi f_c t + \varphi_c) \tag{1.57}$$

where $s_E(t)$ denotes the emitter NIR signal structured as the product of a carrier with NIR frequency f_c, of some hundreds of THz, and phase φ_c and a modulating signal $m_E(t)$. Signal $m_E(t)$, in turn, incorporates AM modulation of either sinusoidal or square wave type in current products with frequency f_m, of some tens of MHz, and φ_m. In current products there are two levels of AM modulation, as shown in Fig. 1.18 and further explained in the next section. The first is AM modulation at NIR frequencies concerning the optical signal $s_E(t)$ used to deliver the modulating signal $m_E(t)$ at the receiver. The second is AM modulation in the high HF or low VHF bands embedded in $m_E(t)$, which delivers information related to round-trip time τ to the receiver, either in terms of phase or time lag. This will be seen in detail in Chap. 3.

The current ToF camera NIR emitters are either lasers or LEDs. Since they cannot be integrated, they are typically positioned in configurations mimicking the presence of a single emitter co-positioned with the optical center of the ToF camera. The geometry of the emitters' position is motivated by making the sum of all the emitted NIR signals equivalent to a spherical wave emitted by a single emitter, called *simulated emitter*, placed at the center of the emitters constellation. The LED configuration of the Mesa Imaging SR4000, shown in Fig. 1.17, is an effective example of this concept.

The arrangement of the actual emitters, such as the one of Fig. 1.17, is only an approximation of the non-feasible juxtaposition of single ToF sensor devices with emitter and receiver perfectly co-positioned and it introduces a number of artifacts, including a systematic distance measurement offset that is larger for close scene points than for far scene points.

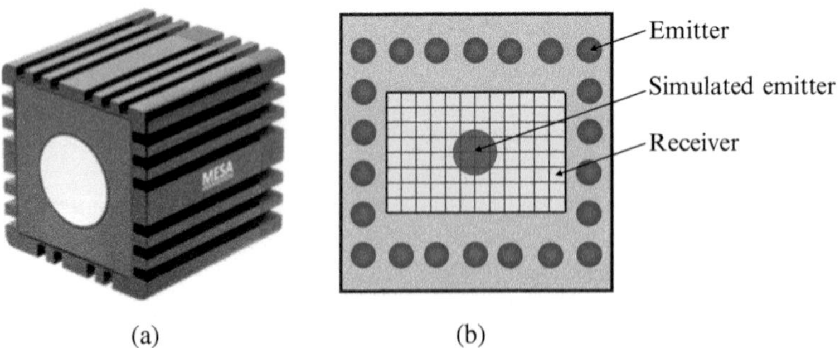

Fig. 1.17 The NIR emitters of the MESA Imaging SR4000: (**a**) the actual depth camera; (**b**) in the scheme the emitters are distributed around the lock-in pixels matrix and mimic a simulated emitter co-positioned with the center of the lock-in pixel matrix

1.4.4.3 ToF Depth Camera Receiver Basics

The heart of ToF camera receivers is a matricial sensor with individual elements, called pixels because of their imaging role, individually and simultaneously capable of independent ToF measurements. In other words each pixel independently computes the delay between the departure of the sent signal $s_E(t)$ and the arrival of the signal $s_R(t)$ back-projected by the scene point P imaged by the pixel. Currently there are three main technological solutions (Chap. 1 of [56]) considered best suited for the realization of such matricial ToF sensors, namely Single-Photon Avalanche Diodes (SPADs) assisted by appropriate processing circuits, standard photo diodes coupled to dedicated circuits and the In-Pixel Photo-Mixing devices. The latter technology includes the lock-in CCD sensor of [38], the photonic mixer device (PMD) [52, 62], and other variations [11, 12]. Chapter 3 will only recall the main characteristics of the In-Pixel Photo-Mixing devices, since so far it is the only one adopted in commercial products [2, 5, 7]. The reader interested in an in-depth treatment of such a technology can refer to [38] and to Chap. 4 of [56]. Current solutions about matricial ToF sensors based on SPADs and standard photo diodes are reported in Chaps. 2 and 3 of [56].

Figure 1.18 offers a system interpretation of the basic functions performed by each pixel of a sensor based on photo-mixing device technology, which are

(a) photoelectric conversion
(b) correlation or fast shutter
(c) signal integration by charge storage on selectable time intervals

The logic of system analysis and circuit effectiveness are opposite. For analysis purposes it is useful to recognize and subdivide the various operations as much as possible. On the contrary, multifunctional components are the typical choice for circuit effectiveness. This chapter and Chap. 3 present ToF depth cameras from a system perspective and the reader must be aware that it does not always coincide with the circuit block description.

Each sensor pixel receives as input the optical NIR signal back-projected by the scene point P imaged by the pixel itself, which can be modeled as

$$s_R(t) = m_R(t)\ \cos(2\pi f_c t + \varphi_c') + n_R(t) \tag{1.58}$$

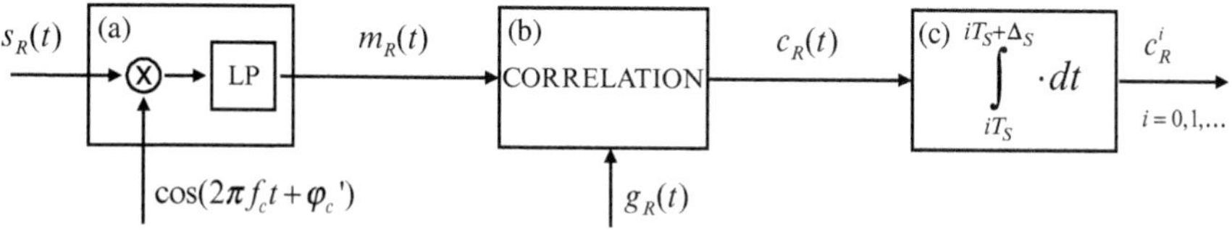

Fig. 1.18 System interpretation of the operation of a single pixel of a sensor based on In-Pixel Photo-Mixing devices technology

where $m_R(t)$ denotes the transformations of the modulating signal $m_E(t)$ actually reaching the receiver, since direct and reflected propagation typically affect some parameters of the transmitted signal $m_E(t)$ (for instance amplitude attenuation is inevitable) and $n_R(t)$ is the background wide-band light noise at the receiver input.[1]

The photoelectric conversion taking place at the pixel in the scheme of Fig. 1.18 is modeled as a standard front-end demodulation stage (a) with a carrier $\cos(2\pi f_c t + \varphi_c')$ at NIR frequency f_c followed by a low pass filter (LP). The input of stage (a) is the optical signal $s_R(t)$ and the output is the baseband electrical signal $m_R(t)$. Stage (a) is, however, only a simple model for the light detection operations taking place at pixel level; details about the actual circuits converting optical signal $s_R(t)$ into electrical signal $m_R(t)$ can be found in [38] and in Chap. 4 of [56]. Stage (b) represents the correlation between baseband signal $m_R(t) + n(t)$ and reference signal $g_R(t)$. Details about the most common types of $g_R(t)$ will be given in Chap. 3. Stage (c) models the charge accumulation process as an integrator operating on time intervals of selectable lengths Δ_S starting at uniformly spaced clock times iT_s, $i = 1, 2, \ldots$ where T_s is the sampling period. Details about indirect ToF measurement methods used to compute phase increment $\Delta\varphi$ or round-trip time delay τ from sampled data will be presented in Chap. 3. Chapter 3 also covers the imaging characteristics of ToF depth cameras and gives an operating paradigm of current commercial ToF depth cameras.

1.5 Book Overview

This introduction provides the basic notions for understanding the working principles of structured light and ToF depth cameras. Readers interested in a more organic and extensive coverage of computer vision are referred to textbooks on this topic, such as [15, 20, 22, 24, 26, 27, 32, 45, 48, 55, 57, 61].

The first part of this book, consisting of the next two chapters, describes in more detail the operation of structured light and ToF depth cameras. Chapter 2 describes structured light systems and their practical issues while Chap. 3 addresses CWAM ToF sensors. In both chapters, we adopt a general system approach emphasizing principles over circuit and implementation details and focus on the characteristics and limitations of the data provided by these technologies. The design of algorithms using depth data described in the rest of the book is directly related to the characteristics of such depth cameras.

---

[1] It is worth to notice that the phase φ_c of the carrier at the transmitter side is generally different from the phase φ_c' at the receiver. Both φ_c and φ_c' are usually unknown, especially in the case of a non-coherent process which is the typical practical solution. However, the system does not need to be aware of the values of φ_c and φ_c' and it is inherently robust to the lack of their knowledge.

The second part of this book focuses on processing the depth data produced by depth cameras in order to obtain accurate geometric information. The extraction of 3D information with the corresponding color data first requires the calibration of all the deployed acquisition devices and their accurate registration. Both photometric and geometric calibration of consumer depth cameras are discussed in detail in Chap. 4, which introduces the calibration of single devices (standard cameras and depth cameras) and then addresses joint calibration between multiple depth and standard cameras. Joint calibration of color and depth cameras is the starting point for super-resolution and data fusion algorithms exploiting color data to improve the information of current depth cameras treated in Chap. 5. Given the relatively recent appearance of consumer depth cameras it is no surprise that the current quality of their depth data may not be comparable with those of images provided by today's digital cameras or video cameras. For this reason, the idea of combining low resolution and high noise depth data with high resolution and low noise images is intriguing. Chapter 5 presents methods for improving the characteristics of an original depth data stream with the assistance of one or two standard cameras. When two standard cameras, acting as a stereo vision system, are used to assist a depth camera, there are two independent sources of depth information. This problem can be formulated as a data fusion task, i.e., combining two or more depth descriptions of the same scene captured by different systems, with the purpose of delivering a unique output stream with characteristics improved with respect to those of the original inputs. Various approaches for this task are presented, based on both fast local schemes, suited to real-time operation, and on more refined global optimization procedures.

The third part of the book addresses various applications where consumer depth cameras give significant contributions. Scene segmentation, i.e., the recognition of the regions corresponding to different objects, is addressed first in Chap. 6, since it is used as pre-processing for many other applications. This is a classical problem not completely solved by way of images after decades of research. The use of depth data together with color can drastically simplify the segmentation task and delivers segmentation tools which outperform segmentation techniques based on a single clue only, either depth or color. Scene segmentation can be divided in three main categories, corresponding to different tasks. The first concerns *video-matting*, namely the separation of foreground objects from the background, a key application for the film industry and many other fields. The second deals with standard segmentation, i.e., the subdivision of the acquired data into the various objects in the scene. The third concerns semantic segmentation, where a set of labels describing the scene is also associated with the segmented regions. Chapter 7 introduces the problem of 3D reconstruction from depth cameras data. Various solutions to adapt the classic 3D modeling pipeline to the peculiar characteristics of these data are discussed in detail. Chapter 8 presents body and hand pose estimation methods, a driving application for the development of consumer depth cameras. This chapter first introduces the basic principles and models for pose representation, then presents pose estimation methods based on a single frame. Finally, it addresses tracking methods to estimate the pose among multiple frames by propagating the information in the temporal domain. The critical issues differentiating hand pose estimation from body pose estimation are also mentioned in the same chapter. Gesture recognition is the subject of Chap. 9, which is divided in two main parts. The first addresses static gesture recognition and the second deals with dynamic gestures. Many different feature descriptors and machine learning techniques for these tasks are discussed in detail.

References

1. Faro, http://faro.com Accessed March 2016
2. Iee, http://www.iee.lu Accessed March 2016
3. Intel RealSense, www.intel.com/realsense Accessed March 2016
4. Leica, http://hds.leica-geosystems.com Accessed March 2016
5. Mesa imaging, http://www.mesa-imaging.ch Accessed March 2016
6. OpenCV, http://opencv.org Accessed March 2016
7. Pmd technologies, http://www.pmdtec.com/ Accessed March 2016
8. Riegl, http://www.riegl.com/ Accessed March 2016
9. Velodyne lidar, http://www.velodynelidar.com Accessed March 2016
10. Zoller and Frolich, http://www.zf-laser.com/ Accessed March 2016
11. J. Andrews, N. Baker, Xbox 360 system architecture. IEEE Micro **26**(2), 25–37 (2006)
12. C.S. Bamji, P. O'Connor, T. Elkhatib, S. Mehta, B. Thompson, L.A. Prather, D. Snow, O.C. Akkaya, A. Daniel, A.D. Payne, T. Perry, M. Fenton, V.-H. Chan, A 0.13 um cmos system-on-chip for a 512 × 424 time-of-flight image sensor with multi-frequency photo-demodulation up to 130 mhz and 2 gs/s adc. IEEE J. Solid-State Circuits **50**(1), 303–319 (2015)
13. Y. Bar-Shalom, *Tracking and Data Association* (Academic Press Professional, Inc., San Diego, CA, 1987)
14. F. Bernardini, H.E. Rushmeier, The 3d model acquisition pipeline. Comput. Graphics Forum **21**(2), 149–172 (2002)
15. G. Borenstein, *Making Things See: 3D Vision with Kinect, Processing, Arduino, and MakerBot* (Maker Media, O'Reilly Media Inc., Sebastopol, 2012)
16. J.Y. Bouguet, Camera calibration toolbox for matlab. http://www.vision.caltech.edu/bouguetj/calib_doc/. Accessed March 2016
17. J.Y. Bouguet, B. Curless, P. Debevec, M. Levoy, S. Nayar, S. Seitz, Overview of active vision techniques, in *Proceedings of ACM SIGGRAPH Workshop, Course on 3D Photography* (2000)
18. D. Claus, A.W. Fitzgibbon, A rational function lens distortion model for general cameras, in *Proceedings of IEEE Conference on Computer Vision and Pattern Recognition* (2005)
19. B. Curless, M. Levoy, A volumetric method for building complex models from range images, in *Proceedings of ACM SIGGRAPH* (New York, 1996), pp. 303–312
20. B. Cyganek, *An Introduction to 3D Computer Vision Techniques and Algorithms* (Wiley, New York, 2007)
21. D. Dardari, A. Conti, U. Ferner, A. Giorgetti, M.Z. Win, Ranging with ultrawide bandwidth signals in multipath environments. Proc. IEEE **97**(2), 404–426 (2009)
22. E.R. Davies, *Computer and Machine Vision*, 4th edn. (Academic, Boston, 2012)
23. J. Davis, D. Nehab, R. Ramamoorthi, S. Rusinkiewicz, Spacetime stereo: a unifying framework for depth from triangulation, in *Proceedings of IEEE Conference on Computer Vision and Pattern Recognition* (2003)
24. O. Faugeras, *Three-Dimensional Computer Vision: A Geometric Viewpoint* (MIT Press, Cambridge, 1993)

25. M.A. Fischler, R.C. Bolles, Random sample consensus: a paradigm for model fitting with applications to image analysis and automated cartography, in *Readings in Computer Vision: Issues, Problems, Principles and Paradigms*, vol. 1 (M. Kaufmann Publishers, Los Altos, CA, 1987), pp. 726–740
26. D.A. Forsyth, J. Ponce, *Computer Vision: A Modern Approach*. Prentice Hall Professional Technical Reference (Prentice Hall, London, 2002)
27. A. Fusiello, *Visione Computazionale. Tecniche di Ricostruzione Tridimensionale* (Franco Angeli, Milano, 2013)
28. A. Fusiello, E. Trucco, A. Verri, A compact algorithm for rectification of stereo pairs. Mach. Vis. Appl. **12**, 16–22 (2000)
29. C. Harris, M. Stephens, A combined corner and edge detector. in *Proceedings of Alvey Vision Conference* (1988), pp. 147–151
30. R.I. Hartley, In defense of the eight-point algorithm. IEEE Trans. Pattern Anal. Mach. Intell. **19**(6), 580–593 (1997)
31. R.I. Hartley, P. Sturm, Triangulation, in *Procedings of ARPA Image Understanding Workshop* (1994)
32. R.I. Hartley, A. Zisserman, *Multiple View Geometry in Computer Vision* (Cambridge University Press, Cambridge, 2004)
33. J. Heikkila, O. Silven, A four-step camera calibration procedure with implicit image correction, in *Proceedings of IEEE Conference on Computer Vision and Pattern Recognition* (1997)
34. A. Heyden, K. Astrom, Euclidean reconstruction from constant intrinsic parameters. in *Proceedings of International Conference on Pattern Recognition*, pp. 339–343
35. T.S. Huang, O. Faugeras, Some properties of the E matrix in two-view motion estimation. IEEE Trans. Pattern Anal. Mach. Intell. **11**(12), 1310–1312 (1989)
36. K. Konolige, Projected texture stereo, in *Proceedings of IEEE International Conference on Robotics and Automation* (2010)
37. K. Konolige, Sparse sparse bundle adjustment, in *Proceedings of British Machine Vision Conference* (BMVA Press, Aberystwyth, 2010), pp. 102.1–102.11
38. R. Lange, 3D Time-of-flight distance measurement with custom solid-state image sensors in CMOS/CCD-technology, Ph.D. thesis, University of Siegen (2000)
39. M. Levoy, K. Pulli, B. Curless, S. Rusinkiewicz, D. Koller, L. Pereira, M. Ginzton, S. Anderson, J. Davis, J. Ginsberg, J. Shade, D. Fulk, The digital michelangelo project: 3d scanning of large statues, in *Proceedings of ACM SIGGRAPH* (Addison-Wesley Publishing Co., New York, 2000), pp. 131–144
40. H. Li, R. Hartley, Five-point motion estimation made easy, in *Proceedings of International Conference on Pattern Recognition* (2006), pp. 630–633
41. C. Loop, Z. Zhang, Computing rectifying homographies for stereo vision, in *Proceedings of IEEE Conference on Computer Vision and Pattern Recognition*, (1999), p. 131
42. D.G. Lowe, Distinctive image features from scale-invariant keypoints. Int. J. Comput. Vis. **60**(2), 91–110 (2004)
43. B.D. Lucas, T. Kanade, An iterative image registration technique with an application to stereo vision, in *Proceedings of International Joint Conference on Artificial Intelligence* (Morgan Kaufmann Publishers Inc., San Francisco, CA, 1981), pp. 674–679
44. Q.T. Luong, O.D. Faugeras, The fundamental matrix: theory, algorithms, and stability analysis. Int. J. Comput. Vis. **17**, 43–75 (1995)
45. Y. Ma, S. Soatto, J. Kosecka, S.S. Sastry, *An Invitation to 3-D Vision: From Images to Geometric Models* (Springer, Berlin, 2003)
46. P.R.S. Mendonca, R. Cipolla, A simple technique for self-calibration in *Proceedings of IEEE Conference on Computer Vision and Pattern Recognition* (1999), p. 505
47. K. Mikolajczyk, C. Schmid, A performance evaluation of local descriptors. IEEE Trans. Pattern Anal. Mach. Intell. **27**(10), 1615–1630 (2005)
48. S.J.D. Prince, *Computer Vision: Models, Learning, and Inference*, 1st edn. (Cambridge University Press, New York, 2012)

49. L. Robert, O. Faugeras, Relative 3d positioning and 3d convex hull computation from a weakly calibrated stereo pair, in *Proceedings of International Conference on Computer Vision* (1993), pp. 540–544
50. J. Salvi, J. Pagès, J. Batlle, Pattern codification strategies in structured light systems. Pattern Recogn. **37**, 827–849 (2004)
51. D. Scharstein, R. Szeliski, A taxonomy and evaluation of dense two-frame stereo correspondence algorithms. Int. J. Comput. Vis. **47**(1–3), 7–42 (2001)
52. R. Schwarte et al., Pseudo-noise (pn) laser radar without scanner for extremely fast 3d-imaging and navigation, in *Proceedings of Microwave and Optronics Conference* (1997)
53. N. Snavely, Bundler, http://www.cs.cornell.edu/~snavely/bundler/. Accessed March 2016
54. N. Snavely, S.M. Seitz, R. Szeliski, Modeling the world from internet photo collections. Int. J. Comput. Vis. **80**(2), 189–210 (2008)
55. G. Stockman, L.G. Shapiro, *Computer Vision*, 1st edn. (Prentice Hall PTR, Upper Saddle River, 2001)
56. D. Stoppa, F. Remondino (eds.), *TOF Range-Imaging Cameras* (Springer, Berlin, 2012)
57. R. Szeliski, *Computer Vision: Algorithms and Applications* (Springer, New York, 2010)
58. C. Tomasi, T. Kanade, Detection and tracking of point features. Technical report, International Journal of Computer Vision (1991)
59. B. Triggs, P.F. McLauchlan, R.I. Hartley, A.W. Fitzgibbon, Bundle adjustment - a modern synthesis, in *Proceedings of ICCV Workshop, Vision Algorithms: Theory and Practice* (Springer, London, 2000), pp. 298–372
60. M. Trobina, Error model of a coded-light range sensor. Technical report, Communication Technology Laboratory Image Science Group, ETH-Zentrum (1995)
61. E. Trucco, A. Verri, *Introductory Techniques for 3-D Computer Vision* (Prentice Hall PTR, Upper Saddle River, 1998)
62. Z. Xu, *Investigation of 3D-Imaging Systems Based on Modulated Light and Optical RF-Interferometry* (Shaker Verlag GmbH, Aachen, 1999)
63. Z. Zhang, T. Kanade, Determining the epipolar geometry and its uncertainty: a review. Int. J. Comput. Vis. **27**, 161–195 (1998)
64. L. Zhang, B. Curless, S.M. Seitz, Rapid shape acquisition using color structured light and multi-pass dynamic programming, in *Proceedings of IEEE International Symposium on 3D Data Processing, Visualization, and Transmission* (2002), pp. 24–36
65. L. Zhang, B. Curless, S.M. Seitz, Spacetime stereo: shape recovery for dynamic scenes, in *Proceedings of IEEE Conference on Computer Vision and Pattern Recognition* (2003)

Part I
Operating Principles of Depth Cameras

Chapter 2
Operating Principles of Structured Light Depth Cameras

The first examples of structured light systems appeared in computer vision literature in the 1990s [5–8, 20] and have been widely investigated since. The first consumer-grade structured light depth camera products only hit the mass market in 2010 with the introduction of the first version of Microsoft Kinect™, shown in Fig. 2.1, based on the Primesensor™ design by Primesense. The Primesense design also appeared in other consumer products, such as the Asus X-tion [1] and the Occipital Structure Sensor [3], as shown in Fig. 2.2. Primesense was acquired by Apple in 2013, and since then, the Occipital Structure Sensor has been the only structured light depth camera in the market officially based on the Primesense design. Recently, other structured light depth cameras reached the market, such as the Intel RealSense F200 and R200 [2], shown in Fig. 2.3.

As explained in Sect. 1.3, the configuration of structured light systems is flexible. Consumer depth cameras can have a single camera, as in the case of Primesense products and the Intel F200, or two cameras, as in the case of the Intel R200 (and of the so called space-time stereo systems [9, 22], which will be shown later to belong in the family of structured light depth cameras).

This chapter, following the approach of Davis et al. [9], introduces a unified characterization of structured light depth cameras in order to present existing systems as different members of the same family. The first section of this chapter introduces camera virtualization, the concept that the presence of one or more cameras does not introduce theoretical differences in the nature of the measurement process. Nevertheless, the number of cameras has practical implications, as will be seen. The second section provides various techniques for approaching the design of the illuminator. The third section examines the most common non-idealities one must take into account for the design and usage of structured light systems. The fourth section discusses the characteristics of the most common commercial structured light systems within the introduced framework.

P. Zanuttigh et al., *Time-of-Flight and Structured Light Depth Cameras*,
DOI 10.1007/978-3-319-30973-6_2

Fig. 2.1 Microsoft Kinect™ v1

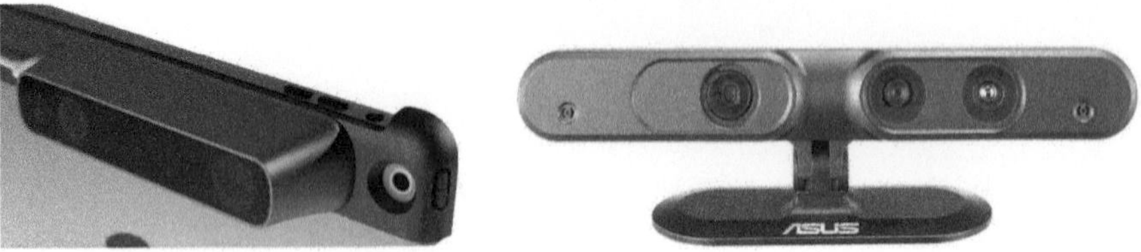

Fig. 2.2 Products based on the Primesense design: structure sensor by Occipital (*left*) and Asus X-tion Pro Live (*right*)

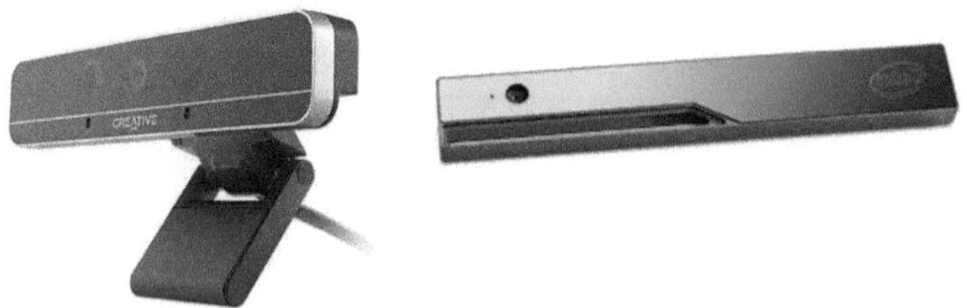

Fig. 2.3 Intel RealSense F200 (*left*) and R200 (*right*)

2.1 Camera Virtualization

In order to introduce the camera virtualization concept, consider Fig. 2.4, in which p_A is projected on three surfaces at three different distances: z_1, z_2 and z_3. For each distance z_i, according to (1.8), the point is framed by C at a different image location $\mathbf{p}_{C,i} = [u_{C,i}, v_C]$ with $v_C = v_A$ and $u_{C,i} = u_A + d_i$ with disparity $d_i = bf/z_i$, in which b is the baseline of the camera-projector system and f is the focal length of the camera and the projector (since they coincide in the case of a rectified system) expressed in pixel units.

The disparity of each pixel p_{C_i} can be expressed as a disparity difference or relative disparity with respect to a selected disparity reference. In particular, if the selected disparity reference is $d_{REF} = d_2$, the values of d_1 and d_3 can be expressed

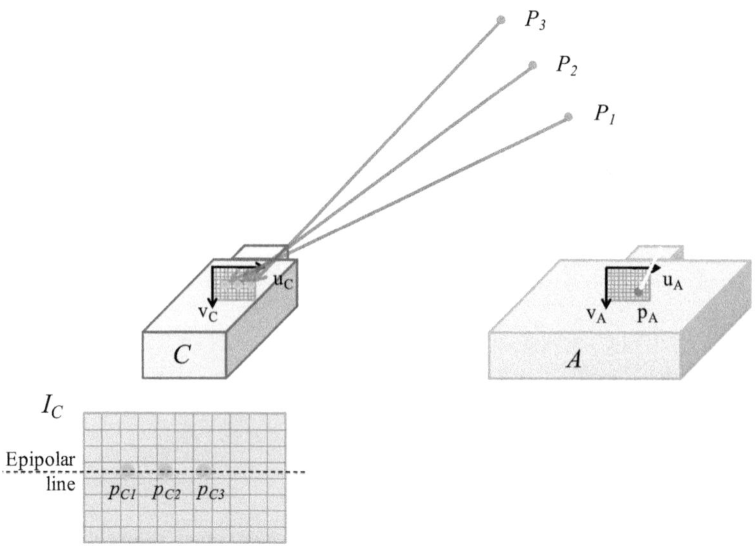

Fig. 2.4 The ray of the projected pattern associated with pixel p_A intersects the scene surface at points P_1, P_2 and P_3 placed at different distances, and is reflected back to pixels p_{C_1}, p_{C_2} and p_{C_3} of the image acquired by the camera

with respect to d_2 as signed difference $d_{REL_1} = d_1 - d_2$ and $d_{REL_3} = d_3 - d_2$. Given the value of $z_{REF} = z_2$ and of d_{REL_1} and d_{REL_3}, the value of z_1 and of z_3 can be computed as

$$\begin{aligned} \Delta z_i &= \frac{1}{\dfrac{1}{z_2} + \dfrac{d_{REL_i}}{bf}} - z_2 \\ z_i &= z_2 + \Delta z_i, \quad i = 1, 3. \end{aligned} \tag{2.1}$$

Equation (2.1) shows how to compute the scene depth z_i for every pixel p_{C_i} of the camera image I_C from a reference depth z_{REF} and relative disparity values d_{REL_i}, taken with respect to the reference disparity value d_{REF} ($d_{REF} = d_2$ in our example).

Note that if the absolute disparity range for the structured light system is $[d_{min}, d_{max}]$, generally with $d_{min} = 0$ (and definitely with $d_{min} \geq 0$) the relative disparity range with respect to the reference disparity d_{REF} becomes $[d_{RELmin}, d_{RELmax}] = [d_{min} - d_{REF}, d_{max} - d_{REF}]$. Also, while relative disparity d_{REL} is allowed to be negative, its absolute counterpart d is strictly non-negative in accordance to the rules of epipolar geometry.

The generalization of the idea behind the above example leads to the so called *camera virtualization* i.e., a procedure hinted in [9], by which a structured light depth camera made by a single camera and an illuminator operates equivalently to a structured light depth camera made by two rectified cameras and an illuminator.

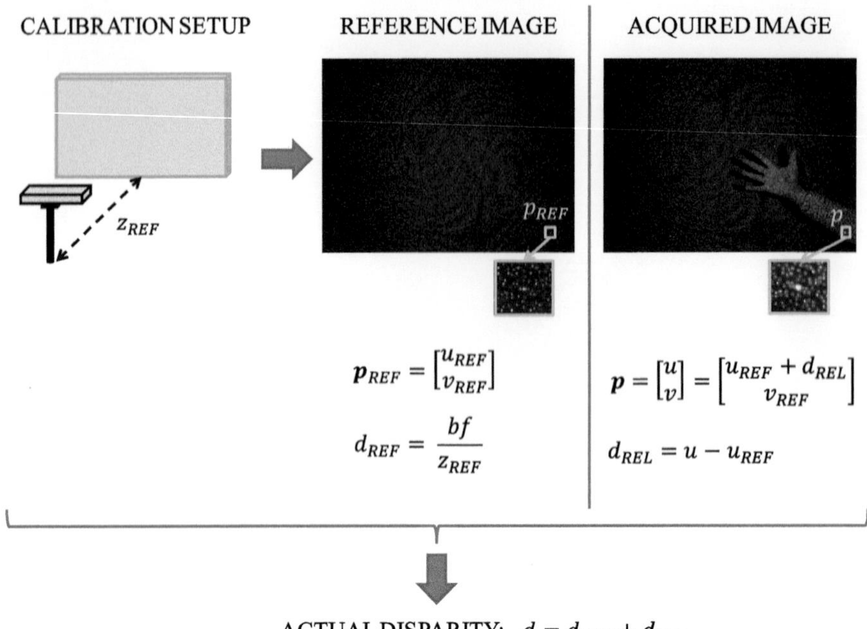

Fig. 2.5 Illustration of the reference image usage: (*left*) The structured light system, in front of a flat surface at known distance z_{REF}; (*middle*) reference image and computation of d_{REF} from z_{REF}; (*right*) generic scene acquired with pixel coordinates referring to the reference image coordinates

Camera virtualization, schematically shown in Fig. 2.5, assumes a *reference image* I_{REF} concerning a plane at known reference distance z_{REF}. The procedure requires one to associate each point p_{REF} with coordinates $\mathbf{p}_{REF} = [u_{REF}, v_{REF}]^T$ of I_{REF} to the corresponding point p of image I_C acquired by camera C and to express its coordinates with respect to those of p_{REF}, as $\mathbf{p} = [u, v]^T = [u_{REF} + d_{REL}, v_{REF}]$. In this way the actual disparity value d of each scene point given by (1.16) can be obtained by adding d_{REF} to the relative disparity value d_{REL} directly computed from the acquired image I_C, i.e.,

$$d = d_{REF} + d_{REL}. \tag{2.2}$$

Furthermore, from (1.17) and (2.2), comparison

$$\begin{aligned} u_C &= u_A + d \\ u_{REF} &= u_A + d_{REF} \end{aligned} \tag{2.3}$$

gives

$$u_C - u_{REF} = d - d_{REF} = d_{REL} \tag{2.4}$$

or

$$u_{REF} = u_C - d_{REL}. \tag{2.5}$$

Equation (2.5) has the same structure of (1.16) and is the desired result, since it indicates that single camera light systems like the one in Fig. 2.4 operate identically to standard stereo systems with a real acquisition camera C as the left camera and a "virtual" camera C' as the right camera co-positioned with the projector. Camera C' has the same intrinsic parameters of camera C.

The same conclusion can be reached less formally but straightforwardly by just noting that the shifting of reference image I_{REF} by its known disparity value d_{REF} gives an image which would be acquired by a camera C' with the same characteristics of C, co-positioned with the projector.

With respect to the reference image, depending whether the projector projects one or more patterns, the reference representation I_{REF} may be made by one or more images.

In order to be stored in the depth camera memory and used for computing the disparity map at each scene acquisition, the reference image I_{REF} can be either acquired by an offline calibration step or just computed by a virtual projection/acquisition based on the mathematical model of the system.

Direct acquisition of reference image I_{REF} represents an implicit and effective way of avoiding non-idealities and distortions due to C and A and their relative placements. A calibration procedure meant to accurately estimate the position of the projector with respect to the camera by a set of acquisitions of a checkerboard with the projected pattern superimposed is provided by Zhang et al. [21].

The advantage of a reference image I_{REF} generated by a close-form method due to a mathematical projection/acquisition model is that it does not need to be stored in memory and may be re-computed on-the-fly as in the case of a temporal pattern in which the coordinates of each pixel are directly encoded into the pattern itself.

Camera virtualization plays a fundamental conceptual role since it decouples the structured light system geometry from the algorithms used on them: in other words, standard stereo algorithms can be applied to structured light systems whether they have one or two cameras, unifying algorithmic methods for passive and active methods independently from the geometric characteristics of the latter.

A natural question prompted by the above observation is, given the complete operational equivalence between single camera and double camera systems, why should one use two cameras instead of one? The reason is that, although the presence of a second physical camera may seem redundant, in practice it leads to several system design advantages. The existence on the market of both single camera and two cameras systems is an implicit acknowledgment of this fact. For instance, the Intel RealSense F200 and the Primesense cameras belong to the first family, while the Intel RealSense R200 camera belongs to the second family.

The usage of two cameras leads to better performance because it simplifies the handling of many system non-idealities and practical issues, such as the distortion of the acquired pattern with respect to the projected one due to camera and projector non-idealities and to their relative alignment. Furthermore, in order to benefit from the virtual camera methodology, the projected pattern should maintain the same geometric configuration at all times. This requirement can be demanding for camera

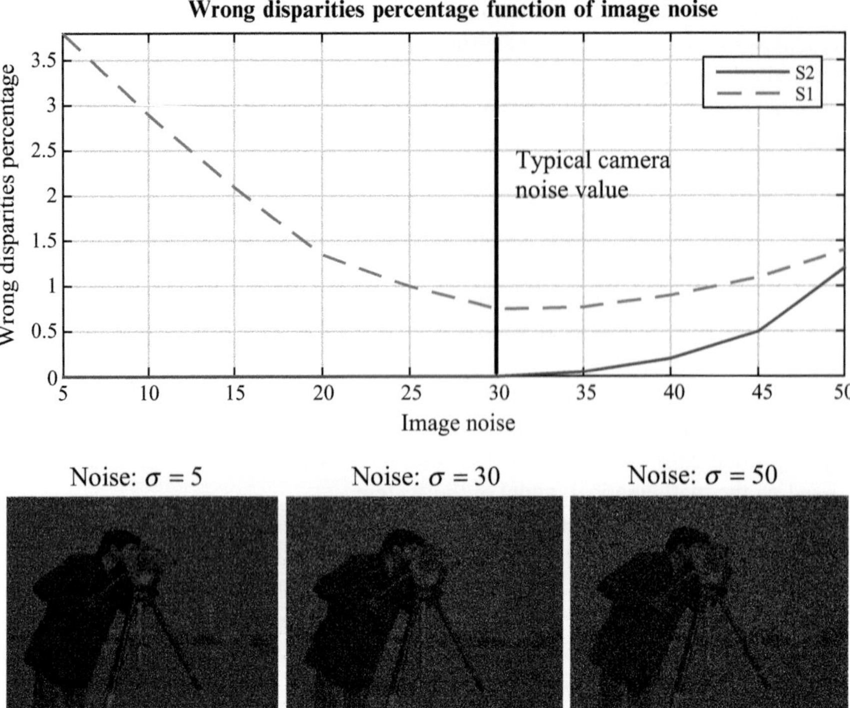

Fig. 2.6 Simulation of the performance of a single camera structured light system projecting the Primesense pattern (S1) and of a double-camera structured light system projecting the Primesense pattern (S2) for a flat scene textured by the "Cameraman" image at various noise levels

systems with an illuminator based on laser technology, because the projected pattern tends to vary with the temperature of the projector. For this reason, an active cooling system is used in the Primesense single camera system design, while it is unnecessary in the two cameras Intel RealSense R200.

Another fundamental weakness of single camera systems is that any ambient illumination at acquisition time leads to a difference between the appearance of the acquired representation and that of the reference representation. This effect, most evident in outdoor scenarios, can be exemplified by the following simulation with a test scene made by a flat wall textured by an image, e.g., the standard "Cameraman" of Fig. 2.6. This scene offers a straightforward depth ground truth which is a constant value everywhere if the structured light system is positioned in a fronto-parallel situation with respect to the wall (i.e., if the optical axis of the rectified system cameras and projector are assumed orthogonal to the wall). At the same time the scene texture helps the selection of matching points in stereo algorithms. With respect to the above scene, let us computationally simulate a structured light system projecting the Primesense pattern with a single acquisition camera, like in commercial products, and a structured light system projecting the

Primesense pattern but carrying two acquisition cameras instead of just one. For simplicity we will call S1 the former and S2 the latter.

As a first approximation, the scene brightness can be considered proportional to the reflectance and illumination made by a uniform component (background illumination) and by a component due to the Primesense pattern. In the case of S1, in order to mimic camera virtualization we consider two acquisitions of a shifted version of "Cameraman", while in S2 we consider only one acquisition per camera, and compare them with respect to the actually projected pattern.

The acquisitions with S1 and S2 are repeated using versions of the "Cameraman" images corrupted by independent additive Gaussian noise with different standard deviations.

Determining which of the two systems performs a better disparity estimation can be easily ascertained from the percentage of non constant, i.e., wrong depth values (in this case produced by a block-matching stereo algorithm with window size 9×9) as a function of the independent additive Gaussian camera noise, as shown in Fig. 2.6 for S1 and S2. The performance of the depth estimation procedure of S1 (red) is worse than the one of S2 (blue), especially for typical camera noise values (green line). In spite of its simplicity, this simulation provides an intuitive understanding of the approximations associated with the presented camera virtualization technique.

In order to cope with the above mentioned illumination issues, single camera structured light systems adopt a notch optical filter on the camera lenses with a band-pass bandwidth tightly matched to that of the projected pattern. Moreover, in the case of extremely high external illumination in the projector's range of wavelengths, a double camera structured light depth camera can be used as a standard stereo system, either by neglecting or switching off the contribution of the active illuminator A.

For both a physical or virtual second camera, the disparity estimation with respect to the reference image I_{REF} corresponds to a computational stereopsis procedure between two rectified images [16, 18]. Given this, one can continue to use *local algorithms*, i.e., methods which consider a measure of the local similarity (covariance) between all pairs of possible conjugate points on the epipolar line and simply select the pair that maximizes it, as observed in Sect. 1.2. The global methods mentioned in Sect. 1.2.1.3 that do not consider each couple of points on their own but exploit global optimization schemes are generally not used with structured light systems.

From now on, in light of all the considerations provided about camera virtualization, this book will only consider structured light depth cameras made by two cameras and an illuminator, with the understanding that any reasoning presented in the remainder of the book also applies to the case of single camera structured light systems.

2.2 General Characteristics

Let us refer for simplicity to the single camera structured light system of Fig. 1.11 in order to recall the operation of structured light systems: each pixel of the projector is associated with a specific local configuration of the projected pattern called *code word*. The pattern undergoes projection by A, reflection by the scene and capture by C. A correspondence algorithm analyzes the code words received by the acquired image I_C in order to compute the conjugate p_c of each pixel p_A of the projected pattern. The goal of pattern design (i.e., code word selection) is to adopt code words effectively decodable even in the presence of non-idealities of the pattern projection/acquisition process pictorially indicated in Fig. 1.11 and explained in the next section. Figure 2.7 shows an example of the data acquired by a single camera structured light system projecting the Primesense pattern.

The data acquired by a structured light depth camera are:

- the images I_C and $I_{C'}$ acquired by cameras C and C', respectively, defined on the lattices Λ_C and $\Lambda_{C'}$ associated with cameras C and C'. The axes that identify Λ_C coincide with u_C and v_C of Fig. 1.11. The axes of $\Lambda_{C'}$ will similarly refer to those of u'_C and v'_C on the virtual or actual camera C'. The values of I_C and $I_{C'}$ belong to interval [0, 1]. Images I_C and $I_{C'}$ can be considered a realization of the random fields $\mathcal{I}_C$ and $\mathcal{I}_{C'}$ defined on Λ_C and $\Lambda_{C'}$, with values in [0, 1]. In single camera structured light systems, like in the case of the Primesense depth camera, C' is a virtual camera, and the image $I_{C'}$ is not available.

The data available at the output of a structured light depth camera are:

- The estimated disparity map, called $\hat{D}_C$, is defined on the lattice Λ_C associated with the C sensor. The values of $\hat{D}_C$ belong to interval $[d_{min}, d_{max}]$, where d_{min} and d_{max} are the minimum and maximum allowed disparity values. Disparity map $\hat{D}_C$ can be considered a realization of a random field $\mathcal{D}_C$ defined on Λ_C, with values in $[d_{min}, d_{max}]$.
- The estimated depth map computed by applying (1.26) to $\hat{D}_C$, called $\hat{Z}_C$, is defined on the lattice Λ_C associated with camera C. The values of $\hat{Z}_C$ belong to the interval $[z_{min}, z_{max}]$, where $z_{min} = bf/d_{max}$ and $z_{max} = bf/d_{min}$ are the

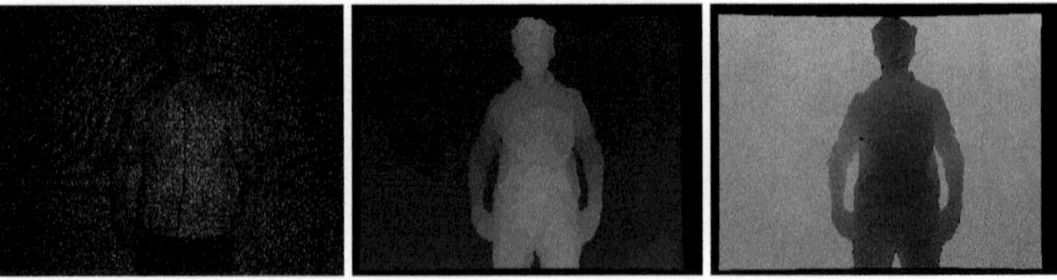

Fig. 2.7 Example of I_C, $\hat{D}_C$ and $\hat{Z}_C$ acquired by the Primesense depth camera

minimum and maximum allowed depth values, respectively. Depth map $\hat{Z}_C$ can be considered as a realization of a random field $\mathscr{Z}_C$ defined on Λ_C, with values in $[z_{min}, z_{max}]$.

2.2.1 Depth Resolution

Since structured light depth cameras are based on triangulation, they have the same depth resolution model as that of standard stereo systems. In particular, their depth resolution Δz can be computed as

$$\Delta z = \frac{z^2}{bf} \Delta d \tag{2.6}$$

where Δd is the disparity resolution. Equation (2.6) shows that the depth resolution is quadratically dependent on the depth of the measured object (i.e., its z coordinate).

Disparity resolution Δd can be 1 in the case of pixel resolution or less than 1 in the case of sub-pixel resolution. Techniques for sub-pixel disparity estimation are well-known in stereo literature [16, 18] and they can also be applied to structured light systems. For example, according to the analysis of Konoldige and Mihelich [11], the Kinect™ v1 uses a sub-pixel refinement process with interpolation factor 8, hence $\Delta d = 1/8$, and according to [17] with baseline $b = 75\,[\text{mm}]$ and focal length approximately $f = 585.6\,[\text{pxl}]$. Figure 2.8 shows a plot of the depth resolution of a system with the same parameters as those of Kinect™ v1 and various sub-pixel precisions.

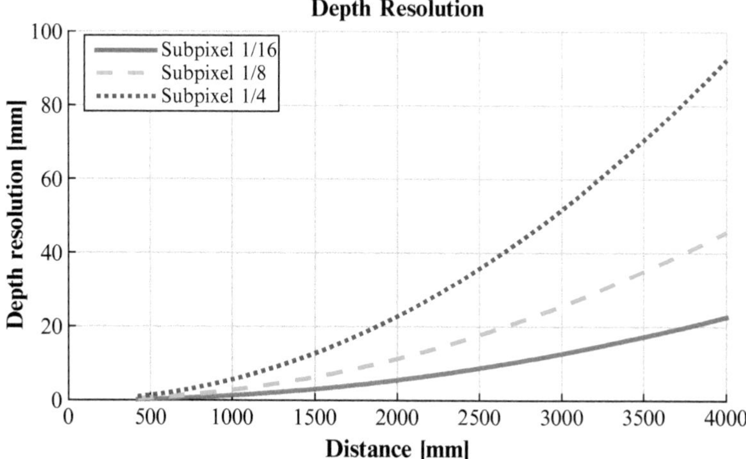

Fig. 2.8 Depth resolution according to (2.6) for systems with baseline $b = 75\,[\text{mm}]$ focal length $f = 585.6\,[\text{pxl}]$ and sub-pixel precisions: $\Delta d = 1/16$, $1/8$ and $1/4$

2.3 Illuminator Design Approaches

The objective of structured light systems is to simplify the correspondence problem through projecting effective patterns by the illuminator A. This section reviews current pattern design methodologies. In addition, the specific design of the illuminator as well as its implementation are at the core of all structured light depth cameras.

A code word alphabet can be implemented by a light projector considering that it can produce n_P different illumination values called *pattern primitives* (e.g., $n_P = 2$ for a binary black-and-white projector, $n_P = 2^8$ for a 8-bit gray-scale projector, and $n_P = 2^{24}$ for a RGB projector with 8-bit color channels). The local distribution of a pattern for a pixel p_A is given by the illumination values of the pixels in a window around p_A. If the window has n_W pixels, there are $n_P^{n_W}$ possible pattern configurations on it. From the set of all possible configurations, N configurations need to be chosen as code words. What is projected to the scene and acquired by C is the pattern resulting from the code words relative to all the pixels of the projected pattern. Let us assume that the projected pattern has $N_R^A \times N_C^A$ pixels p_A^i, $i = 1, \dots, N_R^A \times N_C^A$ where N_R^A and N_C^A are the number of rows and columns of the projected pattern, respectively.

The concept of pattern uniqueness is an appropriate starting point to introduce the various approaches for designing illuminator patterns. Consider an ideal system in which images I_C and $I_{C'}$ are acquired by a pair of rectified cameras C and C' (whether C' is real or virtual is immaterial for the subsequent discussion) and assume the scene to be a fronto-parallel plane corresponding to disparity 0 at infinity and infinite reflectivity. Since the cameras are rectified, let us recall from Sect. 1.2 that conjugate points p and p', i.e., points of I_C and $I_{C'}$ corresponding to the same 3D point P, are characterized by coordinates with the same v-component and u-components differing by disparity d: $\mathbf{p} = [u, v]^T$, $\mathbf{p}' = [u', v']^T = [u - d, v]^T$. The correspondences matching process searches the conjugate of each pixel p in I_C, by allowing d to vary in the range $[d_{min}, d_{max}]$ and by selecting the value $\hat{d}$ for which the local configuration of I_C around $\mathbf{p}$ is most similar to the local configuration of $I_{C'}$ around $\mathbf{p} - [d, 0]^T$ according to a suitable metric.

Images I_C and $I_{C'}$ can carry multiple information channels, for instance encoding data at different color wavelengths (e.g., R, G, B channels) or at multiple timestamps $t = 1, \dots, N$ with N being the timestamp of the most recent frame acquired by cameras C and C'. The local configuration in which the images are compared is a cuboidal window $W(\mathbf{p})$ made by juxtaposing windows centered at $\mathbf{p}$ in the different channels. If there is only one channel (with respect to time), the system is characterized by an instantaneous behavior and is called a *spatial stereo* system, according to [9]. On the contrary, if the matching window is characterized by a single-pixel configuration in the image (e.g., the window is only made by the pixel with coordinate $\mathbf{p}$) and by multiple timestamps, the system is called a *temporal stereo* system. If the matching window has both a spatial and temporal component, the system is called *space-time stereo*. A standard metric to compute the local similarity between I_C in the window $W(\mathbf{p})$ and $I_{C'}$ in the window $W(\mathbf{p}')$ is the Sum of Absolute Differences (SAD) of the respective elements in the two windows, defined as

$$SAD[I_C(W(\mathbf{p})), I_{C'}(W(\mathbf{p}'))] \triangleq \sum_{\mathbf{q}\in W(\mathbf{p}),\mathbf{q}'\in W(\mathbf{p}')} |I_C(\mathbf{q}) - I_{C'}(\mathbf{q}')|. \tag{2.7}$$

Since generally the windows W on which the SAD metric is computed are predefined, the value of the SAD metric is fully specified from $\mathbf{p}$ and d, as emphasized by notation

$$SAD_{I_C,I_{C'},W}(\mathbf{p}, d) \triangleq SAD[I_C(W(\mathbf{p})), I_{C'}(W(\mathbf{p}'))] \tag{2.8}$$

rewritten for simplicity just as $SAD(\mathbf{p}, d)$. For each pixel p one selects the disparity that minimizes the local similarity as $\hat{d}(\mathbf{p}) = \operatorname{argmin} SAD(\mathbf{p}, d)$. A pattern is said to be *unique* if in an ideal system, i.e., a system without any deviation from theoretical behavior, for each pixel p in the lattice of I_C, the value of the SAD metric of the actual estimated disparity d^* coincides with minimum $\hat{d}(\mathbf{p}) = \operatorname{argmin} SAD(\mathbf{p}, d)$, which is unique. The uniqueness U of a pattern is defined as

$$U \triangleq \min_{p\in\Lambda_C} U(\mathbf{p}) \tag{2.9}$$

where $U(\mathbf{p})$ is computed as the second argmin of the SAD metric, excluding the first argmin $\hat{d}(\mathbf{p})$ and the values within one disparity value from it, i.e.,

$$d \in \{d_{min}, \ldots, d_{max}\} \setminus \{\hat{d}(\mathbf{p}) - 1, \hat{d}(\mathbf{p}), \hat{d}(\mathbf{p}) + 1\}. \tag{2.10}$$

Let us further comment on the above definition of uniqueness. For each pixel in the image I_C the uniqueness map $U(\mathbf{p})$ is computed as the cost of the non-correct match that gives the minimum matching error. The higher such cost is, the more robust the pattern is against noise and non-idealities. The minimum uniqueness value across the entire pattern is selected in order to obtain a single uniqueness value for the entire pattern. This enforces the fact that the pattern should be unique everywhere in order to obtain a correct disparity estimation for each pixel, at least in the case of ideal acquisition. The minimum value of uniqueness for a pattern is 0. If a pattern has uniqueness greater than 0, it means that the pattern itself makes the conjugate correspondence detection problem a well-posed problem for each pixel in the pattern, otherwise the correspondence detection is ill-posed, at least for a certain number of pixels in the pattern.

In the above discussion, uniqueness was defined using SAD as a matching cost function, but uniqueness can be defined in terms of other types of metrics or matching costs, such as the Sum-of-Squared-Differences (SSD), Normalized Cross-Correlation (NCC), or the Hamming distance of the Census transform [20].

The choice of matching cost is generally system dependent. For simplicity's sake, in the rest of this book all uniqueness analysis will refer to standard *SAD* defined by (2.7).

2.3.1 Implementing Uniqueness by Signal Multiplexing

The just-defined concept of uniqueness is a function of the number of color channels, the range of values in the image representation, and the shape of the matching window, which may have both a spatial and temporal component. Following the framework of Salvi et al. [15], different choices of these quantities lead to different ways to encode the information used for correspondences estimation, typically within the following four signal multiplexing families:

- wavelength multiplexing;
- range multiplexing;
- temporal multiplexing;
- spatial multiplexing.

Each multiplexing technique performs some kind of sampling in the information dimension typical of the technique, limiting the reconstruction capability in the specific dimension. This concept is instrumental in order to understand the attributes of different structured light depth cameras according to the considered multiplexing techniques.

2.3.1.1 Wavelength Multiplexing

If the projected pattern contains light emitted at different wavelengths, i.e., different color channels, the system is characterized by wavelength-multiplexing. An example is a system with an illuminator projecting red, green, and blue light. Today, there is no commercial structured light depth camera implementing a wavelength multiplexing strategy, but such cameras have been widely studied, for instance in [10] and in [21]. Figure 2.9 shows an image of the projected pattern from [21].

This type of technique makes strong assumptions about the characteristics of the cameras and the reflectance properties of the framed scene. Notably, it assumes the

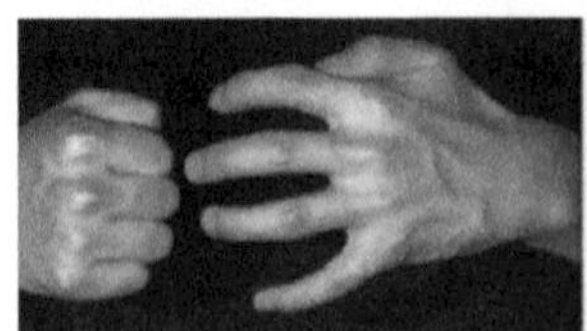 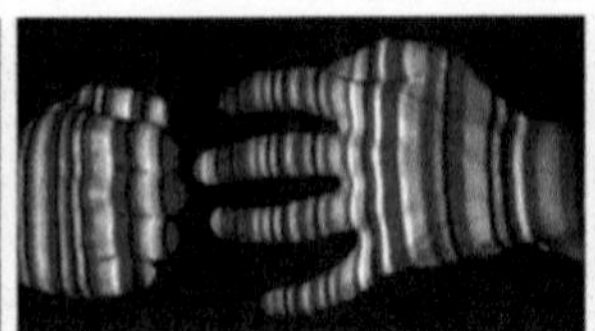 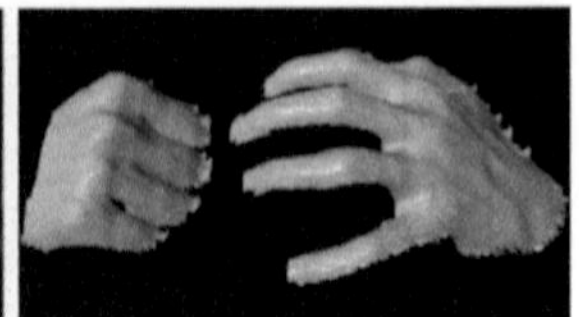

Fig. 2.9 Example of pattern projected by the illuminator of Zhang et al. (*center*) on a scene characterized by two hands (*left*), and the relative depth estimate (*right*). Courtesy of the authors of [21]

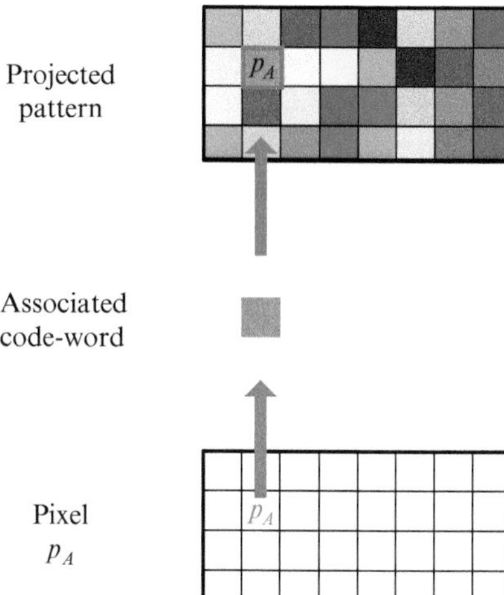

Fig. 2.10 Schematic representation of "direct coding", special case of wavelength multiplexing

camera pixels that collect different channels are not affected by inter-channel cross-talk, which is often present. In addition, the scene is assumed to have a smooth albedo distribution without abrupt reflectivity discontinuities. For an analysis of these assumptions, the interested reader is referred to [21].

Since wavelength-multiplexing approaches sample the wavelength domain, in general they limit the capability of the system to deal with high frequency reflectivity discontinuities at different wavelengths. Therefore, the depth estimates produced by these types of systems tend to be correct for scenes with limited albedo variation and external illumination, but not in other cases.

In *direct coding* [15] each pixel within a scanline is associated with a specific color value, as schematically shown in Fig. 2.10. Hence, direct coding is a special case of wavelength multiplexing where the disparity of a pixel can be estimated by matching windows of size 1.

In general, combining different multiplexing techniques together leads to more robust systems. An example of a system combining wavelength and temporal multiplexing techniques is described in [21].

2.3.1.2 Range Multiplexing

In the case of a single channel, the projected pattern range can be either binary (black or white, as in Primesense products or in the Intel RealSense F200) or characterized by multiple gray level values (as in the Intel RealSense R200). The case of range characterized by multiple gray levels is usually referred as range multiplexing. Figure 2.11 shows the pattern of the Intel RealSense R200, which uses a range multiplexing approach. Figure 2.17 compares the textured pattern produced by the Intel RealSense R200 with the pattern of the Primesense camera, which uses binary dots instead of range multiplexing.

(a) (b)

Fig. 2.11 Pattern projected by the Intel RealSense R200 camera: (**a**) full projected pattern; (**b**) a zoomed version of a portion of the pattern

Even though range multiplexing has interesting properties, it has not received as much attention as other multiplexing methods. In particular, range multiplexing does not require collimated dots, thus avoiding eye safety issues. Projecting grayscale texture allows one to gather more information in the matching cost computation step, in particular with some stereo matching techniques [20].

However, since image acquisition is affected by noise and other non-idealities, the local differences in the images may not exactly reflect the local differences of the projected pattern and such appearance difference may not be the same in the images I_C and $I_{C'}$ of the two cameras. Therefore, the range multiplexing information is at risk of being hidden by the noise and non-idealities of the camera acquisition process. Since range multiplexing samples the range of the projected pattern and of the acquired images, it limits the system's robustness with respect to the appearance non-idealities which may differ for the two cameras, especially in low SNR situations.

A major issue of range multiplexing is that different pixels of the projected texture have different illumination power. Consequently, dark portions of the pattern are characterized by lower power than bright areas. Combined with the fact that the optical power of the emitted pattern decreases with the square of the distance of the framed scene, it is not possible to measure far distances in correspondence to the dark areas of the pattern.

Different from the case of wavelength multiplexing, *direct coding* alone becomes impractical in the case of range multiplexing, as shown by the following example. Consider the case of a disparity range made by 64 disparities, where the projector uses 64 range values and the acquired image range, encoded with 8 bits, has 256 values. If one adopted "direct coding," the system would be robust to non-idealities up to 2 range values. Since the standard deviation of the thermal noise of a typical camera is in the order of 5–10 range values, the noise resilience of "direct coding" in the case of range multiplexing would clearly be inadequate. For this reason, range multiplexing is typically used in combination with other multiplexing techniques, such as *spatial multiplexing*, as in the case of the Intel RealSense R200 camera.

2.3.1.3 Temporal Multiplexing

Temporal multiplexing is a widely investigated technique originally introduced for still scenes [19] and subsequently extended to dynamic scenes in [22], where the illuminator projects a set of N patterns, one after the other. The patterns are typically made by vertical black and white stripes representing the binary values 0 and 1, as shown in Fig. 2.12 for $N = 3$, since in this case there is no need to enforce wavelength or range multiplexing as the system is assumed rectified.

This arrangement at time N ensures a different binary code word of length N for each line pixel p_A; all the rows have the same code words given the vertical symmetry of the scheme. The total number of available code words is 2^N.

In the case of acquisition systems made by a pair of real cameras, this method is typically called *space-time stereo*. For a comprehensive description of temporal multiplexing techniques see [15], while for details on space-time stereo see [9, 22]. The first example of a commercial camera for dynamic scenes based on temporal multiplexing was the Intel RealSense F200 camera.

Figure 2.13 shows an example of projected patterns and acquired images for space-time stereo system.

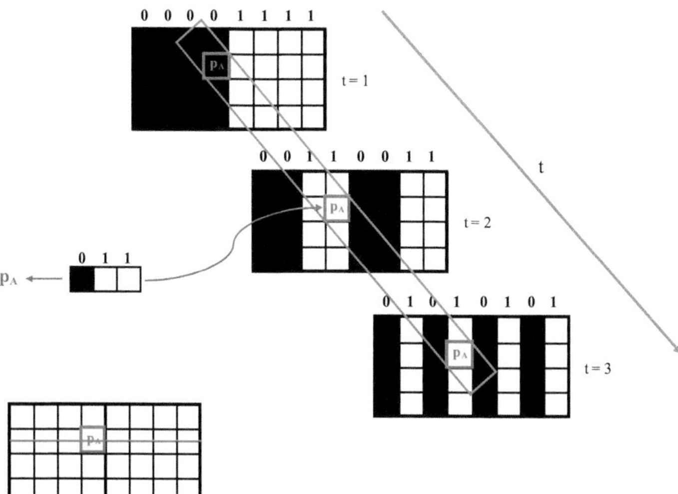

Fig. 2.12 Temporal multiplexing: vertical *black and white stripes* coding a pattern at multiple timestamps (with $N = 3$)

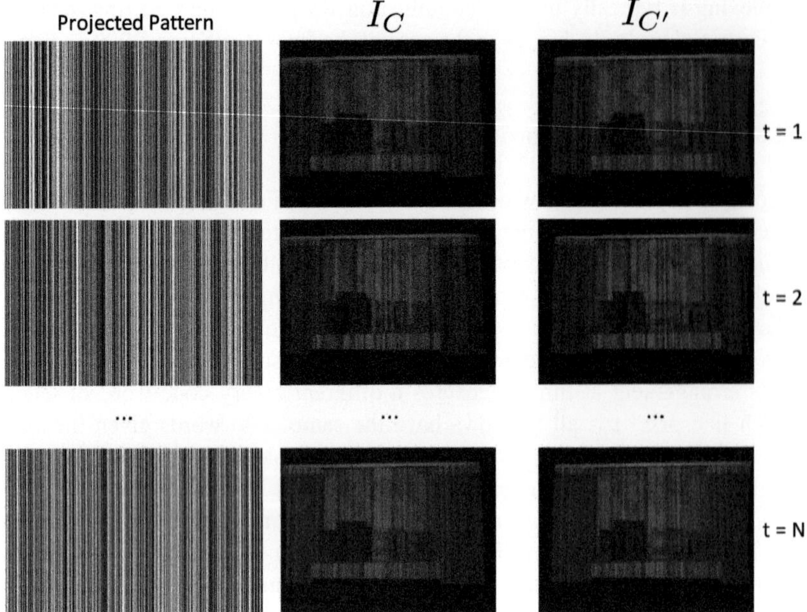

Fig. 2.13 Example of projected patterns and acquired images for temporal multiplexing

The critical component of temporal multiplexing is the method used in order to create the set of projected patterns. As described in [15], there exist several pattern design techniques. The most popular is one based on Gray codes. In Gray codes, each code word differs from the previous one by just one bit. If the length of the code word is N, it is possible to generate exactly 2^N uniquely decodable code words of a Gray code by the scheme pictorially shown in Fig. 2.14. In order to have a pattern with more values than the number of code words, it is possible to stack the patterns side by side. If 2^N is greater than the maximum disparity that the system can measure, the produced code maintains its uniqueness property. Note that this assumes there is one pixel of the projected pattern for each pixel of the acquired images. This hypothesis usually holds for well-engineered systems, but if not verified, one should adjust the reasoning in order to account for the specific properties of the system.

As suggested in [22], projected patterns for temporal multiplexing techniques should be characterized by high spatial frequency. However, as shown in Fig. 2.14, spatial frequencies differ for each pattern and increase for the bottom patterns. In order to address this issue and improve the performance of Gray coded patterns, the authors of [22] suggest shuffling the columns of the projected patterns as shown in Fig. 2.15 for $N = 7$. Note that the same shuffling sequence must be applied to all the patterns in order to preserve the uniqueness property characteristic of the Gray pattern. Another popular variation of Grey coding includes a post-refinement based on phase-shifting.

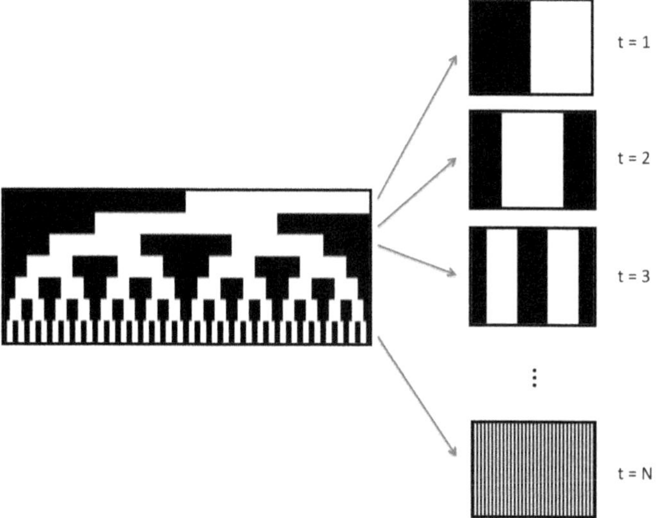

Fig. 2.14 Example of patterns for temporal multiplexing generated with Gray code, with code word length $N = 7$ suited to distinguish between 128 disparity levels

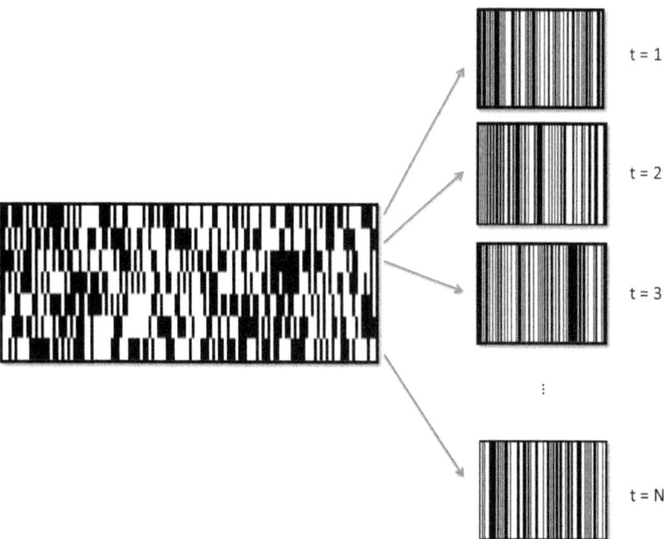

Fig. 2.15 Example of patterns obtained by permuting a Gray code with code word length $N = 7$ suited to distinguish between 128 disparity levels

Once a set of N patterns has been acquired by C and C' it is possible to compute the Gray code representation for each pixel in Λ_C and $\Lambda_{C'}$, by adding a 0 if the acquired image in the pixel is dark and a 1 if the acquired image is bright. This process can be performed by thresholding techniques or by more refined reasoning based on the local configuration of the pattern around the pixel, both in space and time. The computation of the Gray code representation of each pixel of I_C and $I_{C'}$ simplifies the correspondence estimation, performed by associating the pixels of the two images with the same Gray code (provided they satisfy the epipolar geometry constraint, i.e., they lie on the same epipolar line and lead to a disparity value within the valid disparity range). As previously mentioned, if the maximum disparity in the allowed range for the system is smaller than 2^N, the matching problem is well-defined and it is possible to obtain a correct match if the acquisition non-idealities allow one to distinguish between bright and dark measurements.

Temporal multiplexing techniques lead to precise pixel depth estimates (since there is a unique pixel-per-pixel match) and do not suffer the issues of wavelength or range multiplexing. However, they rely on the assumption that during the projection and acquisition of the set of N patterns, the scene remains static. In other words, since temporal multiplexing samples the information in time, it limits the system capability of acquiring depth information of scenes with temporal frequency variations. If the scene is not static during the acquisition of the N projected frames, artifacts occur in the estimated depth map, as shown in Fig. 2.16 for the Intel RealSense F200 camera.

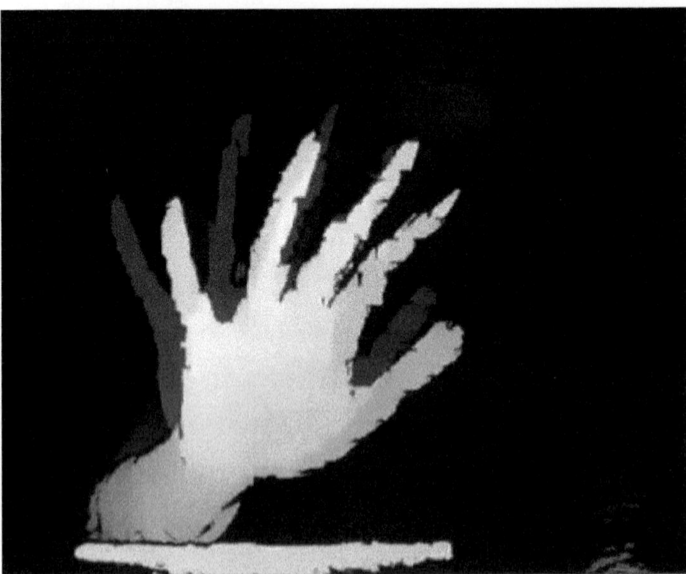

Fig. 2.16 Artifacts in the depth estimate of a moving hand acquired by the Intel RealSense F200 depth camera: the depth of the moving hand should only be the brightest silhouette, however a shadowed hand appears in the estimated depth map

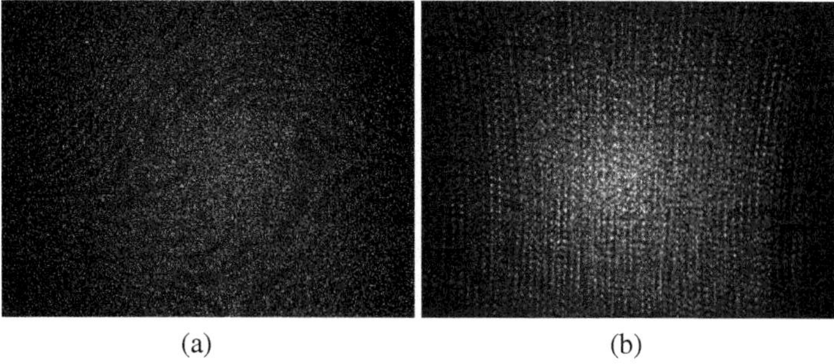

Fig. 2.17 Examples of patterns for spatial multiplexing: (**a**) pattern of the Primesense camera; (**b**) pattern of the Intel RealSense R200 camera

2.3.1.4 Spatial Multiplexing

Spatial multiplexing is the most widely used design approach for structured light illuminators, used for example in the Primesense Carmine camera that has been shipped in many forms, including the Kinect™ v1. While the Primesense design is characterized by a single real camera, other designs such as the Intel RealSense R200 camera and notably the non-commercial systems of Konolige [12] and Lim [13] carry two real cameras.

Similar to time multiplexing systems, in the case of two real cameras, spatial multiplexing is often referred to as active stereo vision or projected texture stereo. Furthermore, when the projected information is made by collimated dots, as in the case of Primesense, the projected light is usually referred to as pattern and as texture otherwise. Two examples of projected patterns for spatial multiplexing systems are shown in Fig. 2.17.

As already seen for time multiplexing, the most critical component design also for spatial multiplexing is the method used to generate the set of projected patterns. A number of techniques can be used for pattern design [15]; among them, those based on De Bruijn patterns received great attention. De Bruijn patterns are unique patterns that can be programmatically generated and eventually refined by imposing non-recurrency, as proposed in [13]. A further refinement can be obtained by stochastic optimization techniques that maximize uniqueness, as proposed in [12].

Let us explore the properties of a De Bruijn pattern and how to build one. Given an alphabet $\mathscr{A}$ with k symbols $\{a_1, \ldots, a_k\}$ and a positive integer n, a De Bruijn sequence $B(n, \mathscr{A})$ is a cyclic sequence of symbols in the alphabet in which all possible sequences of length n of symbols in $\mathscr{A}$ appear exactly once. A light pattern with values equal to those of a De Bruijn sequence is called a De Bruijn pattern. In a De Bruijn pattern, each window of length n within the pattern can be associated with a subsequence of length n of the underlying De Bruijn sequence. As previously recalled, in pattern design it is fundamental to guarantee uniqueness.

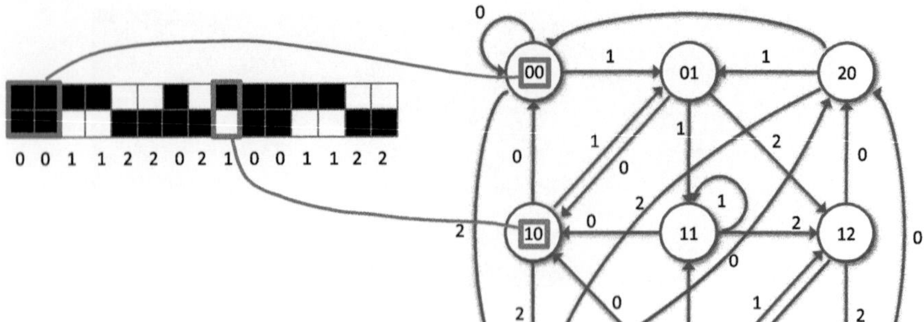

Fig. 2.18 Example of a De Bruijn graph associated with the projected pattern for the case of $B(2, \{0, 1, 2\})$

With De Bruijn sequences, the concept of uniqueness can be translated into the concept of uniqueness of all the sub-strings associated with the different windows in a pattern that satisfies the epipolar constraint, i.e., to all the windows of a pattern that lie on the same row of the pattern and are within the considered disparity range. Since a De Bruijn sequence is a cyclic sequence with the subsequences of length n appearing only once in the sequence itself, the uniqueness of the De Bruijn pattern is ensured [13]. Note that as in the case of temporal multiplexing, one needs to guarantee that patterns are unique only along the epipolar line and that their range is less than the number of disparities. A larger pattern needed to cover the entire field of view, while preserving enough spatial resolution can be obtained by tiling multiple patterns.

De Bruijn patterns can be systematically constructed by associating a graph to the string of the projection pattern and by associating the graph's nodes to each pattern sub-string. Edges connect the nodes when one node corresponds to the next window position of the other node in the scanning process, as pictorially represented in Fig. 2.18.

A De Bruijn graph $G_B(n, \mathscr{A})$ is a directed graph $\{V, E\}$, where V is a set of all possible length-n permutations of symbols in $\mathscr{A}$ and E is the set of directed edges in which the last $n - 1$ symbols of the source node coincide with the first $n - 1$ symbols of the sink node. The label associated with each edge in E is the last symbol of the code word associated with its sink node. A De Bruijn sequence $B(n, \mathscr{A})$ is obtained as a Hamiltonian cycle of $G_B(n, \mathscr{A})$. Examples of De Bruijn sequences are $B(2, \{0, 1\}) = 0011$ and $B(2, \{0, 1, 2\}) = 001122021$, which can be computed respectively with the De Bruijn graphs at the left and right of Fig. 2.19.

De Bruijn sequences characterized by alphabets with more than two symbols can be encoded in a non-binary pattern by associating each symbol in the alphabet to a gray value, or by associating each symbol in the alphabet with one column and multiple rows in a binary configuration, e.g., in the case of the Primesense pattern. The generated patterns can be tiled in order to obtain a projected pattern that satisfies

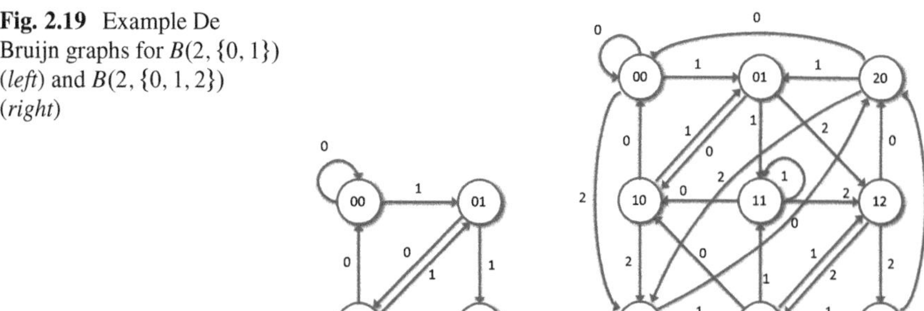

Fig. 2.19 Example De Bruijn graphs for $B(2, \{0, 1\})$ (*left*) and $B(2, \{0, 1, 2\})$ (*right*)

the epipolar constraint, if the horizontal size of each tile is greater than the maximum disparity in the system range. Figure 2.20 shows an example of a pattern generated by the proposed methodology based on De Bruijn sequences.

Fig. 2.20 Example of a pattern generated from a De Bruijn sequence, with alphabet size $k = 7$ and maximum disparity 96

Once a couple of images I_C and $I_{C'}$ from the rectified cameras C and C' are acquired, one can compute the conjugate $p' \in I_{C'}$ of each pixel $p \in I_C$ by standard block matching techniques because the properties of De Bruijn patterns ensure enough information for a unique match in correspondence of the correct disparity value. Of course this is true only as long as non-idealities in the projection and acquisition processes do not affect the pattern uniqueness. A very interesting property of De Bruijn patterns is that the matching window should have at least the size of the alphabet $\mathscr{A}$, which in the previous example is $k = 7$. If a larger window is considered, the pattern does not worsen its uniqueness properties.

Spatial multiplexing techniques allow for independent depth estimates at each frame, avoiding the problems typical of wavelength, range, and time multiplexing. The hypothesis on which spatial multiplexing rests is that the acquired scene may be well approximated by a fronto-parallel surface within the size of the block used in the block matching algorithm for the disparity computation. This is a classical assumption of block matching stereo algorithms which is inevitably violated in the presence of depth discontinuities. For block matching, on one hand one would like to use the smallest possible block size in order to enforce this assumption even on the smallest patches of the scene, hence reducing disparity estimation errors; on the other hand, the use of larger block sizes for block matching leads to better uniqueness performance. Moreover, there is an explicit lower limit in the choice of the block size, the size of the alphabet $\mathscr{A}$ used for generation of the specific De Bruijn sequence associated with the pattern. In other words, since spatial multiplexing techniques exploit the spatial distribution of the projected pattern, they effectively sample the spatial information of the acquired scene, limiting the system's ability to cope with spatial frequency changes of scene depth. As pointed out in [9], this assumption is the counterpart of the static nature of the scene for temporal multiplexing techniques. A practical effect of this sampling of the spatial scene information is that the object contours of the acquired scene appear jagged and do not correspond to the object's actual edges. An example of structured light depth cameras based on spatial multiplexing not leading to pixel-precise depth estimates is offered in Fig. 2.21.

Let us recall from the introduction of this section that according to the terminology adopted in this book, light-coded cameras are structured light depth cameras with the pattern designed by algorithms that generate code words with suitable local pattern distribution characteristics, e.g., De Bruijn patterns. While thinking about algorithmically generated patterns may be more intuitive, Konolige [12] shows that patterns maximizing uniqueness can be also generated by numerical optimization. Even though coded patterns are usually characterized by very predictable uniqueness properties, they do not necessarily have better performance with respect to numerically optimized patterns as shown by Konolige [12]. The analysis of Konolige [12] indicates that the advantages of coded versus non-coded patterns depend on specific system characteristics and concern more system realization issues than fundamental differences in the quality of the uniqueness.

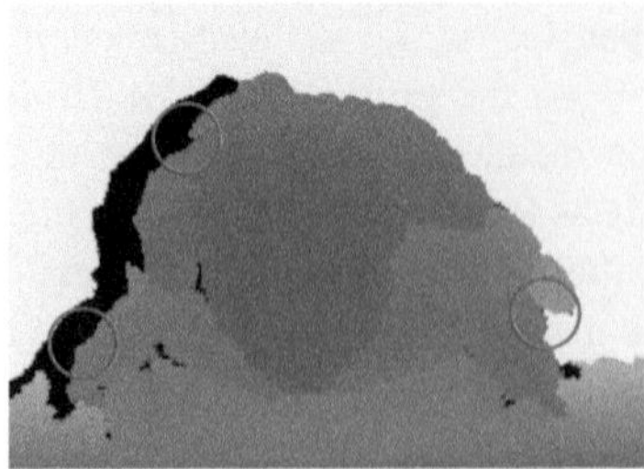

Fig. 2.21 Edge artifacts due to spatial multiplexing: the edges of the depth map acquired by the Intel RealSense R200 (*left*) are jagged with respect to the edges of the actual object (*right*)

2.3.2 Structured Light Systems Non-idealities

After the presentation of different possibilities in pattern design approaches, let us recall that a number of non-idealities might affect the actual implementation of a structured light depth camera, independent from the selected scheme. Some of these non-idealities are related to fundamental properties of optical and imaging systems, e.g., camera and projector thermal noise, while other non-idealities are present in the case of different systems. A list of the most important non-idealities is presented below.

(a) *Perspective distortion.* Since the scene points may have different depth values z, neighboring pixels of the projected pattern may not be mapped to neighboring pixels of I_C. In this case the local distribution of the acquired pattern becomes a distorted version of the relative local distribution of the projected pattern (see the first row of Fig. 2.22).

(b) *Color or gray-level distortion due to scene color distribution and reflectivity properties of the acquired objects.* The projected pattern undergoes reflection and absorption by scene surfaces. The ratio between incident and reflected radiant power is given by the scene reflectance, generally related to the scene color distribution. In the common case of IR projectors, the appearance of the pixel p_C on the camera C depends on the reflectance of the scene surface at the IR frequency used by the projector. For instance, a high intensity pixel of the projected pattern at p_A may undergo strong absorption because of the low reflectance value of the scene point to which it is projected, and the values of its conjugate pixel p_C on I_C may consequently appear much darker. This is an extremely important issue, since it might completely distort the projected code words. The second row of Fig. 2.22 shows how the radiometric power of the projected pattern may be reflected by surfaces of different color.

(c) *External illumination.* The color acquired by the camera C depends on the light falling on the scene's surfaces, which is the sum of the projected pattern and of scene illumination, i.e., sunlight, artificial light sources, etc. This second contribution with respect to code word detection acts as a noise source added to the information signal of the projected light (see third row of Fig. 2.22).

(d) *Occlusions.* Because of occlusions, not all the pattern pixels are projected to 3D points seen by camera C. Depending on the 3D scene geometry, there may not be a one-to-one association between the pattern pixels p_A and the pixels of the acquired image I_C. Therefore, it is important to correctly identify the pixels of I_C that do not have a conjugate point in the pattern in order to discard erroneous correspondences (see fourth row of Fig. 2.22).

(e) *Projector and camera non-idealities.* Both projector and camera are not ideal imaging systems. In particular, they generally do not behave linearly with respect to the projected and the acquired colors or gray-levels.

(f) *Projector and camera noise.* The presence of random noise in the projection and acquisition processes is typically modeled as Gaussian additive noise in the acquired image or images.

Fig. 2.22 Examples of different artifacts affecting the projected pattern (*in the depth maps, black pixels correspond to locations without a valid depth measurement*). *First row:* projection of the IR pattern on a slanted surface and corresponding depth map; observe how the pattern is shifted when the depth values change and how perspective distortion affects the pattern on the slanted surfaces. *Second row:* Primesense pattern projected on a color checker and corresponding color image; observe the dependence of the pattern appearance from the surface color. *Third row:* a strong external illumination affects the acquired scene; the acquired IR image saturates in correspondence of the strongest reflections and the Kinect™ v1, is not able to acquire the depth of those regions. *Fourth row:* the occluded area behind the ear of the stuffed toy is visible from the camera but not from the projector's viewpoint, consequently, the depth of this region cannot be computed

2.4 Examples of Structured Light Depth Cameras

After this presentation of theoretical and practical facts on structured-light depth cameras, we now explore how actual implementations combine the various presented design concepts. This section analyzes the most diffused structured light depth cameras in the market, namely, the Intel RealSense F200, the Intel RealSense R200, and the Primesense camera (AKA Kinect^{TM} v1.)

2.4.1 The Intel RealSense F200

The Intel RealSense F200 [2] has a very compact depth camera that can either be integrated in computers and mobile devices or used as a self-standing device. As shown in Fig. 2.23, the Intel RealSense F200 generally comes with an array of microphones, a color camera, and a depth camera system, made by an IR camera and an IR projector.

The spatial resolution of the depth camera of the Intel RealSense F200 is VGA (640×480), the working depth range is 200–1200 [mm], and the temporal resolution is up to 120 [Hz]. The horizontal Field-of-View (FoV) of the Intel RealSense F200 depth camera is 73° and the vertical FoV is 59°, with a focal length in pixels of approximately 430 [pxl]. Such characteristics are well suited to applications such as face detection [14] or face tracking, gesture recognition, and to applications that frame a user facing the screen of the device. The letter "F" in the name hints at the intended "Frontal" usage of this device.

Figure 2.24 shows the positions of the three most important components of the structured light depth camera, i.e., the IR camera, the IR projector (respectively

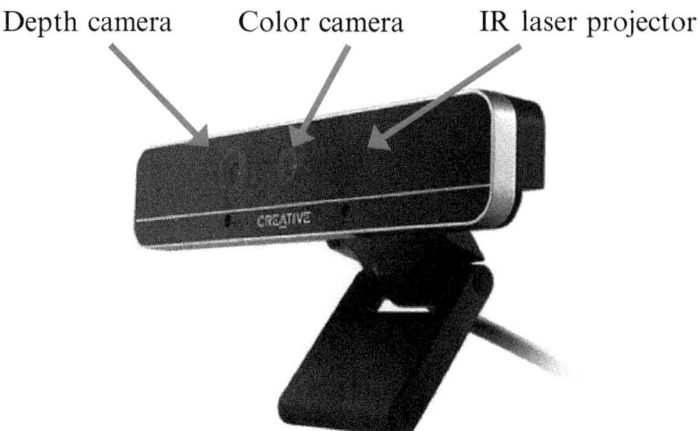

Fig. 2.23 Intel RealSense F200 components: depth camera, color camera and microphone array

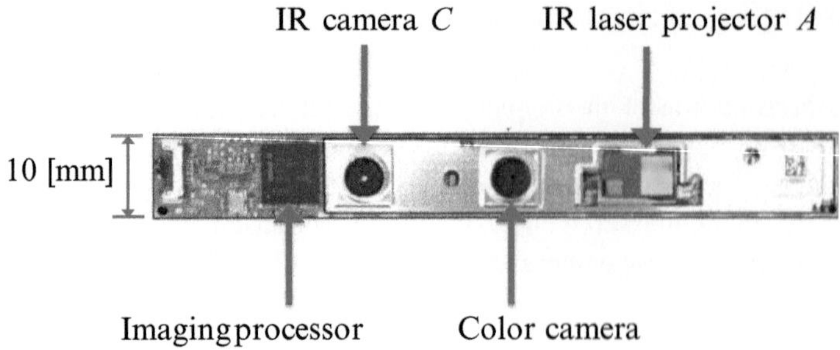

Fig. 2.24 Intel RealSense F200 under the hood

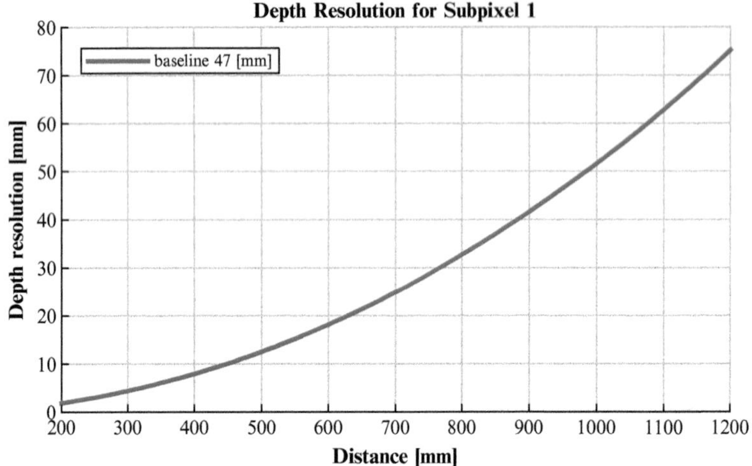

Fig. 2.25 Depth resolution without sub-pixel interpolation vs. measured depth distance of Intel RealSense F200

denoted C and A in the notation of Sect. 1.3) plus a color camera. The presence of a single IR camera indicates that the Intel RealSense F200 exploits the concept of a virtual camera.

Note that the baseline between the IR camera C and the IR projector A is approximately 47 [mm]. Figure 2.25 shows the depth resolution of the Intel RealSense F200 depth camera, without sub-pixel interpolation, as a function of the measured depth, according to (2.6) given the baseline and the focal length in pixels.

The projector of the Intel RealSense F200 is the most interesting component of the depth camera itself. It is a dynamic projector, which projects vertical light stripes of variable width at three different brightness or range levels, an approach similar to Gray code patterns. According to the adopted terminology, the Intel RealSense F200 depth camera uses both temporal and range multiplexing. The impressively high pattern projection frequency in the order of 100 [Hz] makes reverse engineering complex. Figure 2.26 shows the pattern projected by the Intel RealSense F200 (obtained by a very fast camera operating at frame rate 1200 [Hz]).

Fig. 2.26 Patterns projected by the projector of the Intel F200 camera

Figure 2.26 clearly shows that there are at least six layers of independent projected patterns at three range levels, leading to $3^6 = 729$ possible pattern configurations for a set of six frames. Since the number of different configurations is an upper bound for the maximum measurable disparity (corresponding to the closest measurable distance), this characteristic is functional to avoid limitations on the closest measurable depth and to reliably operate in close ranges. Since the Intel RealSense F200 projector does not use spatial multiplexing, there is no spatial sampling and the depth camera operates at full VGA spatial resolution. Figure 2.27 shows that the edge jaggedness typical of spatial multiplexing is not exhibited by the image captured by the Intel RealSense F200 due to its pixel-precise spatial resolution.

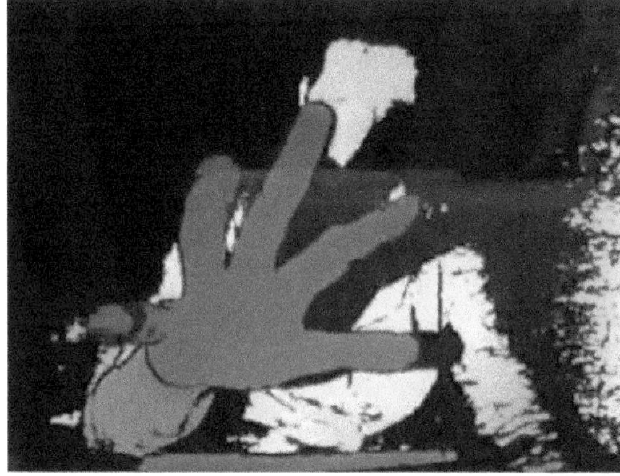

Fig. 2.27 Example of pixel-wise independent depth measurements obtained by the Intel RealSense F200 depth camera. The edges of the framed hand are pixel-precise and do not present edge jaggedness typical of spatial multiplexing techniques

Conversely, the data produced by Intel RealSense F200 exhibit artifacts typical of temporal multiplexing when the scene content moves during the projection of the set of patterns needed for depth estimation. An example of these artifacts is the *ghosting effect* shown by Fig. 2.16. Moreover, the combination of the characteristics of the illuminator design, of the fact that the illuminator produces stripes and not dots, and of the virtual camera approach makes the Intel RealSense F200 depth camera highly sensitive to the presence of external illumination. In facts, as indicated by the

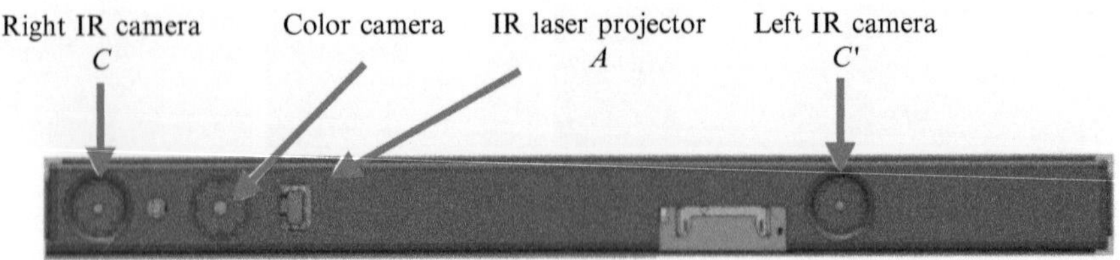

Fig. 2.28 Intel RealSense R200 components: color camera and structured light depth camera made by two IR cameras and one IR projector

official specifications, this structured light system is meant to work indoors, as the presence of external illumination leads to a considerable reduction of its working depth range.

The above analysis suggests that the design of the Intel RealSense F200 depth camera is inherently targeted to a limited depth range allowing for pixel-precise, fast, and accurate depth measurements, particularly well suited for frontal facing applications with maximum depth range of 1200 [mm].

2.4.2 *The Intel RealSense R200*

Like the Intel RealSense F200, the Intel RealSense R200 has a very compact depth camera that can either be integrated in computers and mobile devices or used as a self-standing device. As shown in Fig. 2.28, the Intel RealSense R200 generally comes with a color camera and a depth camera system, made by two IR cameras and not only one, like the Intel RealSense F200, and by an IR projector, respectively denoted as C, C' and A in the notation of Sect. 1.3.

The spatial resolution of the structured light depth camera of the Intel RealSense R200 is VGA (640×480), the working depth range is 510–4000 [mm], and the temporal resolution is up to 60 [Hz]. The horizontal Field-of-View (FoV) of the Intel RealSense R200 depth camera is approximately 56° and the vertical FoV is 43°, with a focal length in pixels of approximately 600 [pxl]. Such characteristics are very well suited for applications such as people tracking and 3D reconstruction, and in general for applications that frame the portion of the world behind the rear part of the device. The letter "R" in the name hints at the intended "Rear" usage of this device.

Figure 2.29 shows the Intel RealSense R200's most important components, namely, the two IR cameras and the IR projector plus the color camera. Since the Intel RealSense R200 carries a pair of IR cameras, there is no need for a virtual camera. The baseline between the left IR camera and the IR projector is 20 [mm] and the baseline between the two IR cameras is 70 [mm]. Since the Intel RealSense R200 does not employ a virtual camera, the baseline value affecting the depth camera resolution is the one between the two IR cameras. Figure 2.30

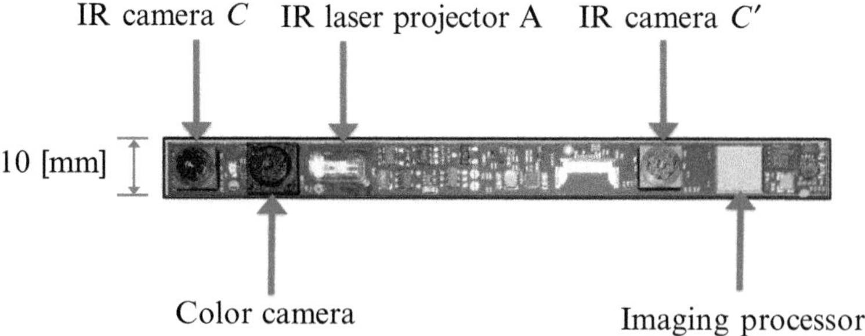

Fig. 2.29 Intel RealSense R200 under the hood

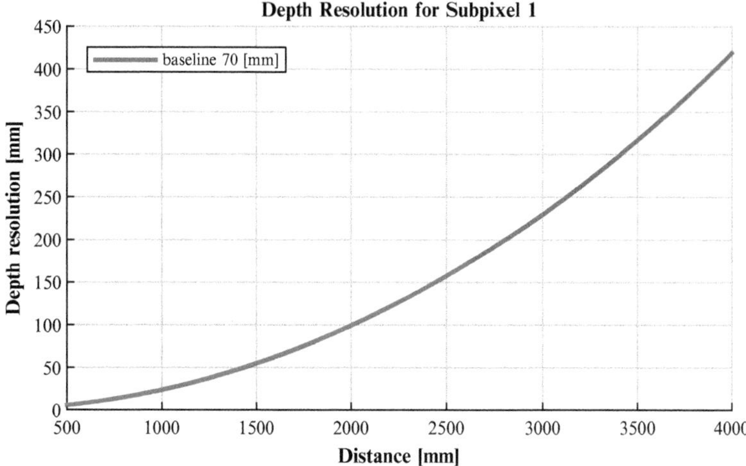

Fig. 2.30 Depth resolution without sub-pixel interpolation vs. measured depth distance of Intel RealSense R200

shows the depth resolution of the Intel RealSense R200 depth camera (without sub-pixel interpolation) as a function of the measured depth, according to (2.6) given the baseline and the focal length in pixels.

Also in this case, the projector of the Intel RealSense R200 is the most interesting component of the depth camera itself. Here, it is a static projector providing texture to the scene. Different from the Primesense camera, the pattern of the Intel RealSense R200's projector is not made by collimated dots. Compared to other cameras, the projector dimensions are remarkably small. In particular, the box length along the depth axis, usually called Z-height, is about 3.5 [mm], a characteristic useful for integration in mobile platforms.

Figure 2.31 shows the pattern projected by the IntelRealSense R200 camera. These images show how the texture is uncollimated and made by elements of different intensity and without a clear structure. The purpose of this texture is to add features to the component of the different reflectance elements of the scene in

Fig. 2.31 Texture projected by the illuminator of the Intel RealSense R200 camera, framed at different zoom levels: (*left*) the full projected pattern; (*center*) a pattern zoom; (*right*) a macro acquisition

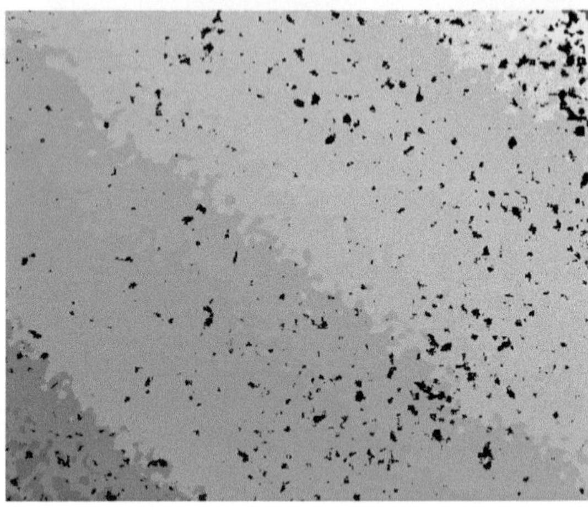

Fig. 2.32 Missing depth estimates, "black holes", in the data produced by the Intel RealSense R200 camera in the acquisition of a scene characterized by almost constant depth values

order to improve uniqueness. Since the projected texture is not collimated, it does not completely dominate the scene uniqueness, with the consequence of possibly missing depth estimates, i.e., of undefined depth values called "black holes" in some areas of the framed scene, as exemplified by Fig. 2.32

The Intel RealSense R200 projects constant illumination that does not vary in time, hence the system is characterized only by range and spatial multiplexing. There is no temporal multiplexing. The estimated depth-maps are therefore characterized by full temporal resolution with an independent depth estimate provided for each acquired frame, and by a subsampled spatial resolution, i.e., the localization of edges in presence of depth discontinuities is bounded by the size of the correlation window used in the depth estimation process. This subsampled spatial resolution leads to coarse estimation of the depth edges, as shown in Fig. 2.33.

The above analysis suggests that the Intel RealSense R200 structured light depth camera is designed to target rear-facing applications, such as objects or environment 3D modeling. The Intel RealSense R200 has an illuminator which projects a texture meant to aid scene reflectance, making this depth camera suitable for acquisitions both indoors and outdoors under reasonable illumination, within nominal range 51–400 [cm]. However, this results in a practical maximum range of about 250 [cm]. Since the projected texture is not made by collimated dots, the depth estimates may

Fig. 2.33 The Intel RealSense R200 camera depth estimation process is based on spatial multiplexing, leading to coarse edges, as clearly shown from the depth map of the leaves of the framed plant

exhibit missing measurements, especially outdoors when the external illumination affects the contribution of the projected texture, and indoors when the scene texture is inadequate to provide uniqueness.

2.4.3 *The Primesense Camera (AKA Kinect™ v1)*

The Primesense camera, AKA Kinect™ v1, is a less compact and more powerful system not suited for integration into mobile devices or computers when compared to the Intel RealSense F200 and R200.[1] As shown in Fig. 2.34, the Primesense system generally comes with a color camera and a structured light depth camera made by an IR camera C and an IR projector A with the notation introduced in Sect. 1.3. While the IR camera of the Primesense system is a high-resolution sensor with 1280×1024 pixels, the depth-map produced by the structured light depth camera is 640×480. In spite of the nominal working depth range being 800–3500 [mm], the camera produces reliable data up to 5000 [mm] and in some cases even at greater distances. The temporal resolution is up to 60 [Hz]. The resolution downscaling not only reduces the sensor acquisition noise by aggregating more pixels, but also improves the effective spatial resolution of the estimated disparity

[1]For completeness, one should recall that the design of the Primesense Capri targeted integration into mobile devices and computers, but it never reached production. This section focuses on the Primesense Carmine, the only product which was commercialized.

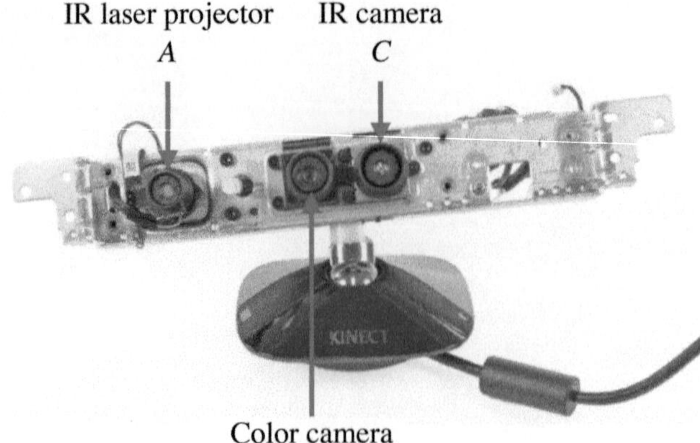

Fig. 2.34 Primesense system components: color camera and depth camera made by an IR camera *C* and an IR projector *A*

map. The horizontal Field-of-View (FoV) of the Primesense structured light depth camera is approximately 58° and the vertical FoV is 44°, with a focal length in pixels of approximately 600 [pxl]. The presence of a high resolution IR camera in the Primesense structured light depth camera gives better performance with respect to the Intel RealSense F200 and R200 in terms of range, spatial resolution, noise, and robustness against external illumination.

The baseline between the IR camera *C* and the IR projector *A* is approximately 75 [mm]. Figure 2.35 shows the depth resolution of the Primesense depth camera, without sub-pixel interpolation and also with an estimated sub-pixel interpolation of $1/8$, according to [11], as a function of the measured depth, according to (2.6) given the baseline and the focal length in pixels.[2]

In the case of the Primesense depth camera as well, the projector is the most interesting component: it is a static projector that produces a pattern made by collimated dots, as shown in Fig. 2.36. The collimated dots pattern appears to be subdivided into 3×3 tiles characterized by the same projected pattern up to holographic distortion. Collimated dots favor long-distance performance. Each tile of the pattern is characterized by a very bright dot at its center, usually called 0-th order, which is an artifact of the collimated laser going through a diffractive optical element.

[2]Even though depth resolution with practical sub-pixel interpolation is reported only for the Primesense structured light depth camera, it is expected to be also present in the Intel RealSense F200 and R200 structured light depth cameras. Since an estimated sub-pixel interpolation value is not available for such structured light depth cameras, it is reported here only for the Primesense structured light depth camera. The practical sub-pixel interpolation value is theoretically better for the Primesense structured light depth camera than for the Intel RealSense F200 and R200 because of the higher resolution of its IR camera.

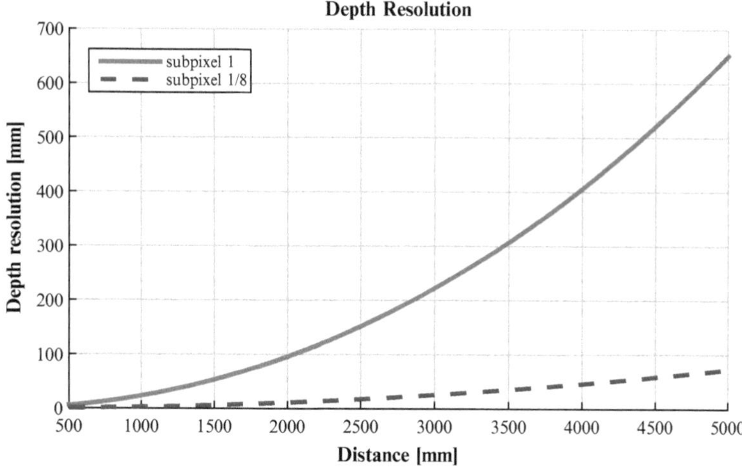

Fig. 2.35 Primesense depth resolution without sub-pixel interpolation and with 1/8 sub-pixel interpolation

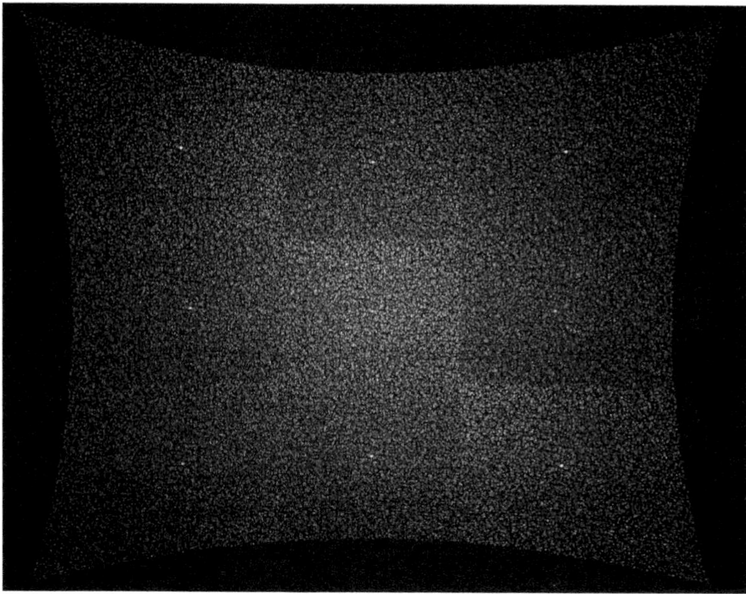

Fig. 2.36 Pattern projected by the Primesense illuminator and acquired by a high-resolution camera

The pattern of the Primesense depth camera has been thoroughly reverse engineered [11]. A summary of the major findings is reported next. A binary representation of the projected pattern is shown by Fig. 2.37. Each one of the 3×3 tiles is made by 211×165 holographic orders (equivalent in diffractive optics to

Fig. 2.37 Binary pattern projected by the Primesense camera reverse engineered by Konoldige and Mihelich [11]. In this representation, there is a single white pixel for each dot of the projected pattern

the concept of pixels in standard DLP projectors), hence the overall tiled pattern is made by $633 \times 495 = 313{,}335$ holographic orders. For each tile only 3861 of these orders are lit (bright spots), for a total of 34,749 lit orders in the tiled pattern. Therefore, on average, there is approximately one lit order for each 3×3 window and approximately 9 of them in a 9×9 window. Figure 2.38 (left) shows the plots of the minimum and average number of dots in a squared window as a function of the window size. The map of the local density for a 9×9 window size is shown in Fig. 2.38 (right).

The uniqueness of the Primesense pattern can be computed according to (2.9). We recall that it is possible to compute a uniqueness value for each pixel and that the overall uniqueness is the minimum of such uniqueness values. The plot of the minimum uniqueness in the pattern, i.e., what has been defined as pattern uniqueness in (2.9), and of the average uniqueness are shown in Fig. 2.39, together with the uniqueness map that can be computed pixel-by-pixel for a squared matching window of size 9×9. This figure shows how the Primesense pattern is a "unique pattern" if one uses a window of at least of 9×9 pixels.

The Primesense pattern only exploits spatial multiplexing without any temporal or range multiplexing. The fact that there is no temporal multiplexing ensures that each frame provides an independent depth estimate. The lack of range multiplexing,

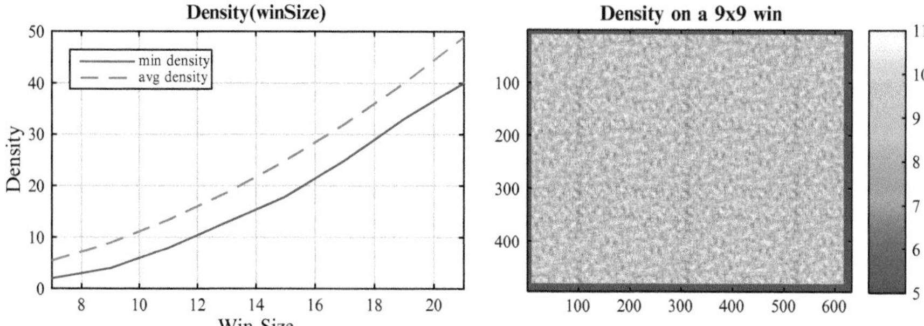

Fig. 2.38 Plot of the minimum and average density of the Primesense pattern as a function of the window size (*left*) and map of the local density for a 9 × 9 window (*right*)

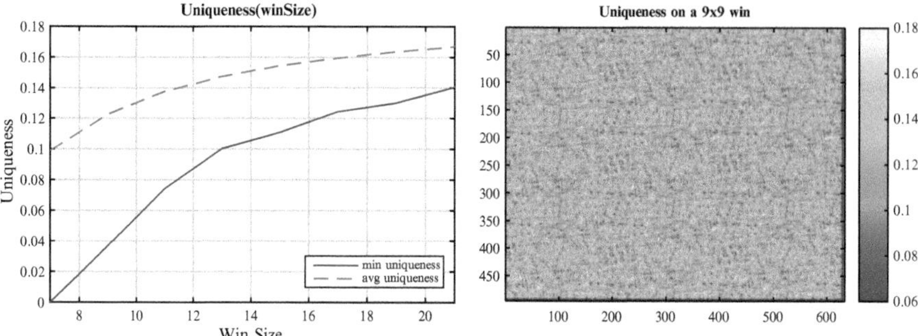

Fig. 2.39 Plot of the minimum and average uniqueness of the Primesense pattern as a function of the window size (*left*) and uniqueness map for a 9 × 9 window (*right*)

as well as the presence of collimated dots, enhances the system's ability to estimate depth at far distances. The adopted spatial multiplexing technique leads to a reduced spatial resolution, i.e., the localization of depth edges is reduced similar to the example in Fig. 2.21.

2.5 Conclusions and Further Reading

Structured light depth cameras encompass many system types. The fundamental components of such systems are their geometry and configuration, i.e., number of cameras, baseline, and position of the projector, and the characteristics of the projected pattern, which should be tailored to the nature of the projector itself. An interesting introductory analysis of structured light depth cameras can be found in [9], and a comprehensive review of the techniques for designing the projected pattern can be found in [15]. Interesting concepts about projected pattern design can be found in [12, 13]. The theory presented in the first sections of this chapter blends together different elements presented in these papers.

Structured light depth cameras are having great success in the mass market. In particular the Primesense system has found many different applications both in user-related products, e.g., gesture recognition for gaming, and in vertical market applications, such as robotics and 3D reconstruction of objects and environments. The Kinect^{TM} depth camera is based on a Primesense proprietary light-coding technique. Neither Microsoft nor Primesense have disclosed the full sensor implementation. Several patents exist, among which [4] covers the technological basis of the depth camera. The interested reader might also refer to current reverse engineering works [11, 17]. The Intel RealSense F200 and R200 cameras are much newer systems targeting various platforms in the mobile market, but provide less information than the Primesense system. The documentation provided by Intel [2] at the moment is the best source of information for them.

Compared to other technologies, such as passive stereo and ToF depth cameras, the predictability properties of the measurements of structured light systems are remarkable since their measurement errors are mainly due to local scene characteristics, such as local reflectivity and geometric configurations. Moreover, the components used to manufacture this type of systems are standard electrical and optical components, such as CCD imaging sensors and Diffractive Optical Elements. Major objective factors like the above mentioned ones have made structured light systems the first and most used consumer-grade depth cameras today.

References

1. Asus, Xtion pro, http://www.asus.com/Multimedia/Motion_Sensor/Xtion_PRO/. Accessed in 2016
2. Intel RealSense, www.intel.com/realsense. Accessed in 2016
3. Occipital Structure Sensor, http://structure.io. Accessed in 2016
4. Patent application us2010 0118123, 2010. Accessed in 2016
5. D. Bergmann, New approach for automatic surface reconstruction with coded light, in *Proceedings of Remote Sensing and Reconstruction for Three-Dimensional Objects and Scenes, SPIE* (1995)
6. K.L. Boyer, A.C. Kak, Color-encoded structured light for rapid active ranging. IEEE Trans. Pattern Anal. Mach. Intell. **9**, 14–28 (1987)
7. B. Carrihill, R.A. Hummel, Experiments with the intensity ratio depth sensor. Comput. Vis. Graph. Image Process. **32**(3), 337–358 (1985)
8. C.S. Chen, Y.P. Hung, C.C. Chiang, J.L. Wu, Range data acquisition using color structured lighting and stereo vision. Image Vis. Comput. **15**(6), 445–456 (1997)
9. J. Davis, D. Nehab, R. Ramamoorthi, S. Rusinkiewicz, Spacetime stereo: a unifying framework for depth from triangulation, in *Proceedings of IEEE Conference on Computer Vision and Pattern Recognition* (2003)
10. T.P. Koninckx, L. Van Gool, Real-time range acquisition by adaptive structured light. IEEE Trans. Pattern Anal.Mach. Intell. **28**(3), 432–445 (2006)

11. K. Konoldige, P. Mihelich, Technical description of kinect calibration. Technical report, Willow Garage, http://www.ros.org/wiki/kinect_calibration/technical (2011)
12. K. Konolige, Projected texture stereo, in *Proceedings of IEEE International Conference on Robotics and Automation* (2010)
13. J. Lim, Optimized projection pattern supplementing stereo systems, in *Proceedings of International Conference on Robotics and Automation* (2009)
14. L. Nanni, A. Lumini, F. Dominio, P. Zanuttigh, Effective and precise face detection based on color and depth data. Appl. Comput. Inform. **10**(1–2), 1–13 (2014).
15. J. Salvi, J. Pagès, J. Batlle, Pattern codification strategies in structured light systems. Pattern Recogn. **37**, 827–849 (2004)
16. D. Scharstein, R. Szeliski, A taxonomy and evaluation of dense two-frame stereo correspondence algorithms. Int. J. Comput. Vis. **47**(1–3), 7–42 (2001)
17. J. Smisek, M. Jancosek, T. Pajdla, 3d with kinect, in *Proceedings of IEEE Workshop on Consumer Depth Cameras for Computer Vision* (2011)
18. R. Szeliski, *Computer Vision: Algorithms and Applications* (Springer, New York, 2010)
19. M. Trobina, Error model of a coded-light range sensor. Technical report, Communication Technology Laboratory Image Science Group, ETH-Zentrum (1995)
20. R. Zabih, J. Woodfill, *Non-parametric Local Transforms for Computing Visual Correspondence*. Lecture Notes in Computer Science (Springer, Heidelberg, 1994)
21. L. Zhang, B. Curless, S.M. Seitz, Rapid shape acquisition using color structured light and multi-pass dynamic programming, in *Proceedings of IEEE International Symposium on 3D Data Processing, Visualization, and Transmission* (2002), pp. 24–36
22. L. Zhang, B. Curless, S.M. Seitz, Spacetime stereo: shape recovery for dynamic scenes, in *Proceedings of IEEE Conference on Computer Vision and Pattern Recognition* (2003)

Chapter 3
Operating Principles of Time-of-Flight Depth Cameras

Time-of-Flight depth cameras (or simply *ToF cameras* in this book) are active sensors capable of acquiring 3D geometry of a framed scene at video rate. Microsoft [6] is currently the major actor in the ToF camera technology arena since it acquired Canesta, a U.S. ToF camera manufacturer, in 2010. This led to the commercialization of the Kinect^{TM} v2 [5]. Commercial products are also available from other manufacturers, such as MESA Imaging [4], PMD Technologies [8] and Sony (which acquired Optrima Softkinetic [9] in 2015). Other companies (e.g., Panasonic [7] and IEE [3]) and research institutions (e.g., CSEM [1] and Fondazione Bruno Kessler [2]) are also actively investigating ToF cameras.

As mentioned in Sect. 1.4.1, this chapter examines the technology adopted in all current commercial ToF camera products. Section 3.1 presents how AM modulation is used within In-Pixel Photo-Mixing devices. Section 3.2 discusses the imaging system supporting ToF sensors. Section 3.3 presents the practical issues driving ToF's performance limits and noise characteristics. Section 3.4 describes how principles introduced in this chapter enter various ToF camera products.

3.1 AM Modulation Within In-Pixel Photo-Mixing Devices

As seen in Sect. 1.4, all the current commercial products adopt homodyne AM modulation with circuitry based on various solutions related to In-Pixel Photo-Mixing devices [17, 23], simply called in-pixel devices in the rest. Figure 3.1 shows a conceptual model of the operation of an homodyne AM transmitter and receiver, which are co-sited in a ToF camera, unlike in typical communication systems. Telecommunication systems convert the signal sent by the transmitter into useful information. In contrast, ToF systems only estimate the round-trip delay of the signal rather than the information encoded inside the signal. The transmitted and received signal from (1.57) and (1.58) reported below for convenience are respectively

P. Zanuttigh et al., *Time-of-Flight and Structured Light Depth Cameras*,
DOI 10.1007/978-3-319-30973-6_3

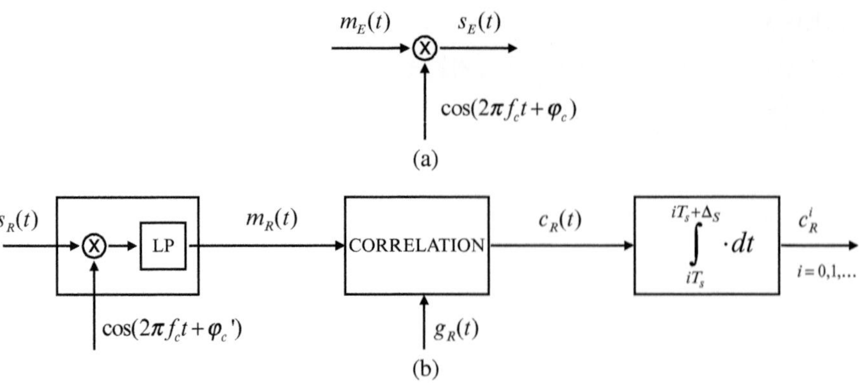

Fig. 3.1 ToF camera: (**a**) transmitter model; (**b**) model of the in-pixel receiver

$$s_E(t) = m_E(t)\cos(2\pi f_c t + \varphi_c) \tag{3.1}$$

and

$$s_R(t) = m_R(t)\cos(2\pi f_c t + \varphi_c') + n_R(t) \tag{3.2}$$

where the carrier frequency f_c corresponds to a NIR wavelength and is in the few hundreds of THz (e.g., $\lambda_c = c/f_c = 860$ [nm] corresponds to $f_c = 348$ [THz]) and $m_E(t)$ is the modulating signal with frequency f_m in the tens of MHz (HF band) and wavelength of a few meters (e.g., $f_m = 16$ [MHz] corresponds to $\lambda_m = 18$ [m]).

It is worth noting that both $s_E(t)$ and $s_R(t)$ are optical signals and that the modulation schemes of Fig. 3.1 are an appropriate description for the operation of the transmitter but not for the photoelectric conversion of the receiver. The schemes of Fig. 3.1 have a conceptual rather than a physical circuit meaning. Indeed, the actual light detection mechanism of the in-pixel devices is such that a baseband voltage signal $m_R(t)$ is generated from the optical input $s_R(t)$, without direct demodulation as in the transmitter side. Therefore the demodulator at the front end of the receiver and the explicit reference to the carrier phase φ_c and φ_c' are only used for modeling simplicity; they hide processes more complex than ones of interest for the purposes of this book.

The electric modulating signal $m_E(t)$ can be either a sine wave of period T_m

$$m_E(t) = A_E[1 + \sin(2\pi f_m t + \varphi_m)] \tag{3.3}$$

with $f_m = 1/T_m$, or a square wave of support $\Delta_m < T_m$ spaced by the modulation period T_m

$$m_E(t) = A_E \sum_{k=0}^{\infty} p(t - kT_m + \varphi_m;\ \Delta_m) \tag{3.4}$$

where

$$p(t;\Delta) = \mathrm{rect}\left(\frac{t-\frac{\Delta}{2}}{\Delta}\right) = \begin{cases} 1 & 0 \le t \le \Delta \\ 0 & \text{otherwise.} \end{cases} \tag{3.5}$$

The pulse $p(t;\Delta)$ in (3.5) is modeled by a rectangle for simplicity, however, this is only a nominal reference signal given the practical difficulty of obtaining sharp rise and fall signals.

At the receiver, after the demodulation of the optical signal $s_R(t)$, the baseband electrical signal $m_R(t)$, of shape similar to that of $m_E(t)$, is correlated with the reference signal $g_R(t)$ with period T_m, obtaining

$$c_R(t) = \int_0^{T_m} m_R(t) g_R(t+t')\,\mathrm{d}t'. \tag{3.6}$$

The signal $c_R(t)$ is sampled according to the "natural sampling" paradigm by the charge accumulator circuit at the back-end of the receiver and can be modeled as a system which at each sampling time iT_s, $i = 0, 1, \ldots,$ returns the integration of $c_R(t)$ in the support Δ_S

$$c_R^i = \int_{iT_S}^{iT_S+\Delta_S} c_R(t)\,\mathrm{d}t. \tag{3.7}$$

Clearly for designing ToF camera sensors there is a countless number of combinations for the value of $m_E(t)$, $g_R(t)$ and Δ_S. The two basic situations of sinusoidal and square modulating signal $m_E(t)$ and related choices of $g_R(t)$ and Δ_S will be discussed next.

3.1.1 Sinusoidal Modulation

In the case of sinusoidal modulation, according to the scheme of Fig. 3.1 the ToF camera transmitter modulates the NIR optical carrier by a modulation signal $m_E(t)$ made by a sinusoidal signal of amplitude A_E and frequency f_m, namely

$$m_E(t) = A_E[1 + \sin(2\pi f_m t + \varphi_m)]. \tag{3.8}$$

Signal $m_E(t)$ is reflected back by the scene surface within $s_E(t)$ and travels back towards the receiver ideally co-positioned with the emitter.

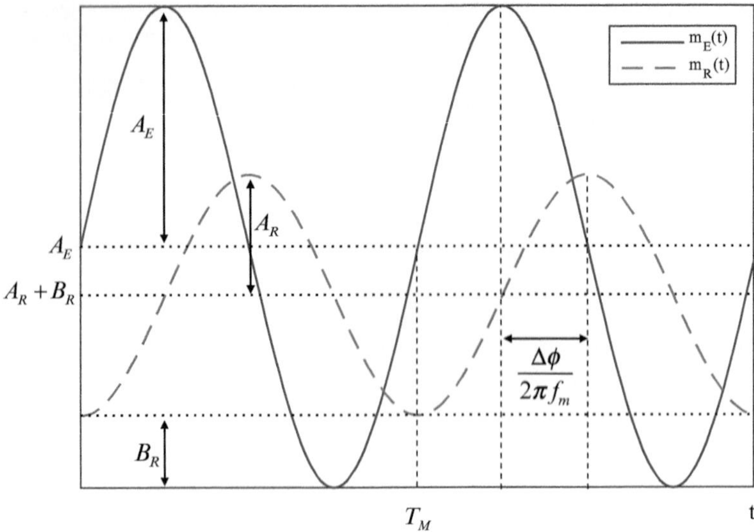

Fig. 3.2 Example of an emitted modulating signal $m_E(t)$ and a received modulating signal $m_R(t)$

The HF/VHF modulating signal reaching the receiver, due to factors such as the energy absorption associated with the reflection, the free-path propagation attenuation (proportional to the square of the distance), and the non-instantaneous propagation of IR optical signals leading to a phase delay $\Delta\varphi$, can be written as

$$\begin{aligned} m_R(t) &= A_R[1 + \sin(2\pi f_m t + \varphi_m + \Delta\varphi)] + B_R \\ &= A_R \sin(2\pi f_m t + \varphi_m + \Delta\varphi) + (A_R + B_R) \end{aligned} \tag{3.9}$$

where A_R is the attenuated amplitude of the received modulating signal and B_R is due to the background light interfering with λ_c and to other artifacts. Figure 3.2 shows an example of emitted and received modulating signal.

For simplicity we will call A_R simply A and $A_R + B_R$ simply $B/2$, obtaining

$$m_R(t) = A \sin(2\pi f_m t + \varphi_m + \Delta\varphi) + \frac{B}{2}. \tag{3.10}$$

Quantity A is called *amplitude*, since it is the amplitude of the useful signal. Quantity B is called *intensity* or *offset*, and it is the sum of the received modulating signal, with a component A_R due to the sinusoidal modulation component at f_m, and an interference component B_R, mostly due to background illumination. It is common to call A and B amplitude and intensity respectively, even though both A and B are signal amplitudes (measured in [V]).

If the correlation signal at the receiver is

$$g_R(t) = \frac{2}{T_m}\left[1 + \cos(2\pi f_m t + \varphi_m)\right] \tag{3.11}$$

the output of the correlation circuit is

$$
\begin{aligned}
c_R(t) &= \int_0^{T_m} m_R(t')g_R(t'+t)\,\mathrm{d}t' \\
&= \frac{2}{T_m}\int_0^{T_m}\left[A\sin(2\pi f_m t' + \varphi_m + \Delta\varphi) + \frac{B}{2}\right]\left[1+\cos(2\pi f_m(t'+t)+\varphi_m)\right]\mathrm{d}t' \\
&= \frac{2}{T_m}\int_0^{T_m} A\sin(2\pi f_m t' + \varphi_m + \Delta\varphi)\,\mathrm{d}t' + \frac{2}{T_m}\int_0^{T_m}\frac{B}{2}\,\mathrm{d}t' + \\
&+ \frac{2}{T_m}\int_0^{T_m} A\sin(2\pi f_m t' + \varphi_m + \Delta\varphi)\cos(2\pi f_m(t'+t)+\varphi_m)\,\mathrm{d}t' + \\
&+ \frac{2}{T_m}\int_0^{T_m}\frac{B}{2}\cos(2\pi f_m(t'+t)+\varphi_m)\,\mathrm{d}t' \\
&= A\sin(\Delta\varphi - 2\pi f_m t) + B.
\end{aligned}
\tag{3.12}
$$

Note that since transmitter and receiver are co-sited, the modulation sinusoidal signal (therefore including its phase φ_m) is directly available at the receiver side.

The unknowns of (3.12) are A, B and $\Delta\varphi$, where A and B are measured in Volts [V] and $\Delta\varphi$ as phase value is a pure number. The most important unknown is $\Delta\varphi$, since as already seen by (1.54) and (1.55), reported below for convenience,

$$\Delta\varphi = 2\pi f_m \tau = 2\pi f_m \frac{2\rho}{c} \tag{3.13}$$

it can deliver distance ρ

$$\rho = \frac{c}{4\pi f_m}\Delta\varphi. \tag{3.14}$$

Unknowns A and B will be shown later to be important for SNR considerations.

In order to estimate the unknowns A, B and $\Delta\varphi$, $c_R(t)$ must be sampled by an ideal sampler, i.e. with $\Delta_S \to 0$ in (3.7), at least 4 times per modulation period T_m [23], i.e., $T_s = T_m/4$. For instance, if the modulation frequency is 30 [MHz], signal $c_R(t)$ must be sampled at least at 120 [MHz]. Assuming a sampling frequency $F_S = 4f_m$, given the 4 samples per period $c_R^0 = c_R(t=0)$, $c_R^1 = c_R(t = 1/F_S)$, $c_R^2 = c_R(t = 2/F_S)$ and $c_R^3 = c_R(t = 3/F_S)$, the receiver estimates values $\hat{A}$, $\hat{B}$ and $\widehat{\Delta\varphi}$ as

$$(\hat{A}, \hat{B}, \widehat{\Delta\varphi}) = \underset{A,B,\Delta\varphi}{\text{argmin}} \sum_{n=0}^{3} \left\{ c_R^n - \left[A \sin\left(\Delta\varphi - \frac{\pi}{2} n \right) + B \right] \right\}^2. \tag{3.15}$$

After some algebraic manipulations of (3.15) one obtains

$$\begin{aligned} \hat{A} &= \frac{\sqrt{\left(c_R^0 - c_R^2\right)^2 + \left(c_R^3 - c_R^1\right)^2}}{2} \\ \hat{B} &= \frac{c_R^0 + c_R^1 + c_R^2 + c_R^3}{4} \\ \widehat{\Delta\varphi} &= \text{atan2}\left(c_R^0 - c_R^2, c_R^3 - c_R^1\right). \end{aligned} \tag{3.16}$$

The final distance estimate $\hat{\rho}$ can be obtained from (3.14) as

$$\hat{\rho} = \frac{c}{4\pi f_m} \widehat{\Delta\varphi}. \tag{3.17}$$

If one takes into account that the sampling is not ideal but actually made by a sequence of rectangular pulses of width Δ_S within the standard natural sampling model, the estimates of A, B, and $\widehat{\Delta\varphi}$ in this case become [24]

$$\begin{aligned} \hat{A}' &= \frac{\pi}{T_S \sin\left(\frac{\pi \Delta_S}{T_S}\right)} \hat{A} \\ \hat{B}' &= \frac{\hat{B}}{\Delta_S} \\ \widehat{\Delta\varphi}' &= \widehat{\Delta\varphi} \end{aligned} \tag{3.18}$$

showing that the phase shift $\Delta\varphi$ is independent from the size of the sampling duration Δ_S, that instead affects both the estimate of A and B. A typical value of Δ_S is $\Delta_S = T_m/4 = 1/(4f_m)$.

3.1.2 *Square Wave Modulation*

In the case of square wave modulation, according to the scheme of Fig. 3.1 the ToF camera transmitter modulates the NIR optical carrier by a square wave $m_E(t)$ of amplitude A_E and frequency $f_m = 1/T_m$ in the HF/VHF band

$$m_E(t) = A_E \sum_{k=0}^{\infty} p(t - kT_m;\ \Delta_m) \tag{3.19}$$

where $\Delta_m \leq T_m$. The phase φ_m of $m_E(t)$ is not explicitly written for notational simplicity. Because of the co-siting of transmitter and receiver, $m_E(t)$ is available also at the receiver and the specific value of φ_m for practical demodulation purposes is irrelevant.

The back-reflected HF/VHF modulating signal reaching the receiver within $s_R(t)$ of Fig. 3.1 can be written as

$$m_R(t) = A \sum_{k=0}^{\infty} p(t - \tau - kT_m;\ \Delta_m) + B \tag{3.20}$$

where A is the attenuated amplitude of the received modulating signal, B is due to the background light interfering with λ_c and τ is the round-trip time. Clearly, A_E is known and A, τ and B are unknown since the first two depend on target distance and material NIR reflectivity, and the latter on the background noise.

In the square wave modulation case there are many ways to estimate A, B and τ, which will be introduced next with a few examples.

Let us first consider the situation exemplified by Fig. 3.3 where $m_E(t)$ is defined in (3.19), $m_R(t)$ is defined in (3.20) and

$$g_R(t) = \sum_{k=0}^{\infty} (-1)^k p(t - 2kT_S;\ 2T_S). \tag{3.21}$$

The following reasoning assumes $\tau < T_S$ and $T_S = T_m/4$ and it can be generalized to $m_E(t)$ and $m_R(t)$ having pulses $p(t; \Delta)$ with a different support Δ_m.

For notational convenience, in Fig. 3.3 the areas of the portions of the useful signal of $m_E(t)$ falling respectively in the first, second and third sampling period are denoted as

$$\begin{aligned} R =& (T_S - \tau)A \\ Q =& T_S A \\ Q - R =& \tau A \end{aligned} \tag{3.22}$$

while the area of the optical noise signal, modeled for simplicity as a constant deterministic signal, in each sampling period is denoted as

$$W = BT_S. \tag{3.23}$$

In this case, from (3.6) and (3.7), again without considering the noise $n_R(t)$, the outputs of the back-end integrator stage are

$$c_R^i = \int_{iT_S}^{iT_S + \Delta_S} m_R(t) g_R(t)\, \mathrm{d}t \tag{3.24}$$

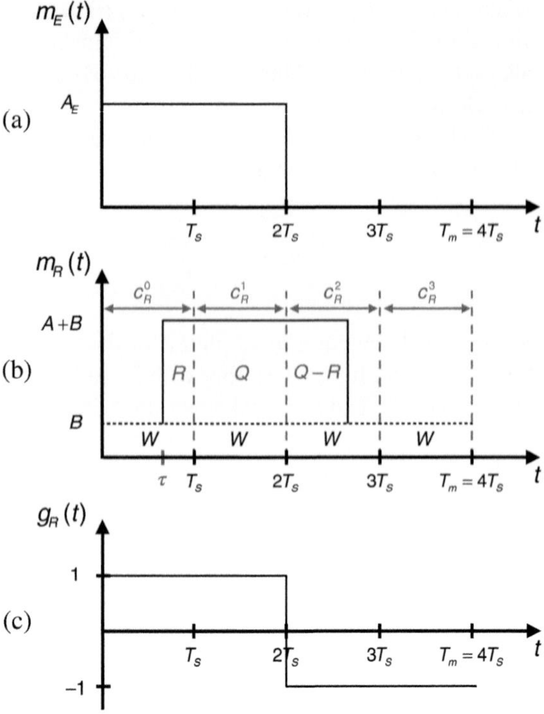

Fig. 3.3 Example of one period of square wave signaling: (**a**) $m_E(t)$, (**b**) $m_R(t)$ and (**c**) $g_R(t)$

where $\Delta_S = T_S$. As Fig. 3.3 schematically indicates, they correspond to the sum of the area of the two components of $m_R(t)$ in each sampling period T_S equal to

$$
\begin{aligned}
c_R^0 &= R + W = Q\left(1 - \frac{\tau}{T_S}\right) + W \\
c_R^1 &= Q + W \\
c_R^2 &= -[Q - R + W] = -\left[Q\frac{\tau}{T_S} + W\right] \\
c_R^3 &= -W.
\end{aligned}
\tag{3.25}
$$

From (3.25) it is straightforward to see that

$$
\begin{aligned}
\hat{\tau} &= \frac{T_S}{2}\left(1 - \frac{c_R^2 + c_R^0}{c_R^1 + c_R^3}\right) \\
\hat{A} &= \frac{1}{T_S}\left(c_R^1 + c_R^3\right) \\
\hat{B} &= -\frac{c_R^3}{T_S}.
\end{aligned}
\tag{3.26}
$$

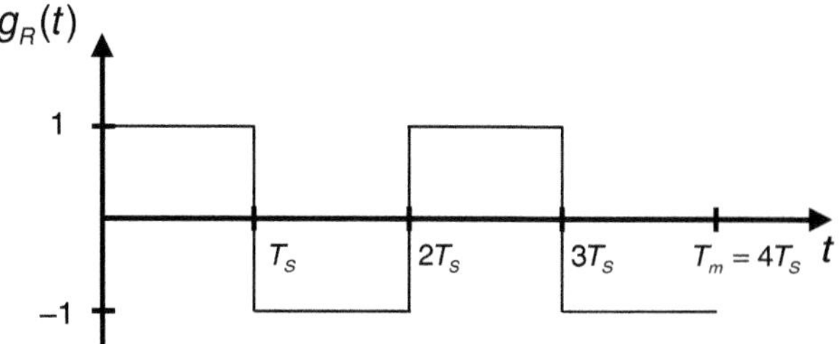

Fig. 3.4 Reference function $g_R(t)$ defined by (3.27)

With $m_E(t)$ and $m_R(t)$ defined as in (3.19) and (3.20) and within assumptions $\tau < T_S$, $\Delta_S = T_S$ and $T_S = T_m/4$ there are many other possible choices for $g_R(t)$, such as

$$g_R(t) = \sum_{k=0}^{\infty} (-1)^k p(t - kT_S;\ T_S) \tag{3.27}$$

shown in Fig. 3.4. This choice of $g_R(t)$ implies different signs of c_R^i for $i = 1, 2$ and the estimates become

$$\begin{aligned} \hat{\tau} &= \frac{T_S}{2}\left(1 - \frac{c_R^2 - c_R^0}{c_R^1 - c_R^3}\right) \\ \hat{A} &= \frac{1}{T_S}\left(c_R^3 - c_R^1\right) \\ \hat{B} &= \frac{c_R^3}{T_S}. \end{aligned} \tag{3.28}$$

With $g_R(t)$ defined as

$$g_R(t) = 1 \tag{3.29}$$

the estimates become

$$\begin{aligned} \hat{\tau} &= \frac{T_S}{2}\left(1 - \frac{c_R^0 - c_R^2}{c_R^1 - c_R^3}\right) \\ \hat{A} &= \frac{1}{T_S}\left(c_R^1 - c_R^3\right) \\ \hat{B} &= \frac{c_R^3}{T_S}. \end{aligned} \tag{3.30}$$

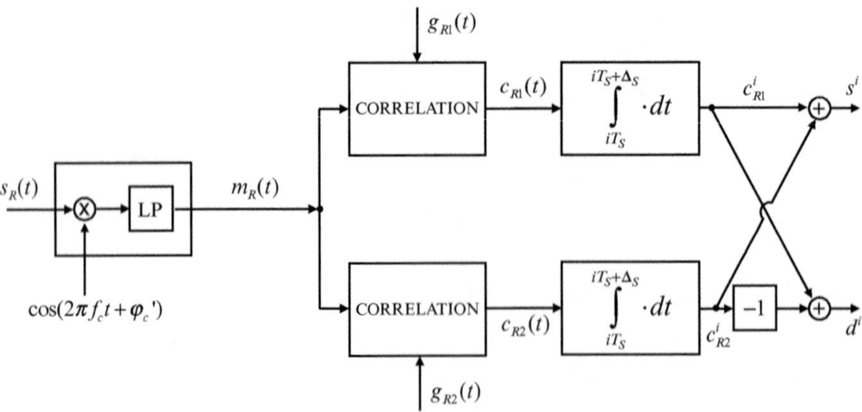

Fig. 3.5 Conceptual model of the differential scheme for the in-pixel receiver

Figure 3.5 shows an alternative scheme for the in-pixel receiver, typically called differential, differing from the scheme of Fig. 3.3 for the presence of two correlators with reference signal $g_{R1}(t)$ and $g_{R2}(t)$ respectively defined as

$$\begin{aligned} g_{R1}(t) &= \sum_{k=0}^{\infty} p(t-(2k)2T_S;\ T_S) \\ g_{R2}(t) &= \sum_{k=0}^{\infty} p(t-(2k+1)2T_S;\ T_S) \end{aligned} \tag{3.31}$$

which operate in parallel. The correlation stage is followed by a subsequent stage where samples c^i_{R1} and c^i_{R2} are added and subtracted obtaining

$$\begin{aligned} s^i &= c^i_{R1} + c^i_{R2} \\ d^i &= c^i_{R1} - c^i_{R2}. \end{aligned} \tag{3.32}$$

At a circuit level, the double correlation and integration stage of Fig. 3.5 is amenable to simple and effective solutions, such as a clock signal of sampling period T_S controlling that the incoming photons contribute to charge c^i_{R1} when the clock signal is high, and to charge c^i_{R2} when the clock signal is low [10]. From area relationships (3.32), which apply also in this case, and from Fig. 3.6 it is readily seen that

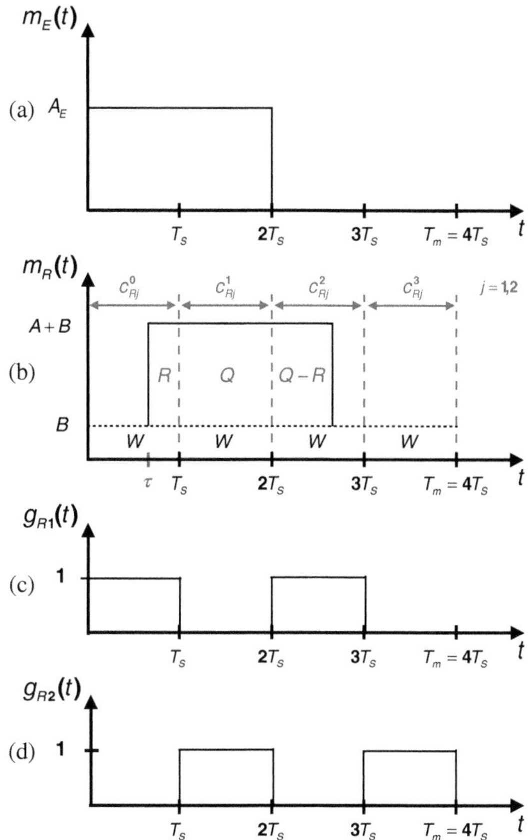

Fig. 3.6 Example of one period of square wave signaling for the differential scheme of the in-pixel receiver: (**a**) $m_E(t)$, (**b**) $m_R(t)$ and correlation reference signals (**c**) $g_{R1}(t)$ and (**d**) $g_{R2}(t)$ defined in (3.31)

$$
\begin{aligned}
c^0_{R1} &= R + W & c^0_{R2} &= 0 & s^0 &= R + W & d^0 &= R + W \\
c^1_{R1} &= 0 & c^1_{R2} &= Q + W & s^1 &= Q + W & d^1 &= -[Q + W] \\
c^2_{R1} &= Q - R + W & c^2_{R2} &= 0 & s^2 &= Q - R + W & d^2 &= Q - R + W \\
c^3_{R1} &= 0 & c^3_{R2} &= W & s^3 &= W & d^3 &= -W
\end{aligned}
\tag{3.33}
$$

from which the unknown parameters can be estimated as

$$
\begin{aligned}
\hat{\tau} &= T_S \frac{d^1 + d^0}{d^1 - d^3} = T_S \frac{d^3 + d^2}{d^3 - d^1} \\
\hat{A} &= \frac{1}{T_S}(d^3 - d^1) \\
\hat{B} &= -\frac{1}{T_S} d^3.
\end{aligned}
\tag{3.34}
$$

In this case both c^i_{R1} and c^i_{R2} have an intrinsic sampling period of $2T_S$, with c^i_{R2} lagged by T_S with respect to c^i_{R1}. Consequently, samples s^i and d^i of (3.32) carry the same information, therefore unknown parameters $\hat{\tau}$, $\hat{A}$ and $\hat{B}$ could also be obtained from samples s^i instead of d^i with some sign changes.

If instead of using $g_{R1}(t)$ and $g_{R2}(t)$ defined in (3.31) as reference signals for the correlations of the receiver of Fig. 3.6, one used $g_{R1}(t)$ and $g_{R2}(t)$ defined in (3.27) and (3.29) respectively, i.e.,

$$\begin{aligned} g_{R1}(t) &= \sum_{k=0}^{\infty}(-1)^k p(t-kT_S;\ T_S) \\ g_{R2}(t) &= 1 \end{aligned} \tag{3.35}$$

as shown in Fig. 3.7, from the area relationships (3.32) one obtains

$$\begin{array}{llll} c^0_{R1} = R+W & c^0_{R2} = R+W & s^0 = 2R+2W & d^0 = 0 \\ c^1_{R1} = -Q-W & c^1_{R2} = Q+W & s^1 = 0 & d^1 = -2Q-2W \\ c^2_{R1} = Q-R+W & c^2_{R2} = Q-R+W & s^2 = 2Q-2R+2W & d^2 = 0 \\ c^3_{R1} = -W & c^3_{R2} = W & s^3 = 0 & d^3 = -2W. \end{array} \tag{3.36}$$

The estimates obtained from samples s^i and d^i, $i = 0, 1, 2, 3$ in this case are

$$\begin{aligned} \hat{\tau} &= \frac{T_S}{2}\left(1+\frac{s^2-s^0}{d^3-d^1}\right) \\ \hat{A} &= \frac{d^3-d^1}{2T_S} \\ \hat{B} &= -\frac{1}{2T_S}d^3. \end{aligned} \tag{3.37}$$

Equations (3.36) and (3.37) show that the sampling period of s^i and d^i is $2T_S$ and not T_S as for c^i_{R1} and c^i_{R2} and that s^i and d^i carry different information. Equation (3.37) shows that A and B can be estimated only from d^i but the estimate of τ requires both s^i and d^i.

As a final consideration let us note that if one was only interested in estimating the two parameters τ and A, relying on theoretical or statistical considerations for the noise estimate, the number of measurements per modulation period could be halved. Within our model such a situation corresponds to $B = 0$ and would require the assumptions $\tau < T_S$, $\Delta_S < T_S$ and $T_S = T_m/2$, equivalent to two measurements per modulation period T_m instead of four as in the previous cases. Our simple model requires the assumption $B = 0$ in order to be extended to the case of two samples per period, even though the noise component B cannot be zero and it can be estimated

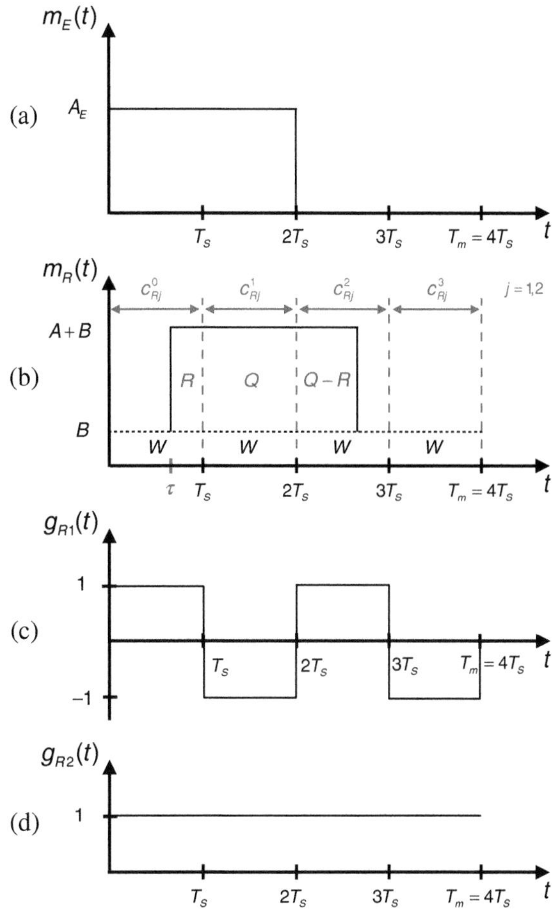

Fig. 3.7 Example of one period of square wave signaling for the differential scheme of the in-pixel receiver: (**a**) $m_E(t)$; (**b**) $m_R(t)$ and correlation reference signals (**c**) $g_{R1}(t)$ and (**d**) $g_{R2}(t)$ defined as in (3.27) and (3.29)

not from the values of the sample c^i_{R1} and c^i_{R2} but from circuital considerations and measurements. Practically, this can be the case in which B is considered constant within the temporal scale of the receiver sampling period, hence B can be estimated only once and then removed for a subsequent set of measurements. The first example of the differential scheme of the in-pixel receiver with $g_{R1}(t)$ and $g_{R2}(t)$ defined as in (3.31) but with $T_S = T_m/2$, corresponds to the signals of Fig. 3.8.

From area relationships (3.22), which still apply, and Fig. 3.8 the samples within each period are

$$\begin{aligned} c^0_{R1} &= R \quad c^0_{R2} = 0 \qquad s^0 = R \qquad d^0 = R \\ c^1_{R1} &= 0 \quad c^1_{R2} = Q-R \quad s^1 = Q-R \quad d^1 = -Q+R. \end{aligned} \tag{3.38}$$

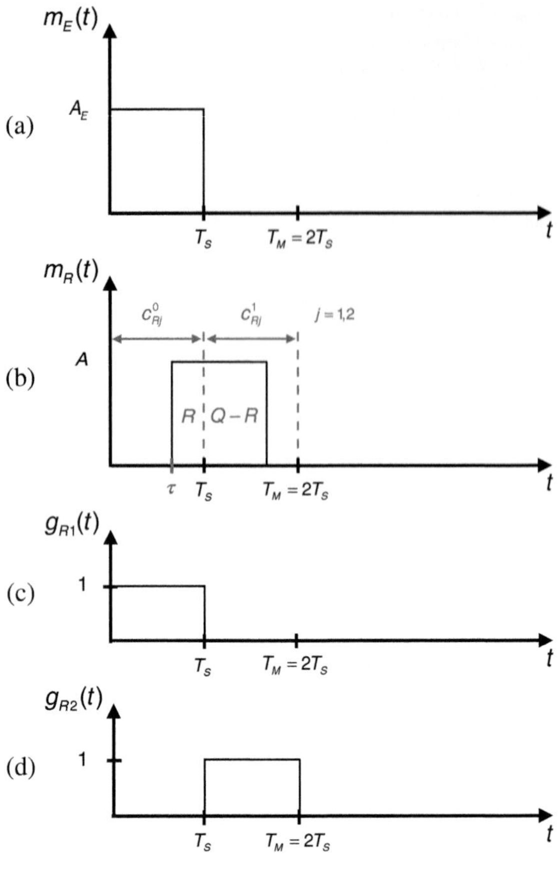

Fig. 3.8 Example of one period of square wave signaling for the differential scheme of the in-pixel receiver with assumptions $B = 0$, $\Delta_S = T_S$ and $T_S = T_m/2$: (**a**) $m_E(t)$, (**b**) $m_R(t)$ defined by in (3.19) and (3.20) and (**c**) $g_{R1}(t)$ and (**d**) $g_{R2}(t)$ defined by (3.31)

From (3.38) the unknown parameters can be estimated as

$$\begin{aligned} \hat{\tau} &= T_S \frac{d^1}{d^1 - d^0} \\ \hat{A} &= \frac{1}{T_S}(d^0 - d^1). \end{aligned} \tag{3.39}$$

The second example of a differential in-pixel receiver corresponding to $g_{R1}(t)$ and $g_{R2}(t)$ defined by (3.35) within assumptions $B = 0$ and $T_S = T_m/2$ is illustrated by Fig. 3.9. From area relationships (3.32) and Fig. 3.9, the samples within each period assume values

$$\begin{aligned} &c_{R1}^0 = R \quad && c_{R2}^0 = R \quad && s^0 = 2R \quad && d^0 = 0 \\ &c_{R1}^1 = -Q + R \quad && c_{R2}^1 = Q - R \quad && s^1 = 0 \quad && d^1 = -2(Q - R). \end{aligned} \tag{3.40}$$

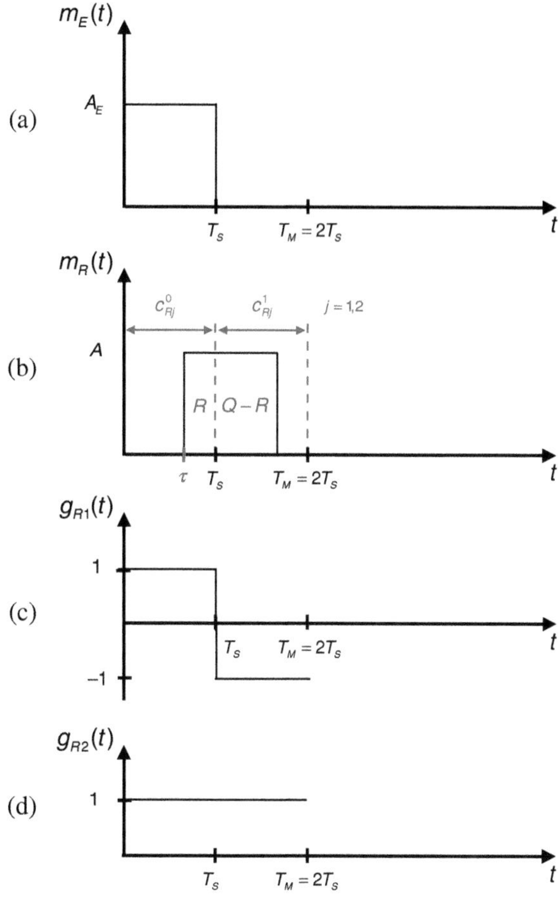

Fig. 3.9 Example of one period of square wave signaling for the differential scheme of the in-pixel receiver with assumptions $B = 0$, $\Delta_S = T_S$ and $T_S = T_m/2$: (**a**) $m_E(t)$, (**b**) $m_R(t)$ defined by in (3.19) and (3.20) and (**c**) $g_{R1}(t)$ and (**d**) $g_{R2}(t)$ defined by (3.27) and (3.29)

Equation (3.40) shows that the sampling interval of s^i and d^i coincides with the modulation interval $T_m = 2T_S$. Let us denote the sum of samples s^i and d^i within one modulation period T_m respectively as

$$\begin{aligned} S &= s^0 + s^1 = s^0 \\ D &= d^0 + d^1 = d^1. \end{aligned} \tag{3.41}$$

The unknown parameters in this case become

$$\begin{aligned} \hat{\tau} &= \frac{T_S}{2}\left(1 + \frac{d^1 + s^0}{d^1 - s^0}\right) = \frac{T_S}{2}\left(1 + \frac{D + S}{D - S}\right) \\ \hat{A} &= -\frac{1}{T_S}(d^1 - s^0) = -\frac{1}{T_S}(D - S) \end{aligned} \tag{3.42}$$

3.2 Imaging Characteristics of ToF Depth Cameras

This section examines the imaging characteristics of ToF depth cameras. A good starting point is recalling from Sect. 1.4.4 and Fig. 3.1 that the components of ToF cameras are:

(a) a transmitter made by an array of LEDs which generates a sinusoidal or square wave modulating signal in the high HF or low VHF bands, of tens of MHz, embedded in an optical NIR signal, in hundreds of THz;
(b) a suitable optics diffusing the optical signal generated by the transmitter to the scene;
(c) a suitable optics collecting the NIR optical radiation echoed by the scene and imaging it onto the receiver matricial ToF sensor. This component includes an optical band-pass filter with center-band tuned to the NIR carrier frequency of the transmitter in order to improve the SNR;
(d) a matricial ToF sensor of $N_R \times N_C$ pixels estimating simultaneously and independently the distance between each ToF sensor pixel p_T and the imaged scene point P;
(e) suitable circuitry providing the needed power supply and control signals to transmitter and receiver;
(f) a user interface for inputting the ToF depth camera parameters and reading the output data.

ToF depth cameras, in spite of their complexity due to the components listed above, can be modeled as pin-hole imaging systems since their receiver has the optics (c) and the sensor (d) made by a $N_R \times N_C$ matrix of lock-in pixels.

Figure 3.10, whose drawing is reminiscent of the NIR emitters configuration of the MESA SR4000 but whose model applies to any ToF depth camera NIR constellation, shows the signaling process at the basis of the relationships between the various scene points and the respective sensor pixels. Namely, the NIR signal sent by the constellation of emitters is conveyed to the scene (blue arrow) by the transmitter optics, is echoed by different portions of the scene, travels back to the camera and through the optics (red arrow) and is finally received by the different lock-in pixels of the ToF sensor.

All the pin-hole imaging system notation and concepts introduced in Sect. 1.1 apply to ToF depth cameras. The notation will be used with pedix T in order to recall that it refers to a ToF depth camera. The CCS of the ToF camera will be called the *$3D-T$ reference system*. The position of a scene point P with respect to the $3D-T$ reference system will be denoted as P_T and its coordinates as $\mathbf{P}_T = [x_T, y_T, z_T]^T$. Coordinate z_T of P_T is called the *depth* of point P_T and the z_T-axis is called the *depth axis* (Fig. 3.11).

The coordinates of a generic sensor pixel p_T of lattice Λ_T with respect to the 2D-T reference system are represented by vector $\mathbf{p}_T = [u_T, v_T]^T$, with $u_T \in [0, \dots, N_C]$

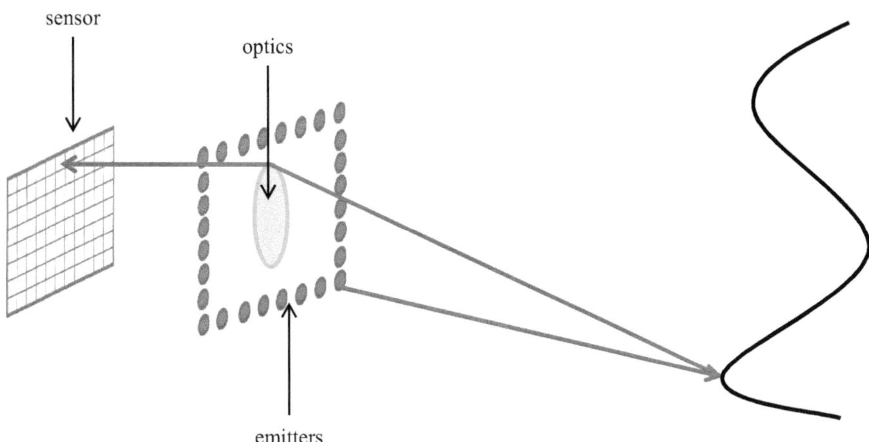

Fig. 3.10 Basic ToF depth camera structure and signaling: propagation towards the scene (*blue arrow*), reflection or echoing (*from the black surface on the right*), back-propagation (*red arrow*) towards the camera through the optics (*green*) and reception (*red sensor*)

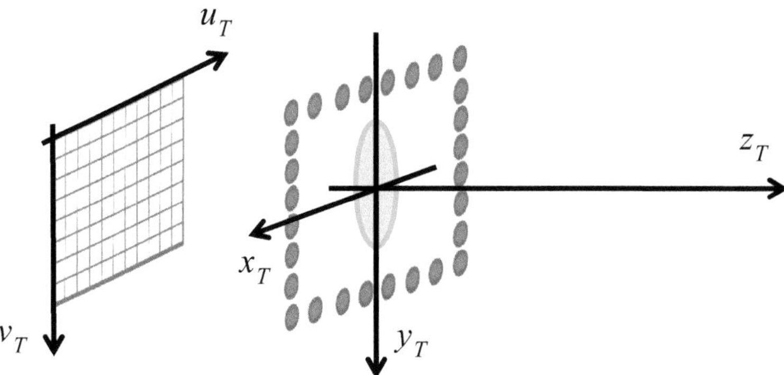

Fig. 3.11 2D T-reference system (with axes $u_T - v_T$) and 3D T-reference system (with axes $x_T - y_T - z_T$)

and $v_T \in [0, \ldots, N_R]$. Therefore the relationship between the $3D$ coordinates $\mathbf{P}_T = [x_T, y_T, z_T]^T$ of a scene point P_T and the $2D$ coordinates $\mathbf{p}_T = [u_T, v_T]^T$ of the pixel p_T receiving the NIR radiation echoed by P_T is given by the perspective projection equation, rewritten for clarity's sake as

$$z_T \begin{bmatrix} u_T \\ v_T \\ 1 \end{bmatrix} = \mathbf{K}_T \begin{bmatrix} x_T \\ y_T \\ z_T \end{bmatrix} \tag{3.43}$$

where the ToF camera intrinsic parameters matrix $\mathbf{K}_T$ is defined as in (1.5).

Because of lens distortion, coordinates $\mathbf{p}_T = [u_T, v_T]^T$ of (3.43) are related to the coordinates $\hat{\mathbf{p}}_T = [\hat{u}_T, \hat{v}_T]^T$ actually measured by the ToF camera by a relationship of type $\hat{\mathbf{p}}_T = [\hat{u}_T, \hat{v}_T]^T = \Psi(\mathbf{p}_T)$, where $\Psi(\cdot)$ is a distortion transformation as described in Sect. 1.1.2. Model (1.14) supported by the camera calibration procedure [19] is also widely used for ToF cameras, as well as other more complex models, e.g., [14].

As already explained, each sensor pixel p_T directly estimates the radial distance $\hat{r}_T$ from its corresponding scene point P_T. With minor and neglectable approximation due to the non-perfect collinearity between emitters, pixel p_T and $3D-T$ reference system origin, the measured radial distance $\hat{r}_T$ can be expressed as

$$\hat{r}_T = \sqrt{\hat{x}_T^2 + \hat{y}_T^2 + \hat{z}_T^2} = \left|\left|[\hat{x}_T^2, \hat{y}_T^2, \hat{z}_T^2]^T\right|\right|_2. \tag{3.44}$$

From radial distance $\hat{r}_T$ measured at pixel p_T with distorted coordinates $\hat{\mathbf{p}}_T = [\hat{u}_T, \hat{v}_T]^T$ the $3D$ coordinates of $\mathbf{P}_T$ can be computed according to the following steps:

1. Given the lens distortion parameters, estimate the non-distorted $2D$ coordinates $\mathbf{p}_T = [u_T, v_T]^T = \Psi^{-1}(\hat{\mathbf{p}}_T)$, where $\Psi^{-1}(\cdot)$ is the inverse of $\Psi(\cdot)$.
2. The estimated depth value $\hat{z}_T$ can be computed from (3.43) and (3.44) as

$$\hat{z}_T = \frac{\hat{r}_T}{||\mathbf{K}_T^{-1}\,[u_T, v_T, 1]^T||_2} \tag{3.45}$$

   where $\mathbf{K}_T^{-1}$ is the inverse of $\mathbf{K}_T$.
3. The estimated coordinates values $\hat{x}_T$ and $\hat{y}_T$ can be computed by inverting (3.43), i.e., as

$$\begin{bmatrix} \hat{x}_T \\ \hat{y}_T \\ \hat{z}_T \end{bmatrix} = K_T^{-1} \begin{bmatrix} u_T \\ v_T \\ 1 \end{bmatrix} \hat{z}_T. \tag{3.46}$$

Let us next summarize the operation of a ToF camera as an imaging system for the case of sinusoidal modulation (similar considerations apply to the case of square wave modulation). Each ToF camera sensor pixel, at each period T_m of the modulation sinusoid, collects four samples c_R^0, c_R^1, c_R^2 and c_R^3 of the NIR signal reflected by the scene. Every N periods of the modulation sinusoid, where N is a function of the integration time, each ToF sensor pixel estimates an amplitude value $\hat{A}$, an intensity value $\hat{B}$, a phase value $\widehat{\Delta\varphi}$, a radial distance value $\hat{r}_T$ and the $3D$ coordinates $\hat{\mathbf{P}}_T = [\hat{x}_T, \hat{y}_T, \hat{z}_T]^T$ of the corresponding scene point.

Since amplitude $\hat{A}$, intensity $\hat{B}$ and depth $\hat{z}_T$ are estimated at each sensor pixel, ToF depth cameras handle them in matricial structures, and return them as 2D maps or depth maps. Therefore, every N periods of the modulation sinusoid

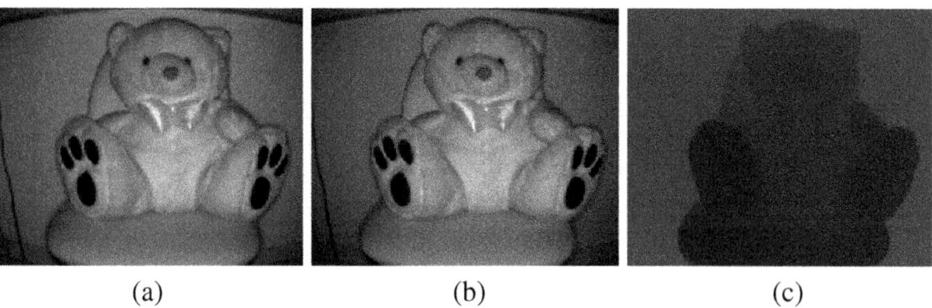

(a) (b) (c)

Fig. 3.12 Example of (**a**) $\hat{A}_T$, (**b**) $\hat{B}_T$ and (**c**) $\hat{Z}_T$

(corresponding to several tens of times per second), a ToF depth camera provides as output the following data:

- an *amplitude map* $\hat{A}_T$, i.e., a matrix obtained by juxtaposing the amplitudes estimated at all the ToF sensor pixels. It is defined on lattice Λ_T and its values, expressed in volts [V], belong to the pixel non-saturation interval. Map $\hat{A}_T$ can be modeled as realization of a random field $\mathscr{A}_T$ defined on Λ_T, with values expressed in volts [V] in the pixel non-saturation interval;
- an *intensity map* $\hat{B}_T$, i.e., a matrix obtained by juxtaposing the intensity values estimated at all the ToF sensor pixels. It is defined on lattice Λ_T and its values, expressed in volts [V], belong to the pixel non-saturation interval. Map $\hat{B}_T$ can be modeled as the realization of a random field $\mathscr{B}_T$ defined on Λ_T, with values expressed in volts [V] in the pixel non-saturation interval;
- a *depth map* $\hat{Z}_T$, i.e, a matrix obtained by juxtaposing the depth values estimated at all the ToF sensor pixels. It is defined on lattice Λ_T and its values, expressed in [mm], belong to the interval $[0, r_{MAX} = c/(2f_m))$. Map $\hat{Z}_T$ can be considered as the realization of a random field $\mathscr{Z}_T$ defined on Λ_T, with values expressed in [mm] in the interval $[0, r_{MAX})$.

By mapping amplitude, intensity and depth values to interval [0, 1], the three maps $\hat{A}_T$, $\hat{B}_T$, and $\hat{Z}_T$ can be represented as images as shown in Fig. 3.12 for a sample scene. For the scene of Fig. 3.12, images $\hat{A}_T$ and $\hat{B}_T$ are very similar because the scene illumination is constant.

3.3 Practical Implementation Issues of ToF Depth Cameras

The previous sections highlight the conceptual steps needed to measure the distances of a scene surface by a ToF depth camera, but they do not consider a number of non-idealities described below, which must be taken into account in practice.

3.3.1 Phase Wrapping

The first fundamental limitation of ToF sensors comes from the fact that the estimate of $\widehat{\Delta\varphi}$ is obtained from an arctangent function, which has co-domain $[-\pi/2, \pi/2]$. Therefore, the estimates of $\widehat{\Delta\varphi}$ can only assume values in this interval. Since the physical delays entering the phase shift $\Delta\varphi$ of (3.13) can only be positive, it is possible to shift the arctan(·) co-domain to $[0, \pi]$ in order to have a larger interval available for $\widehat{\Delta\varphi}$. Moreover, the usage of atan2(·, ·) allows one to extend the co-domain to $[0, 2\pi]$. From (3.17) it is immediate to see that the estimated distances are within range $[0, c/(2f_m)]$. If for instance $f_m = 30$ [MHz], the interval of measurable distances is [0–5] [m].

Since $\widehat{\Delta\varphi}$ is estimated modulo 2π from (3.17) and the distances greater than $c/(2f_m)$ correspond to $\widehat{\Delta\varphi}$ greater than 2π, they are incorrectly estimated. In practice the distance returned by (3.17) corresponds to the remainder of the division between the actual $\Delta\varphi$ and 2π, multiplied by $c/(2f_m)$, a well-known phenomenon called *phase wrapping* since it refers to a periodic wrapping around 2π of phase values $\widehat{\Delta\varphi}$. Clearly, if f_m increases, the interval of measurable distances becomes smaller, and vice-versa. Possible solutions to overcome phase wrapping include the usage of multiple modulation frequencies or of non-sinusoidal wave-forms (e.g., chirp wave-forms).

3.3.2 Harmonic Distortion

The generation of perfect sinusoids of the needed frequency is not straightforward. In practice [12], actual sinusoids are obtained as low-pass filtered versions of square waveforms emitted by LEDs. Moreover, the sampling of the received signal is not ideal, but it takes finite time intervals Δ_S, as shown in Fig. 3.13. The combination of these two factors introduces an harmonic distortion in the estimated phase-shift $\widehat{\Delta\varphi}$ and consequently in the estimated distance $\hat{\rho}$. Such harmonic distortion leads to a systematic offset component dependent on the measured distance. A metrological characterization of this harmonic distortion effect is reported in [21, 29]. Figure 3.14 shows that the harmonic distortion offset exhibits a kind of oscillatory behavior which can be up to some tens of centimeters, clearly reducing the accuracy of distance measurements. As reported in Chap. 4, this systematic offset can be fixed by a lookup table correction.

3.3.3 Photon-Shot Noise

Because of the light-collecting nature of the receiver, the acquired samples c_R^0, c_R^1, c_R^2 and c_R^3 are affected by photon-shot noise, due to dark electron current and photon-generated electron current as reported in [13]. Dark electron current can be reduced

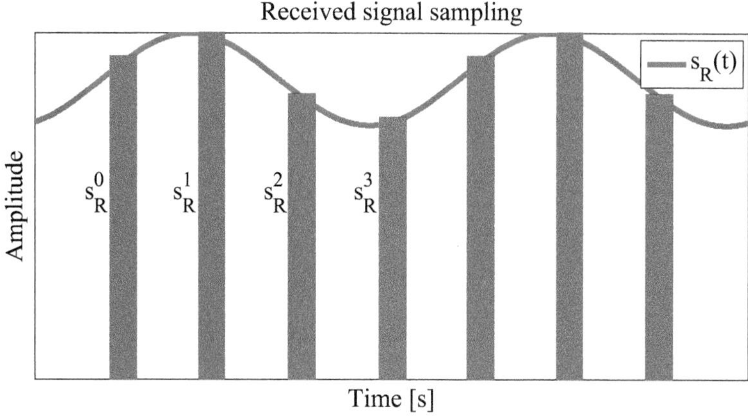

Fig. 3.13 Pictorial illustration of non instantaneous sampling of the received signal $s_R(t)$

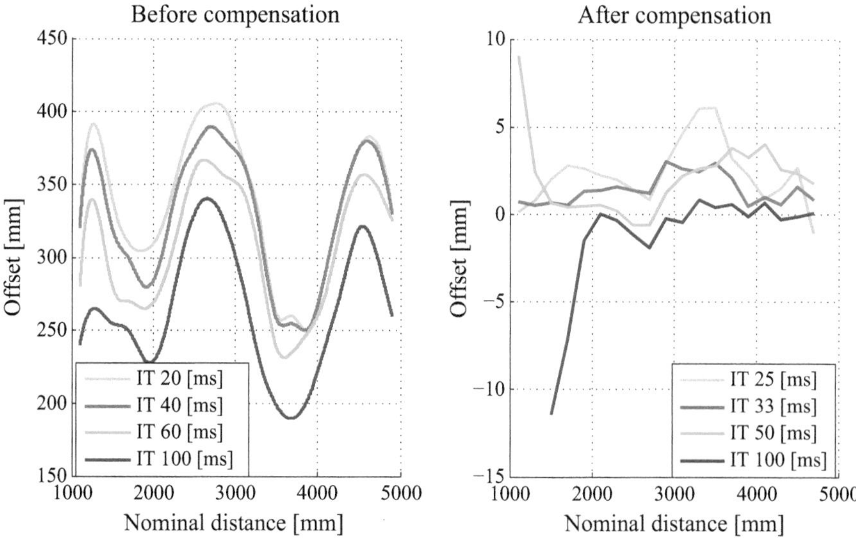

Fig. 3.14 *Left:* systematic distance measurements offset due to harmonic distortion before compensation (from [21]). *Right:* systematic distance measurements offset after compensation (courtesy of MESA Imaging)

by lowering the sensor temperature or by technological improvements. Photon-generated electron current, due to light-collection, cannot be completely eliminated. Photon-shot noise is statistically characterized by a Poisson distribution. Since $\hat{A}$, $\hat{B}$, $\widehat{\Delta\varphi}$ and $\hat{\rho}$ are computed directly from the corrupted samples c_R^0, c_R^1, c_R^2 and c_R^3, their noise distribution can be computed by propagating the Poisson distribution through (3.16)–(3.17). A detailed analysis of error and noise propagations can be found in [24].

Quite remarkably, the probability density function of the noise affecting estimate $\hat{\rho}$, in the case of sinusoidal modulation, according to [13, 24] can be approximated by a Gaussian[1] with standard deviation

$$\sigma_\rho = \frac{c}{4\pi f_m \sqrt{2}} \frac{\sqrt{B}}{A}. \tag{3.47}$$

Standard deviation (3.47) determines the precision, or repeatability, of the distance measurement and is directly related to f_m, A and B. In particular, if the received signal amplitude A increases, the precision improves. This suggests that the precision improves as the measured distance decreases and the reflectivity of the measured scene point increases.

Equation (3.47) also indicates that as the interference intensity B of the received signal increases, precision worsens. This means that precision improves as the scene background IR illumination decreases. Note that B may increase because of two factors: an increment of the received signal amplitude A or an increment of the background illumination. While in the second case the precision gets worse, in the first case there is an overall precision improvement, given the squared root dependence of B in (3.47). Finally, observe that B cannot be 0 as it depends on carrier intensity A.

If modulation frequency f_m increases, precision improves. The modulation frequency is an important parameter for ToF sensors, since f_m is also related to phase wrapping and the maximum measurable distance. In fact, if f_m increases, the measurement precision improves, while the maximum measurable distance decreases (and vice-versa). Therefore, there is a trade-off between distance precision and range. Since f_m is generally a tunable parameter, it can be adapted to the distance precision and range requirements of the specific application.

3.3.4 *Saturation and Motion Blur*

Averaging over multiple periods is effective against noise but it introduces dangerous side effects, such as *saturation* and *motion blur*. Saturation occurs when the received quantity of photons exceeds the maximum quantity that the receiver can collect. This phenomenon is particularly notable in presence of external IR illumination (e.g., direct solar illumination) or in the case of highly reflective objects (e.g., specular surfaces). The longer the integration time, the higher the quantity of collected photons and the more likely the possibility of saturation. Specific solutions

[1] An explicit expression of the Gaussian probability density function mean is not given in [13, 24]. However, the model of Muft and Mahony [24] provides implicit information about the mean which is a function of both A and B, and contributes to the distance measurement offset. For calibration purposes the non-zero mean effect can be included in the harmonic distortion.

have been developed in order to avoid saturation, i.e., in-pixel background light suppression and automatic integration time setting [12, 13].

Motion blur is another important phenomenon accompanying time averaging. It is caused by imaged objects moving during integration time, as in the case of standard cameras. Time intervals of the order of 1–100 [ms] make object movement likely unless the scene is perfectly still. In the case of moving objects, the samples entering (3.50) do not concern a specific fixed scene point at subsequent instants as is expected in theory, but different scene points at subsequent instants, causing distance measurement artifacts. The longer the integration time, the higher the likelihood of motion blur but better the distance measurement precision. Integration time is another parameter to set according to the characteristics of the specific application.

3.3.5 Multipath Error

In the model presented in Sect. 3.1.1, we assumed that the signal $s_E(t)$ transmitted from the source is reflected back by the scene in a single ray. In a more realistic scenario, the signal transmitted from the source will encounter multiple objects in the environment that produce reflected, diffracted, or scattered copies of the transmitted signal, as shown in Fig. 3.15. These additional copies of the transmitted signal, called *multipath* signal components, are summed together at the receiver, leading to a combination of the incoming light paths and thus to a wrong distance estimation. Since the radial distance of a scene point P from the ToF camera is computed from the time-length of the shortest path between P and the camera, the multipath effect leads to over-estimation of the scene points' distances.

Figure 3.15 shows an optical ray (red) incident to a non-specular surface reflected in multiple directions (green and blue). The ideal propagation scenario of Fig. 3.1, with co-positioned emitters and receivers, considers only the presence of the green ray of Fig. 3.15, i.e., the ray back reflected in the direction of the incident ray and

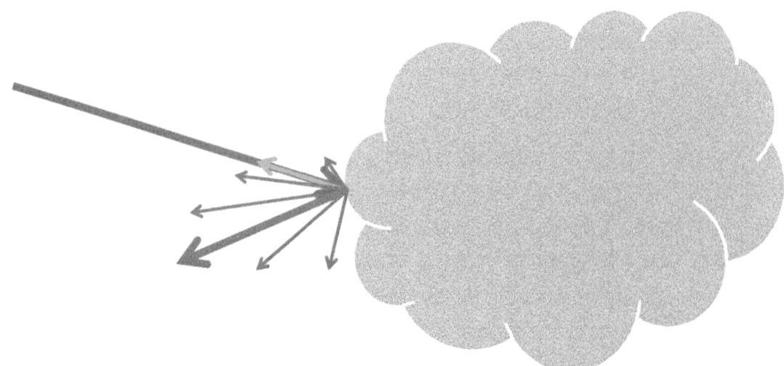

Fig. 3.15 Scattering effect

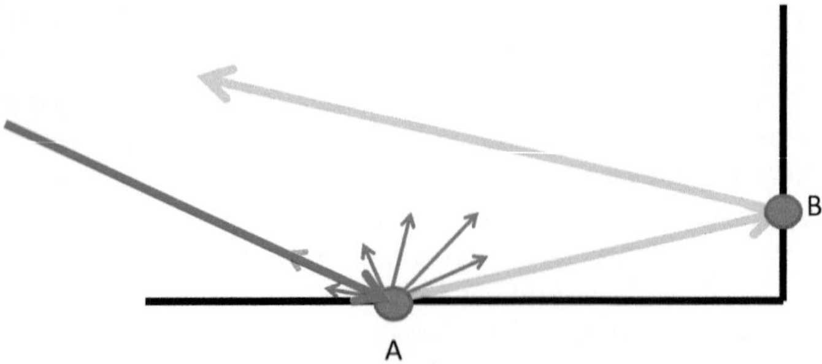

Fig. 3.16 Multipath phenomenon: the incident ray (*red*) is reflected in multiple directions (*blue and orange rays*) by the surface at point *A*. The *orange ray* reaches then *B* and travels back to the ToF sensor

disregards the presence of the other (blue) rays. In practical situations, however, the presence of the other rays may not always be negligible. In particular, the ray specular to the incident ray direction with respect to the surface normal at the incident point (thick blue ray) generally is the reflected ray with greatest radiometric power.

All the reflected (blue) rays may first hit other scene points and then travel back to the ToF sensor, therefore affecting distance measurements of other scene points. For instance, as shown in Fig. 3.16, an emitted ray (red) may be first reflected by a point surface *A* with a scattering effect. One of the scattered rays (orange) may then be reflected by another scene point *B* and travel back to the ToF sensor. The distance measured by the sensor pixel relative to *B* is therefore a combination of two paths, namely path to ToF camera-*B*-ToF camera and path ToF camera-A-B-ToF camera. The coefficients of such a combination depend on the optical amplitude of the respective rays.

There are multiple sources of the multipath effect, and most of them are related to the properties of the scene. In general, all materials reflect incoming light in all directions, so a normal scene will produce indirect reflections everywhere and each camera pixel will measure the superposition of infinite waves. Fortunately, most of the time the indirect reflections are order of magnitudes weaker than direct reflections, and the camera can easily resolve the reflected signal. When the object is highly reflective or transparent, however, the camera pixel will receive multiple signals with different phase and attenuation, leading to incorrect measurements. One of the most visible effects of multipath interference is relative to concave corners, which often appear rounded in ToF depth maps. This happens because each point belonging to one side of the corner will receive light reflected by any point of the other side and reflects parts of it towards the camera, resulting in an over-estimation of the distances of the points on the corner surface. The interference of different waves is not necessarily related only to the scene; the optics and other internal components of the ToF camera may scatter and reflect small amounts of the received signal as well.

In order to model the multipath error, in the case of sinusoidal modulation we can rewrite (3.2), by considering N incoming waves

$$s_R(t) = \sum_{i=1}^{N} (a)_i sin(2\pi f_m t + \Delta\varphi_i) + B_i. \tag{3.48}$$

Since the sum of sinusoidal functions is still a sinusoid, and it is difficult to estimate the contribution of each independent ray, in practice only two components are considered: a first direct signal, and a second indirect signal that takes into account all the additional reflections. With these assumptions, (3.48) can be rewritten as

$$s_R(t) = [Asin(2\pi f_m t + \Delta\varphi) + B] + [A_{MP} sin(2\pi f_m t + \Delta\varphi_{MP}) + B_{MP}] \tag{3.49}$$

where the second component takes into account the multipath signals.

In the literature, there are several works that propose solutions to multipath interference, Whyte et al. [30] reviews current state of the art techniques used to correct for this error. When a single frequency is used in the presence of multipath, the relationship among modulation frequency, measured phase, and amplitude is non-linear. By exploiting modulation frequency diversity, it is possible to iteratively reconstruct the original signal using two or more modulation frequencies [11, 15, 22], and find a closed-form solution by using four modulation frequencies [16]. Another solution to address multipath is to use coded waves [12, 20] where the signal in (3.8) is replaced by a binary sequence or more particular custom codes, and the received signal is estimated by means of sparse deconvolution. The general idea is that the combination of pure sinusoidal signals is still a sinusoid and this creates a unicity problem at the receiver. The use of different signals instead allows one to recognize when the received signal has been corrupted by the scene.

3.3.6 *Flying Pixels*

Another problem similar to multipath is the *flying pixel* effect. Since the pixels of any imaging sensor don't have infinitesimal size but some physical size, as shown in Fig. 3.17, each pixel receives the radiation reflected from all the points of the corresponding small scene patch and the relative distance information. If the scene patch is a flat region with constant reflectivity, the approximation that there is a single scene point associated with the specific pixel does not introduce any artifacts. However, if the scene patch corresponds to a discontinuity of the scene reflectivity, the values of $\hat{A}_T(p_T)$ and $\hat{B}_T(p_T)$ estimated by the correspondent pixel p_T average different reflectivity values.

A worse effect occurs if the scene patch associated with p_T corresponds to a depth discontinuity. In this case, assume that a portion of the scene patch is at a closer distance, called z_{near}, and another portion at further distance, called z_{far}.

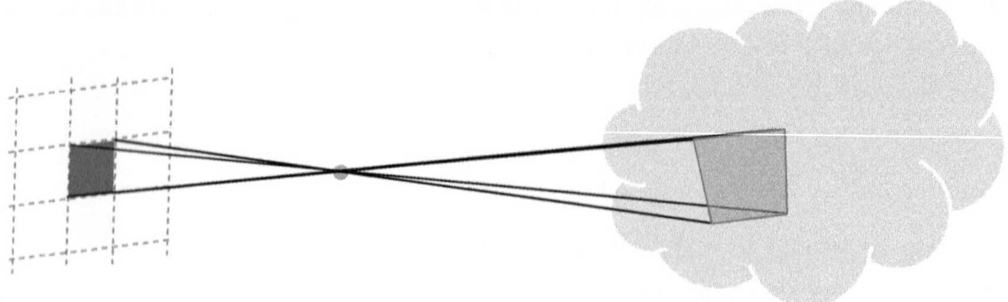

Fig. 3.17 Finite size scene patch (*right*) associated with a pixel (*left*) of any imaging sensor

The resulting depth estimate $\hat{Z}_T(p_T)$ is a convex combination of z_{near} and z_{far}, where the combination coefficients depend on the percentage of area at z_{near} and at z_{far} respectively reflected on p_T. The presence of flying pixels leads to severe depth estimation artifacts, as shown by the example of Fig. 3.18, where foreground and background are blended together.

The most effective solutions to this problem tackle the detection and eventual correction of these points as shown in [26]. More recent works aim at providing a confidence value for each pixel, based on analysis of intensity and amplitude of the received signal [25].

3.3.7 Other Noise Sources

There are several other noise sources affecting the distance measurements of ToF sensors, notably *flicker* and *kTC noise*. The receiver's amplifier introduces a Gaussian-distributed thermal noise component. Since the amplified signal is quantized in order to be digitally processed, this introduces another source of error, customarily modeled as random noise. Quantization noise can be controlled by the number of used bits and it is typically neglectable with respect to the other noise sources. All noise sources, except photon-shot noise, may be reduced by adopting high quality components. A comprehensive description of various ToF noise sources can be found in [12, 13, 23, 24].

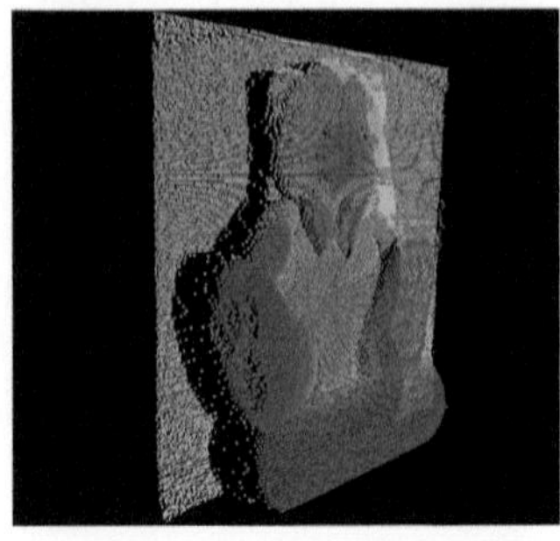

Fig. 3.18 An example of flying pixels at the depth edge between object and wall

Averaging distance measurements over several modulation periods T_m is a classical provision to mitigate the noise effects. If N is the number of periods, in the case of sinusoidal modulation the estimated values $\hat{A}$, $\hat{B}$ and $\widehat{\Delta\varphi}$ become

$$\hat{A} = \frac{\sqrt{\left(\frac{1}{N}\sum_{n=0}^{N-1} c_R^{4n} - \frac{1}{N}\sum_{n=0}^{N-1} c_R^{4n+2}\right)^2 + \left(\frac{1}{N}\sum_{n=0}^{N-1} c_R^{4n+1} - \frac{1}{N}\sum_{n=0}^{N-1} c_R^{4n+3}\right)^2}}{2}$$

$$\hat{B} = \frac{\sum_{n=0}^{N-1} c_R^{4n} + \sum_{n=0}^{N-1} c_R^{4n+1} + \sum_{n=0}^{N-1} c_R^{4n+2} + \sum_{n=0}^{N-1} c_R^{4n+3}}{4N}$$

$$\widehat{\Delta\varphi} = \operatorname{atan2}\left(\frac{1}{N}\sum_{n=0}^{N-1} c_R^{4n} - \frac{1}{N}\sum_{n=0}^{N-1} c_R^{4n+2}, \frac{1}{N}\sum_{n=0}^{N-1} c_R^{4n+1} - \frac{1}{N}\sum_{n=0}^{N-1} c_R^{4n+3}\right) \tag{3.50}$$

where

$$\begin{aligned} c_R^{4n} &= c_R\left(\frac{4n}{F_S}\right) \\ c_R^{4n+1} &= c_R\left(\frac{4n+1}{F_S}\right) \\ c_R^{4n+2} &= c_R\left(\frac{4n+2}{F_S}\right) \\ c_R^{4n+3} &= c_R\left(\frac{4n+3}{F_S}\right). \end{aligned} \tag{3.51}$$

This provision reduces but does not completely eliminates noise effects. The averaging intervals used in practice are typically between 1 [ms] and 100 [ms]. For instance, when $f_m = 30$ [MHz] (the modulating sinusoid period is 33.3×10^{-9} [s]), the averaging intervals concern a number of modulating sinusoid periods from 3×10^4 to 3×10^6. The averaging interval length is generally called *integration time*, and its proper tuning is extremely important in ToF measurements. Long integration times lead to repeatable and reliable ToF distance measurements at the expense of motion blur effects.

3.4 Examples of ToF Depth Cameras

3.4.1 Kinect™ v2

With the introduction of the Xbox One gaming device in November 2013, Microsoft also presented a second version of the Kinect™, called Kinect™ v2 for simplicity.

Kinect^{TM} v2 with respect to Kinect^{TM} v1 is a completely different product, since it employs a ToF depth camera while Kinect^{TM} v1 employed a structured light depth camera. A release of Kinect^{TM} v2 suited for connection to a standard PC was made available together with the Software Development Kit in 2014. As with the Kinect^{TM} v1, the Kinect^{TM} v2 includes the depth sensing element, a video camera and an array of microphones.

A high level description of the operating principles of Kinect^{TM} v2 can be found in [5], while more details are given in Microsoft patents. The ToF depth camera was developed from former products by Canesta, a ToF depth camera producer acquired by Microsoft in 2010. Some innovative details introduced in order to overcome some of the issues of Sect. 3.3 are worth noting. Kinect^{TM} v2 is able to acquire a 512×424 pixels depth map (the largest resolution achieved by a ToF depth camera at the time of writing this book) at 50 fps with a depth estimation error typically smaller than 1 % of the measured distances. The emitted light is modulated by a square wave (see Sect. 3.1.2) instead of a sinusoid as in most previous ToF depth cameras. The receiver ToF sensor is a differential pixels array, i.e., each pixel has two outputs and the incoming photons contribute to one or the other according to the current state of a clock signal. The clock signal is the same square wave used for the modulation of the emitter. Let us denote with U the signal corresponding to the photons arriving when the clock is high and L the signal corresponding to the low state of the clock. The difference $(U - L)$ depends on both the amount of returning light and on the time it takes to come back, and allows one to estimate the time lag used to compute the distance. Square wave modulation helps against harmonic distortion issues.

Another well-known critical trade-off is between precision and the maximum measurable range given by (3.17) and (3.47), i.e., by increasing f_m the measurement precision increases but the measurable range gets smaller. Kinect^{TM} v2 deals with this issue by using multiple modulation frequencies which are 17, 80 and 120 [MHz]. Multiple modulation frequencies allow one to extend the acquisition range, overcoming limits due to phase wrapping. Indeed, the correct measurement can be disambiguated by identifying the measurement values consistent with respect to all three modulation frequencies, as visually exemplified by Fig. 3.19.

Another improvement introduced by the Kinect v2 is the capability of simultaneously acquiring two images with different shutter times, namely 100 and 1000 [μs] and selecting whichever one leads to the best pixel by pixel result; this is made possible by the non-destructive pixel reading feature of its sensor.

3.4.2 MESA ToF Depth Cameras

MESA Imaging, which was founded in 2006, is a spin-off of the Swiss Center for Electronics and Microtechnology (CSEM) acquired by Heptagon in 2014. It was one of the first companies to commercialize ToF depth cameras and its main product, the SwissRanger, is now in its 4th generation (Fig. 3.20). Different from

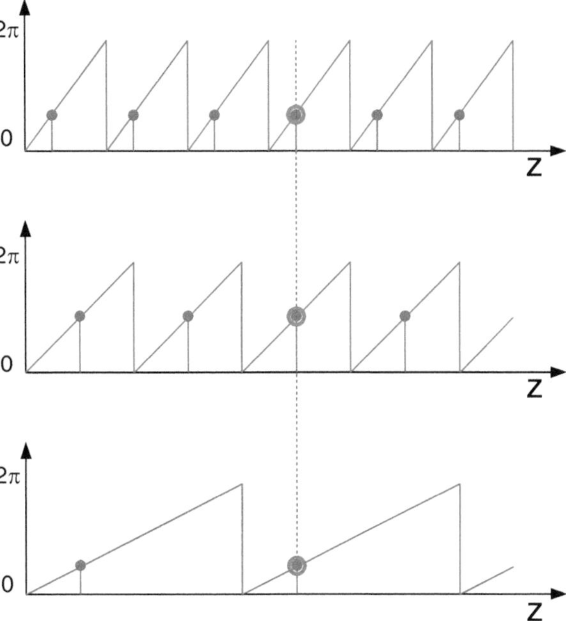

Fig. 3.19 Disambiguation of phase wrapping errors by multiple modulation frequencies. Notice how by looking at each single plot there are ambiguities on the actual distance value (represented by the *multiple dots*), but by comparing all the plots it is possible to disambiguate the various measurements

the Kinect™ v2, the SwissRanger is an industrial grade product developed for measurement applications rather than for interfaces or gaming. The SwissRanger uses CWAM with sinusoidal modulation according to the principles presented in this chapter. For a detailed description see [4]. The modulation frequency can be chosen among 14.5, 15, 15.5, 29, 30 and 31 [MHz]. Typical SwissRanger operation

(a) (b) (c)

Fig. 3.20 Mesa Imaging ToF depth cameras: (**a**) MESA Imaging SR3000™; (**b**) MESA Imaging SR4000™; (**c**) MESA Imaging SR4500™

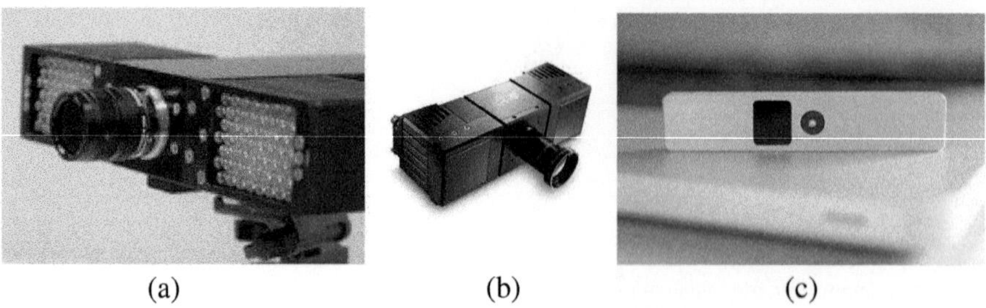

(a) (b) (c)

Fig. 3.21 PMD ToF depth cameras: (**a**) PMD PhotonICs; (**b**) PMD CamCube; (**c**) PMD CamBoard pico

is in the [0, 5] [m] range with nominal accuracy of 1 [cm] at 2 m and in the longer range of [0, 10] [m] with accuracy of 1.5 [cm] at 2 m. Notably, one can use up to three SwissRanger cameras together without interference issues.

3.4.3 PMD Devices

PMD technologies is another early ToF depth camera producer (Fig. 3.21). This spin-off of the Center for Sensor Systems (ZESS) of the University of Siegen (Germany) was founded in 2002. In 2005, it launched the Efector camera, its first commercial product. The company then introduced the Efector 3D in 2008, a 64×48 pixels ToF depth camera developed for industrial use. In 2009 the company launched the CamCube, a 204×204 pixels ToF depth camera characterized by the highest resolution ToF sensor until the introduction of Kinect™ v2. The initial focus of the company was on industrial applications but recently it entered other fields, including automotive, gesture recognition and consumer electronics (it is taking part in Google's Project Tango). Recent products include the CamBoard, a 200×200 single board 3D ToF depth camera, and the PhotonICs 19k-S3 chip for camera developers and system integrators. PMD depth cameras operate according to the CWAM modulation principles introduced in Sect. 3.1.

3.4.4 ToF Depth Cameras Based on SoftKinetic Technology

SoftKinetic is a Belgian company, founded in 2007 and acquired by Sony in 2015, which has produced two generations of ToF depth cameras, the DS311 and the newer DS325 (Fig. 3.22). The latter, introduced in 2013, is a short range ToF depth camera with an optimal working range of 0.1–1 [m] targeted to gesture recognition applications; it can acquire data up to around 2 [m] but beyond 1 [m] its accuracy dramatically decreases. This ToF depth camera is based on the company's patented

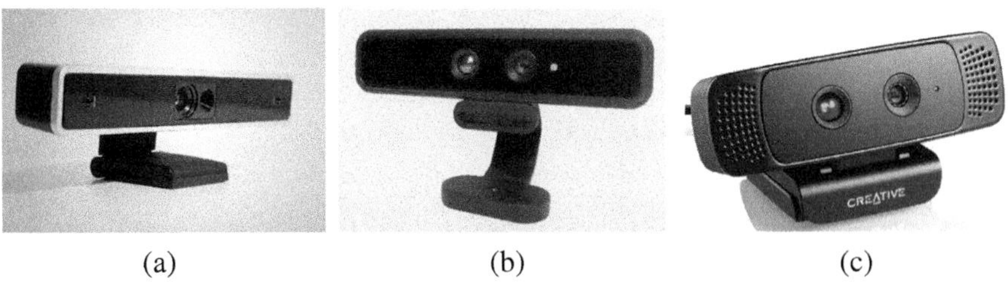

(a) (b) (c)

Fig. 3.22 SoftKinetic ToF depth cameras: (**a**) SoftKinetic DS311; (**b**) SoftKinetic DS325; (**c**) Creative Senz3D™

CMOS pixel technology, called *Current Assisted Photonic Demodulation (CAPD)*. This technique uses a driving current to move electrons towards two different detecting junctions as a result of an alternating current, with a result similar to the differential pixels of the Kinect v2. SoftKinetic adopts CWAM modulation with a square wave modulating signal similar to the Kinect™ v2. The transmitter uses a laser illuminator and the receiver has a resolution of 320×240 pixels. The DS325 can acquire data at up to 60 [fps] within a nominal range 15 [cm]–1 [m], thus being particularly suited for hand gesture recognition applications and computer interfaces. SoftKinetic made available an updated driver allowing one to acquire data up to 4 [m], but the range increase decreases the resolution or the frame rate. The accuracy is about 1.4 [cm] at 1 [m] and the built-in calibration is not very accurate, making the DS325 more suited to gesture recognition applications than to 3D reconstruction purposes. The camera is also sold by Creative under the Senz3D™ name: the DS325 and the Senz3D™ essentially share same hardware with a different case and exterior appearance.

3.5 Conclusions and Further Reading

This chapter introduces the operating principles of continuous wave homodyne amplitude modulation within In-Pixel Photo-Mixing devices in a unified treatment for both sinusoidal and square wave modulation cases.

A comprehensive review of current ToF technology, not only the CWAM type, can be found in [27]. The doctoral dissertation of Robert Lange [23] remains a very readable and accurate description of classical CWAM ToF operation and technology.

ToF depth cameras' image formation can be approximated by the pin-hole model, typical of standard cameras, recalled in Sect. 1.1.1. More extensive treatments on topics such as image formation, projective geometry and camera calibration can be found in [14, 18, 19, 28]. More on ToF camera calibration will be seen in Chap. 4.

The various implementation issues affecting the practical operation and performance of ToF depth cameras are considered in Sect. 3.3. In-depth analysis of ToF noise sensor sources and practical provisions against noise can be found in

[12, 13, 21, 24]. A comprehensive treatment of multipath interference in CWAM ToF depth cameras and related issues, such as flying pixels, can be found in [17]. More specific multipath interference issues are treated in [11, 15, 16, 30].

Section 3.4 reviews current ToF depth camera products which exemplify the operating principles of Sect. 3.1.

Let us finally observe that ToF cameras can be considered as special Multiple-Input and Multiple-Output (MIMO) communication systems, where the emitters array is the input array and the lock-in matrix of the ToF receiver sensor the output array. In principle, this framework would allow one to approach multipath as customarily done in communication systems. However, the number of input and output channels of a ToF camera vastly exceeds the complexity of the MIMO systems used in telecommunications, in which the number of inputs and outputs rarely exceeds the tens of units. Consider that already for the early ToF depth cameras like the MESA SR4000, there were 24 input channels associated with the transmitter's emitters, and 176×144 output channels associated with the lock-in pixels of the receiver's sensor. Even though the current multipath analysis methods used for MIMO systems cannot be applied to ToF depth cameras, the application of communications systems techniques for characterizing ToF depth cameras operations and improving their performance appears an attractive possibility.

References

1. Csem (2016), http://www.csem.ch. Last accessed in 2016
2. Fbk (2016), http://www.fbk.eu. Last accessed in 2016
3. Iee (2016), http://www.iee.lu. Last accessed in 2016
4. Mesa Imaging (2016), http://www.mesa-imaging.ch. Last accessed in 2016
5. Microsoft Kinect™(2016), http://www.xbox.com/en-US/kinect. Last accessed in 2016
6. Microsoft® (2016), http://www.microsoft.com. Last accessed in 2016
7. Panasonic d-imager (2016), http://www.panasonic-electric-works.com. Last accessed in 2016
8. Pmd Technologies (2016), http://www.pmdtec.com/. Last accessed in 2016
9. Softkinetic (2016), http://www.softkinetic.com/. Last accessed in 2016
10. J. Andrews, N. Baker, Xbox 360 system architecture. IEEE Micro **26**(2), 25–37 (2016)
11. A. Bhandari, A. Kadambi, R. Whyte, C. Barsi, M. Feigin, A. Dorrington, R. Raskar, Resolving multipath interference in time-of-flight imaging via modulation frequency diversity and sparse regularization. Opt. Lett. **39**(6), 1705–1708 (2014)
12. B. Buttgen, P. Seitz, Robust optical time-of-flight range imaging based on smart pixel structures. IEEE Trans. Circuits Syst. I: Regul. Pap. **55**(6), 1512–1525 (2008)
13. B. Buttgen, T. Oggier, M. Lehmann, R. Kaufmann, F. Lustenberger, Ccd/cmos lock-in pixel for range imaging: challenges, limitations and state of the art, in *1st Range Imaging Research Day* (2005)
14. D. Claus, A.W. Fitzgibbon, A rational function lens distortion model for general cameras, in *Proceedings of IEEE Conference on Computer Vision and Pattern Recognition* (2005)
15. A.A. Dorrington, J.P. Godbaz, M.J. Cree, A.D. Payne, L.V. Streeter, Separating true range measurements from multi-path and scattering interference in commercial range cameras, *Proceedings of SPIE* **7864**, 786404–786404-10 (2011)
16. J.P. Godbaz, M.J. Cree, A.A. Dorrington, Closed-form inverses for the mixed pixel/multipath interference problem in amcw lidar, *Proceedings of SPIE* **8296**, 829618–829618-15 (2012)

17. J.P. Godbaz, A.A. Dorrington, M.J. Cree, Understanding and ameliorating mixed pixels and multipath interference in amcw lidar, in *TOF Range-Imaging Cameras*, ed. by F. Remondino, D. Stoppa (Springer, Berlin/Heidelberg, 2013), pp. 91–116
18. R.I. Hartley, A. Zisserman, *Multiple View Geometry in Computer Vision* (Cambridge University Press, Cambridge , 2004)
19. J. Heikkila, O. Silven, A four-step camera calibration procedure with implicit image correction, in *Proceedings of IEEE Conference on Computer Vision and Pattern Recognition* (1997)
20. A. Kadambi, R. Whyte, A. Bhandari, L. Streeter, C. Barsi, A. Dorrington, R. Raskar, Coded time of flight cameras: sparse deconvolution to address multipath interference and recover time profiles. ACM Trans. Graph. **32**(6), 167:1–167:10 (2013)
21. T. Kahlmann, H. Ingensand, Calibration and development for increased accuracy of 3d range imaging cameras. J. Appl. Geod. **2**, 1–11 (2008)
22. A. Kirmani, A. Benedetti, P.A. Chou, Spumic: simultaneous phase unwrapping and multipath interference cancellation in time-of-flight cameras using spectral methods, in *Proceedings of IEEE International Conference on Multimedia and Expo*, pp. 1–6 (2013)
23. R. Lange, 3D Time-Of-Flight distance measurement with custom solid-state image sensors in CMOS/CCD-technology, Ph.D. thesis, University of Siegen, (2000)
24. F. Mufti, R. Mahony, Statistical analysis of measurement processes for time-of flight cameras, in *Proceedings of SPIE* (2009)
25. M. Reynolds, J. Doboš, L. Peel, T. Weyrich, G.J. Brostow, Capturing time-of-flight data with confidence, in *Proceedings of IEEE Conference on Computer Vision and Pattern Recognition* (2011)
26. A. Sabov, J. Krüger, Identification and correction of flying pixels in range camera data, in *Proceedings of ACM Spring Conference on Computer Graphics* (New York, 2010), pp. 135–142
27. D. Stoppa, F. Remondino, *TOF Range-Imaging Cameras* (Springer, Berlin, 2013)
28. R. Szeliski, *Computer Vision: Algorithms and Applications* (Springer, New York, 2010)
29. C. Uriarte, B. Scholz-Reiter, S. Ramanandan, D. Kraus, Modeling distance nonlinearity in tof cameras and correction based on integration time offsets, in *Progress in Pattern Recognition, Image Analysis, Computer Vision, and Applications* (Springer, Berlin/Heidelberg, 2011)
30. R. Whyte, L. Streeter, M.J. Cree, A.A. Dorrington, Review of methods for resolving multi-path interference in time-of-flight range cameras, in *Proceedings of IEEE Sensors* (2014), pp. 629–632

Part II
Extraction of 3D Information from Depth Cameras Data

Chapter 4
Calibration

Color imaging instruments, such as photo and video cameras, and depth imaging instruments, such as ToF and structured light depth cameras, require preliminary calibration in order to be used for measurement purposes.

Calibration must account both for geometric and photometric properties, and should be accurate and precise for reliable measurements. Geometric calibration accounts for the internal characteristics, called *intrinsic parameters*, and the spatial positions of the considered instruments, called *extrinsic parameters*. Photometric calibration accounts for the relationship between the light emitted from a scene point and the light information acquired by the sensor.

Each sensor can be used alone or combined with other imaging sensors, as will be seen in Chap. 5, therefore, depth estimates require a calibration of the overall system. In this book we are not interested in the calibration of each internal single component of ToF and structured light depth cameras, since we adopt a user's perspective and consider depth cameras as single entities, disregarding their internal composition. In our view, a depth camera is a device providing depth information from a certain reference system. For the internal structure of structured light depth cameras and ToF depth cameras, and their working principles, the reader is sent to Chaps. 2 and 3 respectively.

The first part of this chapter formalizes calibration for generic imaging devices, in order to provide a unified framework for calibrating the specific instruments considered in this book, i.e. standard cameras, structured light depth cameras, and ToF depth cameras. The second part considers the calibration of standard cameras, providing practical solutions to the problem of calibrating one or more standard cameras together. The third part discusses how to calibrate structured light depth cameras and ToF depth cameras in order to exploit the data associated with them. The last part accounts for the calibration of heterogeneous imaging systems, where multiple depth cameras and standard cameras are combined together, paving the way for depth sensor fusion addressed in Chap. 5.

P. Zanuttigh et al., *Time-of-Flight and Structured Light Depth Cameras*,
DOI 10.1007/978-3-319-30973-6_4

4.1 Calibration of a Generic Imaging Device

4.1.1 Measurement's Accuracy, Precision and Resolution

Consider any instrument, for instance a generic imaging sensor, measuring a given quantity a. Let us denote with $\hat{a}$ the measured value and with a^* the actual quantity to be measured, i.e., the true value, or *ground-truth*. Both $\hat{a}$ and a^* assume values in the interval $[a_{min}, a_{max}]$. *Accuracy* and *precision* of the measurement are two important concepts describing the quality of the measurement itself.

Accuracy, also called bias in the literature, describes the closeness of the agreement between the measured value $\hat{a}$ and the true value a^*. This property is especially important when the user is interested in estimating the true value, and is less important when measurement repeatability is the critical issue.

Precision, often referred to as variability, describes the closeness of agreement between measured quantity values $\hat{a}_i$, obtained by replicating measurements of the same quantity a under specified conditions. A measurement's precision is usually expressed numerically by way of standard deviation or percentages of coefficient variations. Some measurements require high precision, i.e. small variability in the measurements, and may tolerate a constant offset with respect to the true value, which in most cases is easily detectable and removable during the calibration procedure.

Figure 4.1 exemplifies the two concepts in the case of multiple measurements of a certain quantity a, distributed according to a Gaussian probability density function. Accuracy is the difference between the average $\hat{a}$ and the true value a^*, while precision is the measurement's variance.

The lack of precision or accuracy in the measurement data have two different effects quantifiable by two types of error: *systematic errors* reduce a measurement's accuracy and *stochastic errors* reduce a measurement's precision.

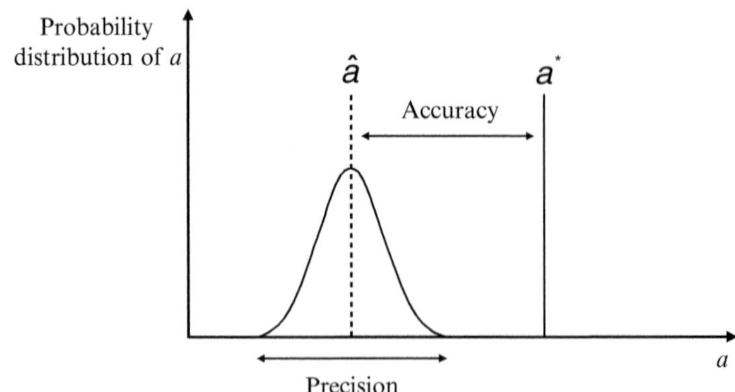

Fig. 4.1 Measurement of a quantity a: *Accuracy* describes how close the measurement is to the true value. *Precision* concerns the similarity of the measurements among themselves

Systematic errors always occur with the same value when acquisition and environmental conditions remain the same. They are typically predictable and constant, and if they can be identified, they can also usually be removed. If a set of measurements contains a systematic error, increasing the number of measurements does not improve the accuracy. When the systematic error is caused by the environment interfering with measurements, its reduction is more difficult and requires a more refined calibration procedure. If the systematic error of a certain measurement increases with distance from the device, the measured property is called *myopic*. The identification of myopic properties of a certain device allows one to use them to fit calibration models or lookup tables to reduce the systematic error.

Stochastic errors are associated with different measurements of the same quantity and can be characterized by a mean and variance value. The variance, in general, can be reduced by averaging multiple measurements, while the mean can be incorporated with systematic errors. The central limit theorem ensures that stochastic errors tend to be normally distributed when many independent measurements are added together.

In addition to accuracy and precision, measurements are also characterized by *resolution*, which is the smallest variation of the measured quantity that can be detected. Resolution is a property of the instrument, and higher resolution implies smaller minimum difference between two measurements. Increasing the resolution of the measuring instrument usually improves both accuracy and precision of the measurements.

4.1.2 General Calibration Procedure

The calibration of an instrument, measuring a quantity a, is the process of setting or correcting the relationship between the measured quantity $\hat{a}$ and the actual quantity a^* within a valid interval $[a_{min}, a_{max}]$. The purpose of the calibration procedure is to ensure the correct behavior of the device during its utilization, in order to provide accurate measurements. Using the concepts introduced previously, the calibration procedure relies on the analysis of errors in the estimation of $\hat{a}$ with respect to a^*, in order to increase the measurement accuracy and precision. In particular, stochastic errors are usually reduced by repeating the calibration procedure multiple times and by averaging multiple measurements. Systematic errors, instead, intuitively represent the residual "cost" associated with the calibration routine.

The relationship between $\hat{a}$ and a^* can be derived by two approaches:

- By means of a parametric function: $a^* = f(\hat{a}, \theta)$, where θ denotes a set of parameters characterizing the instrument operation, and $f(\cdot)$ a suitable function relating $\hat{a}$ and θ to a^*. In this case, the calibration becomes a parameter estimation problem and the optimal parameters value $\hat{\theta}$ can be obtained from a set of J matches between ground-truth and measured quantities $(a_1^*, \hat{a}_1), (a_2^*, \hat{a}_2), \ldots, (a_J^*, \hat{a}_J)$.

For instance, it is common to estimate the calibration parameters $\hat{\theta}$ by a Mean Squared Error (MSE) approach

$$\hat{\theta} = \underset{\theta}{\operatorname{argmin}} \frac{1}{N} \sum_{j=1}^{J} \left(a_j^* - f(\hat{a}_j, \theta)\right)^2 . \tag{4.1}$$

Since in this case the calibration is computed as a parameter estimation problem, indeed it is addressed as *parametric calibration* or *model-based calibration*. Equation (4.1) accounts for both kinds of errors. The first part of the equation

$$\frac{1}{J} \sum_{j=1}^{J} (\cdot) \tag{4.2}$$

represents a way to deal with stochastic errors, by averaging over N measurements, and the function to minimize

$$\left(a_j^* - f(\hat{a}_j, \theta)\right)^2 \tag{4.3}$$

represents the systematic error.

- By using a "brute force" approach. Since the set of possible attribute values is well known ($a^* \in [a_{min}, a_{max}]$), one may uniformly sample the attribute interval obtaining J ground truth sampled values $a_1^*, a_2^*, \ldots, a_J^*$. One can then perform a measurement with respect to each ground truth value $a_j^*, j = 1, \ldots, J$, therefore obtaining a set of relative measurements $\hat{a}_1, \hat{a}_2, \ldots, \hat{a}_J$ and establishing a relationship between ground-truth $a_1^*, a_2^*, \ldots, a_J^*$ and measurements $\hat{a}_1, \hat{a}_2, \ldots, \hat{a}_J$, by a lookup table with J couples $(a_1^*, \hat{a}_1), (a_2^*, \hat{a}_2), \ldots, (a_J^*, \hat{a}_J)$ as entries. This calibration approach is usually called *non-parametric calibration* or *model-less calibration*. The number of interval samples J affects the systematic error; the higher the number of intervals, the more accurate the estimated relationship is but the more laborsome the calibration process becomes. Stochastic errors are reduced by measuring multiple times the quantities $\hat{a}_j$ before their association with a_j^*.

Each one of the two calibration methods above has advantages and disadvantages. Model-based calibration, on one hand, requires a model which may not always be available and might be difficult to obtain. On the other hand, modeling the measurement process generally has the advantage of reducing the number of parameters to be estimated to a small set (e.g., eight parameters for the case of a standard camera calibration), thus reducing the number J of required calibration measurements with respect to the model-less calibration case. The availability of an analytical correspondence between each possible ground-truth attribute a^* and its relative measurement $\hat{a}$ is another advantage.

Model-less calibration clearly has the advantage of not needing any model, but has the drawback of requiring a large number J of calibration measurements in order to reduce the sampling effects. This approach is more prone to over-fitting than the previous one, because J is generally much greater than the number of parameters to be estimated. There is not an analytical correspondence between each possible ground-truth value a^* and its measured value $\hat{a}$. Moreover a direct although not analytical correspondence is available only for the set of sampled ground truth values. For the other values, the correspondences can be approximated by interpolation with inevitable sampling artifacts.

4.1.3 Supervised and Unsupervised Calibration

Calibration methods can be divided into two general categories, *supervised* and *unsupervised* methods, depending on whether prior knowledge of the scene and user intervention are required in the calibration process.

Supervised methods for an imaging device or sensor require a calibration object with known geometric properties to be imaged by the sensor, so that the mapping of the calibration object to the sensor can be constrained and retrieved by matching the known 3D points P_i to the 2D points p_i seen by the sensor. There are different techniques for calibrating an imaging device in a supervised manner, depending on the dimensions of the calibration objects, and they can be mainly grouped into three categories:

- **3D object** based, in which the calibration object is a 3D structure of known geometry such as orthogonal or parallel planes [13];
- **2D plane** based, in which the calibration object is a planar pattern shown at different orientations. The most common example is the planar checkerboard [69];
- **1D line** based, in which the calibration object is a set of collinear points, such as a line moving around a fixed point [70].

Figure 4.2 shows an example for each of the three categories of calibration objects.

Among these three categories, there is no single calibration method that is best for all scenarios, but a few recommendations are worth being mentioning [45, Chap. 2]. Highest accuracy can usually be obtained by 3D objects, but this requires very accurate calibration objects and hence high costs. Planar objects are usually

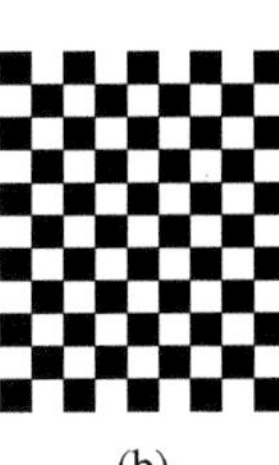

(a) (b) (c)

Fig. 4.2 Example of calibration objects: (**a**) orthogonal planes (3D object); (**b**) planar checkerboard (2D plane); (**c**) line moving around a fixed point (1D line)

more affordable and are a good compromise between accuracy and ease of use: planar checkerboards built with various accuracy degrees are very popular among computer vision researchers and practitioners. Mono dimensional calibration tools, although not commonly used, are useful in camera network calibration, where 2D or 3D objects are unlikely to be simultaneously visible by all cameras.

Unsupervised methods, on the other hand, do not use any calibration object.[1] The scene rigidity provides the information needed to calibrate the camera. This concept is often seen together with 3D reconstruction as explained in Sect. 1.2.3 in connection with the problem of reconstructing the 3D geometry of a scene without prior knowledge of the calibration parameters. For standard cameras, unsupervised calibration usually refers to the concept of *auto-calibration*, introduced in Sect. 4.2.3, in which camera parameters are directly determined from a set of uncalibrated images. These techniques usually require a large number of parameters and the need to solve rather difficult mathematical problems. A widely used simplification is to add more constraints to the unsupervised methods, in order to reduce the order of the problem and improve convergence.

Unsupervised calibration usually cannot reach accuracy values comparable to those of supervised procedures, since the number of parameters to estimate is higher. However, there are cases where pre-calibration using external tools is not available (e.g., for a 3D reconstruction from a video sequence) and self-calibration is the only choice. In addition, since unsupervised methods make no or little usage of prior knowledge and do not require user interaction, they enable the concept of online-calibration (or dynamic calibration) in which the camera is calibrated during the vision task.

Self-calibration methods are appealing since they make no or little usage of prior knowledge and reduce user interaction, different from supervised methods which typically exploit constraints to determine the camera parameters and assume additional aid provided by the user.

4.1.4 Calibration Error

Measurement errors cannot be completely eliminated, only reduced within known ranges. Once calibration is performed, an error remains between the estimated and actual values, commonly called *calibration error*. Estimating the calibration error is very important since it may become the bottleneck for accuracy and precision of specific applications.

The calibration error can be computed in different ways for the case of model-based and model-less calibration upon a set of M measurements $\hat{a}_k$, $k = 1, \ldots, M$ with relative ground-truth values a_k^*, $k = 1, \ldots, M$. Note that this set of measure-

[1]Since no calibration apparatus are needed for unsupervised calibration, this method can be considered as zero-dimensional.

ments is different from the set of J measurements adopted in the calibration process described by (4.1), in order to adopt a rigorous cross-validation approach. In the case of model-based calibration, calibration error e can be computed from the M error estimates as

$$e = \frac{1}{M}\sum_{k=1}^{M}\left|a_k^* - f(\hat{a}_k, \hat{\theta})\right|. \tag{4.4}$$

In model-less calibration, there is no model to use in (4.4). However, given a measured quantity $\hat{a}_k$, its corresponding ground-truth value a_k^* can be approximated by interpolation: $\mathscr{F}\left[\hat{a}_k, (a_1^*, \hat{a}_1), \ldots, (a_N^*, \hat{a}_N)\right]$ where $\mathscr{F}[\cdot]$ is a suitable interpolation method (e.g., in imaging applications nearest-neighbor, bilinear, bicubic or splines interpolation). Calibration error e in the case of model-less calibration can be computed from the M error estimates as

$$e = \frac{1}{M}\sum_{j=1}^{M}\left|a_j^* - \mathscr{F}(\hat{a}_j, (a_1^*, \hat{a}_1), \ldots, (a_N^*, \hat{a}_N))\right|. \tag{4.5}$$

In both cases, the goal of a calibration method is to give a consistent procedure, meaning that several calibrations of the same device should deliver similar results, providing the lowest possible calibration error.

Once the relationship between measured and actual attribute values is known, one may define a way to map the measured values to the corresponding actual values in order to compensate for the systematic differences between these quantities.

The compensation may increase the accuracy but not the precision of the instrument measurements, since it cannot account for error randomness. In the calibration process itself, and whenever feasible in actual measurements, precision can be improved by averaging over multiple measurements. However, when the measurements are not performed for calibration purposes, especially in the case of dynamic scenes, averaging may not be possible to use for reducing stochastic errors. As a simple example, consider the case of a depth camera capable of operating only at 30 [fps], to be deployed in a specific application requiring real time interaction with dynamic scenes. In this case multiple frames cannot be averaged over time since this process would introduce blurring effects.

An important reliability feature for any measurement device, and for depth cameras in particular, is their ability to remain calibrated over time. Different aspects can influence camera behavior depending on its sensor family. Therefore, the tolerances guaranteeing the correct usage of each sensor are different.

For standard cameras, for example, the sensor temperature can influence the image quality: the warmer the sensor, the noisier the images. Temperature can also affect the focal length of the camera, and there exist methods to estimate and compensate these focal length changes [50, 59].

Stereo cameras and structured light systems instead are vulnerable with respect to the spatial displacement of each component, meaning the sensor's quality is strongly affected by the application of mechanical forces that modify the system geometry. A possible source of *decalibration* is the connection and disconnection of the cable to and from the camera, or strong vibrations of the sensor. Some alterations of the calibrated configuration are detectable and techniques exist to dynamically calibrate the system, some others are not even detectable and require manual recalibration of the system. For stereo cameras, for example, if the cameras move along the direction of the baseline, it is impossible to detect that the geometry has changed by only looking at the two images, and the decalibration effect is that the 3D reconstruction is scaled with respect to the real scene. A rotation of one camera around its x axis, instead, is easily detectable and fixable by comparing the epipolar lines of the two cameras. Figure 4.3 shows an example of a decalibrated stereo camera, both for a horizontal and vertical translation of the right camera. In the first case, the decalibration cannot be detected, as it results only in an incorrect disparity estimate, while in the second case, the epipolar constraints on the two images are no longer valid and decalibration can be detected.

As seen in Chap. 2, structured light depth cameras are conceptually equivalent to stereo cameras, with the addition that for some architectures the knowledge of the projector's position is fundamental. In these cases, the entire system is even more fragile than passive stereo systems with respect to cameras and projector displacements.

For ToF cameras, mechanical changes in the sensor structure are less dangerous, but photometric properties are more sensitive to changes. For example, a sensor temperature variation may cause a frequency shift of the laser which matches the passband frequency of the optics filters, with the effect of higher signal attenuation possibly causing wrong depth estimates. To mitigate this, ToF cameras usually carry dedicated hardware to keep temperature constant.

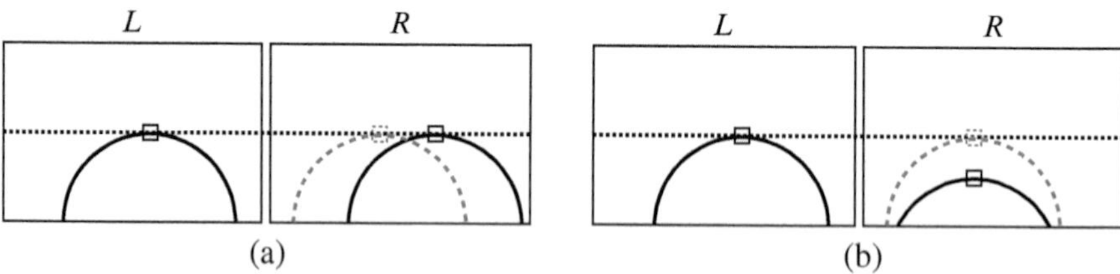

Fig. 4.3 Decalibration of a stereo camera: (**a**) Horizontal translation (not detectable); (**b**) Vertical translation (detectable). *Solid black lines* show the actual acquisition, *dashed red lines* show the acquisition before decalibration

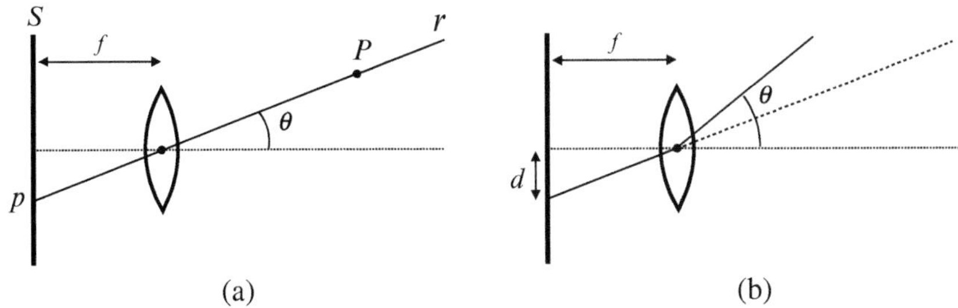

Fig. 4.4 Geometric model of a standard cameras: (**a**) Pin-hole model; (**b**) General projection (*dashed line* is the perspective projection)

4.1.5 Geometric Calibration

The most common geometric model for standard cameras is the pin-hole camera model introduced in Sect. 1.1.1 reported for convenience in Fig. 4.4a where S denotes the *image plane*, C the center of projection, also called *nodal point* or *optical center*, with the associated *optical axis*, and f the distance of the nodal point from the image plane. The point p in the image plane is associated with the *optical ray* r and represents the projection of the scene point P and of all the points belonging to the ray r through C and P. Although very simple, this model is at the base of almost every geometric consideration related to cameras in computer vision and computer graphics. The perspective projection of a pin-hole camera can be described by the following formula:

$$d = f \tan(\theta) \tag{4.6}$$

where θ is the vertical angle between the optical axis and the incoming ray r, and d is the vertical distance between the image point and the principal point. A similar expression holds for the horizontal angle and distance from the principal point. This model holds for standard lenses with field of view smaller than 70°. Ultra-wide angle or fisheye lenses with field of views up to 180° are usually designed in compliance with a more general projection model, which is still function of f and θ.

Perhaps the most common models are the equidistance projection $d = f\theta$ or the stereographic projection $d = 2f \tan(\theta/2)$. Figure 4.4b, where the dashed line is the perspective projection, represent a general non perspective projection law, also applicable to perspective projection in the case of a non-ideal lens [30].

In standard cameras, lens distortion causes a systematic error due to the deviation of the optical rays from perspective projection of the pin-hole model. Although distortion can assume different configurations [65], the most encountered effects are radially symmetric. Most cameras tend to pull points radially toward the optical center, a phenomenon referred to as *pin-cushion* distortion (Fig. 4.5a). Another type of lens non-ideality tends to push points away from the optical center along

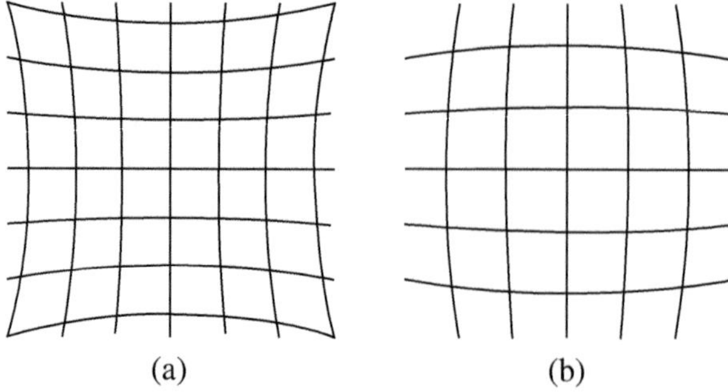

Fig. 4.5 Example of radial distortion: (**a**) Barrel distortion; (**b**) Pincushion distortion

the radial direction and is called *barrel* distortion (Fig. 4.5b). Misalignment in the elements of compound optics produce a tangential distortion typically less impactful than the radial distortion. However, model (1.14) allows one to easily handle radial and tangential distortion together.

According to the notation of Sect. 4.1, the attribute a of geometric camera calibration is the geometry (origin and orientation) of the optical ray. The actual quantity a^* is related to the measured quantity $\hat{a}$ through all the intrinsic parameters of matrix **K**, distortion parameters **d**, and extrinsic parameters **R** and **t**. The geometric calibration of a standard camera concerns the estimation of the relationship between the actual and the measured optical ray relative to each camera sensor pixel, or equivalently, the relationship between a generic 3D point P and its projection p in the image plane. From Eqs. (1.9) and (1.14) this relationship between the point coordinates **P** and **p** can be expressed as

$$\mathbf{p} = \Psi^{-1}(\mathbf{MP}) \tag{4.7}$$

where $\Psi(\cdot)$ denotes the distortion transformation and $\mathbf{M} = \mathbf{K}[\mathbf{R} \mid \mathbf{t}]$ is the projection matrix (1.9). The output of geometric calibration is therefore given by:

1. Intrinsic parameters **K**, i.e. f_x, f_y, c_x and c_y;
2. Distortion coefficients **d**, e.g., in the Heikkila model [22] k_1, k_2, k_3, d_1 and d_2;
3. Extrinsic parameters **R** and **t**, i.e. rotation and translation of the CCS with respect to the WCS.

Once the camera is calibrated, its distortion parameters are known and the image undistortion procedure, e.g. (1.14), can be applied to all the images acquired by the camera in order to compensate for its radial and tangential distortion, i.e., for its systematic error in the measurement of the optical rays' orientation. The undistortion procedure also allows one to directly obtain the projection or image point p of P by means of Eq. (1.9), simplifying Eq. (4.7) in $\mathbf{p} = \mathbf{MP}$.

There are different techniques to practically estimate the calibration parameters and they can be divided into two main categories: *metric* and *non metric*. In Sect. 4.2.1 we will describe the most common ways to calibrate the geometry of a standard camera.

A geometric calibration is called metric if it relies on known metric information about the scene, such as the standard technique from Zhang [69]. The parameters are usually estimated through homographies between 3D and 2D points. In this case, all the parameters are usually calibrated together, implicitly coupling internal (intrinsic and distortion coefficients) and external parameters, resulting in high errors on the camera internal parameters if the assumption of perspective projection (pin-hole model) is violated.

Non metric techniques instead rely on qualitative constraints and enforce other geometric properties such as collinearity, parallelism, and orthogonality [33]. Most of these methods do not require additional metric information of the scene and exploit the fact that under perspective projection, straight lines must always project to straight lines in the image, parallel lines must be rectified to parallel, and orthogonal lines to orthogonal. Collinearity, parallelism, and orthogonality impose enough constraints to retrieve internal parameters and distortion coefficients, decoupling them from the extrinsic parameters. Non metric techniques are usually employed for calibration of ultra-wide angle or fisheye lenses, in which the perspective projection model does not hold.

A detailed analysis of calibration parameters and relationships is beyond the scope of this book and can be found in classical computer vision readings, such as [6, 11, 21, 22, 69].

4.1.6 Photometric Calibration

The relationship between the amount of light reflected or emitted at a scene point P and the brightness of the corresponding pixel p in the image plane is influenced by many factors, such as lighting conditions, objects materials and colors, and the quality of the camera lenses, among many other elements. Figure 4.6a shows the simplest model that relates the image *irradiance* E, i.e., the amount of light falling on the image surface (measured in $[W \times m^{-2}]$), and the scene *radiance* L, i.e., the amount of light radiated from a surface (measured in $[W \times m^{-2} \times rs^{-1}]$).

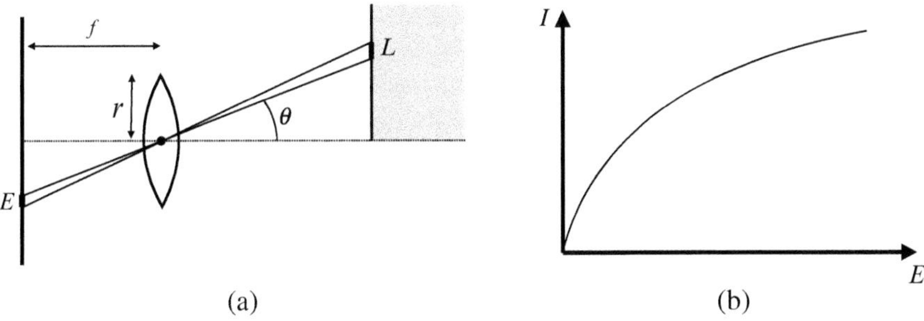

Fig. 4.6 Radiometric image formation: (**a**) Relationship between scene radiance and image irradiance; (**b**) Camera response function

The fundamental model of radiometric image formation can be described by this linear relationship in L:

$$E = \frac{L\pi r^2 \cos^4(\theta)}{f^2} \tag{4.8}$$

where f is the focal length, r is the radius of the camera lens, and θ is the angle between the optical center and the considered ray. The reader interested in the derivation of (4.8) is sent to [63, Chap. 2].

The function relating image irradiance E and pixel intensity I is usually a nonlinear transformation accounting for the photo-electric conversion implemented by the electronics behind the pixel called the *camera response function* (CRF). Even though most computer vision algorithms assume a linear relationship between scene irradiance and the brightness measured by the imaging system, in practice, this is seldom the case. Almost always, there exists a non-linear mapping between scene irradiance and measured brightness, of the type shown in Fig. 4.6b.

Another important photometric characteristic of a non ideal lens, such as any real optics, is the image brightness intensity fall-off from the center to the fringe [63]. From (4.8), we can indeed see that the radial fall-off of the irradiance E follows a $\cos^4(\theta)$ law. This spatial variation is commonly called *vignetting effect* and is usually minimal at the center and higher toward the periphery. Figure 4.7a shows an example of vignetting. In most compound lenses, vignetting increases with the aperture size [51]. If no correction is applied, this forces one to use cameras at narrow apertures, where the vignetting effect is negligible. Images taken in such settings however are more noisy than those taken at wider apertures. Most digital cameras today have built-in functions to compensate for this effect. The use of microlenses over the image pixel sensor can reduce the effect of vignetting even more.

In addition to vignetting, another photometric artifact requiring particular attention is *chromatic aberration*, which produces color distortions, also called color

(a) (b)

Fig. 4.7 Photometric artifacts on a standard camera: (**a**) image affected by vignetting; (**b**) color fringes due to chromatic aberration

fringes, especially in wide-angle color cameras. Each color in the optical spectrum is reflected differently by the optics, because lenses have different refractive indices at different light wavelengths. Artifacts are more visible in the edges or where there is a strong color change, introducing wrong colors. For black and white images, chromatic aberration produces blurred edges. Figure 4.7b shows a checkerboard where color fringes are visible at the checkers' boundary. Although effects of chromatic aberration result in permanent errors in the acquired images, different filtering techniques have been proposed to correct for such artifacts [10, 41].

For depth estimation, photometric calibration, although not as fundamental as geometric calibration, plays an important role in photometric stereo algorithms, in which the surface normals of the scene are estimated by analyzing the lighting conditions and used to reconstruct the 3D geometry of the scene. This technique was originally introduced by Woodham [66] and made popular by Horn with the introduction of *shape from shading* methods [25], where data coming from a single image with different lighting conditions are used to estimate the depth map of the scene.

4.2 Calibration of Standard Cameras

Let us recall that "standard camera" in this book refers to a digital camera or a digital video camera, since image formation principles are the same for both these imaging devices. The attributes measured by a standard camera are the attributes measured by all its sensor pixels.

The camera operation as pin-hole system, according to the perspective projection introduced in Sect. 1.1.1, and the hypothesis of infinitesimal size pixels, can be characterized by two different processes: the association of the camera sensor pixels p to the corresponding scene point P by the associated optical ray, as shown in Fig. 4.4, and the measurement of the scene color by the photo-electric conversion associated with each camera sensor pixel. The calibration of standard cameras can be therefore divided in two parts:

- geometric calibration, i.e., the estimate of the relationship between sensor pixels and relative scene points or optical rays;
- photometric calibration, i.e., the estimate of the relationship between the actual scene point color and the color measured by the camera at the corresponding pixel.

A comprehensive treatment of geometric and photometric camera calibration can be found in [55, 58, 69].

4.2.1 Calibration of a Single Camera

The calibration of a single camera has been extensively studied in the photogrammetry and computer vision community [6, 11, 21, 22, 45, 69], and even now new solutions continue to be proposed. The quantity measured by a standard camera is the color set associated with scene points. A calibration procedure for a single camera aims at providing a set of quantities that allow one to associate each camera pixel to the corresponding ray. The parameters needed for this association are:

- the intrinsic parameters matrix $\mathbf{K}$
- the distortion coefficients $\mathbf{d}$

Once the camera is calibrated, after inverting (1.3) it is possible to know for each pixel $\mathbf{p} = (u, v)^T$ what the equation of the ray associated with it is, given by (1.25) and rewritten next for reader's convenience

$$\begin{bmatrix} x/z \\ y/z \\ 1 \end{bmatrix} = \mathbf{K}^{-1} \begin{bmatrix} u \\ v \\ 1 \end{bmatrix} = \mathbf{K}^{-1}\tilde{\mathbf{p}} \tag{4.9}$$

where $\mathbf{P} = [x/z, y/z, 1]^T$ represents the coordinates with respect to the CCS of the 3D point belonging to the ray connecting the sensor point p with the center of projection O. The more general case, where the reference coordinate system does not coincide with the CCS, can be derived from the knowledge of $\mathbf{R}$ and $\mathbf{t}$ from (1.6).

This section provides a summary of geometric and photometric calibration techniques for a single camera that are most frequently used in computer vision. The geometric calibration of a single camera used to be performed with 3D objects of known size, such as the one of Fig. 4.2a, but the techniques introduced by Zhang [69] resorting to a planar calibration object, such as a black and white checkerboard soon became most popular due to their ease of use. The estimation of the relationship concerning measured and actual correspondences between camera sensor pixels and 3D scene points by a planar checkerboard can be summarized in three steps:

1. Detection of feature points in the acquired images;
2. Estimation of intrinsic (and extrinsic) parameters;
3. Refinement of all the parameters including intrinsic, extrinsic and distortion coefficients, through nonlinear optimization.

We will refer to the technique [69], which requires a still camera to observe the planar pattern moved to different positions and orientations, as shown in the top sequence of Fig. 4.8. This technique is equivalent to keeping the planar pattern still and moving the camera so that it frames the pattern from different positions and orientations. The minimum number of positions/orientations theoretically is 2, but practically, to improve the quality of the calibration, a minimum of 10 different positions/orientations is recommended. It is a good practice to acquire the

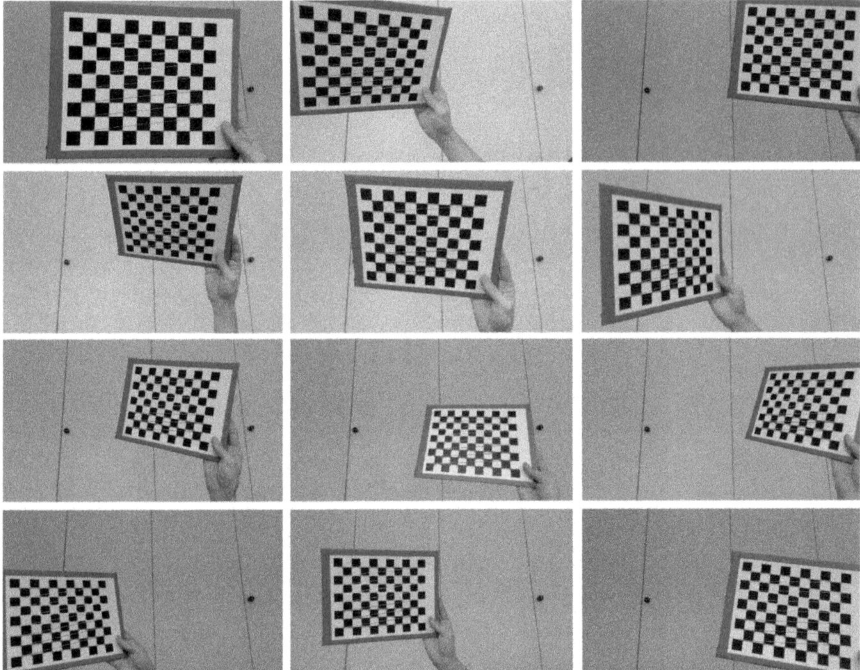

Fig. 4.8 Acquisition of planar checkerboard for camera calibration

checkerboards at distances included in the camera working space. In the acquisition process, the pattern or the camera can be moved without needing to know the absolute pose. Any planar pattern can be used, as long as the metric of the object is known and the homographies of the plane can be estimated. In the black and white checkerboard with checkers of known size, the easily detectable corners of the checkers are the feature points required in the first step. Given M images of the planar checkerboard with J corners each,[2] we obtain a set of points p_k^j, where $k = 1, \ldots, M$ and $j = 1, \ldots, J$. To generate a good estimate of the distortion coefficients, the acquisitions should guarantee that the set of feature points covers the whole camera field of view, for reasons that will be clear soon.

The second step involves the estimation of intrinsic and extrinsic parameters,[3] obtainable by the homography between the model plane and its image. We can define the WCS as the reference system with origin at the checkerboard's top left corner, x and y axis on the checkerboard plane and z axis orthogonal to the checkerboard plane as shown in Fig. 4.9. The 3D coordinates of the checkers'

[2]The calibration procedure remains valid even if not all the feature points are visible in every image but in order to simplify the notation we assume that all the feature points are visible in every image.

[3]It is simpler to think that the planar checkerboard is fixed and the camera moves, even though the converse situation leads to the same image set and is typically more convenient.

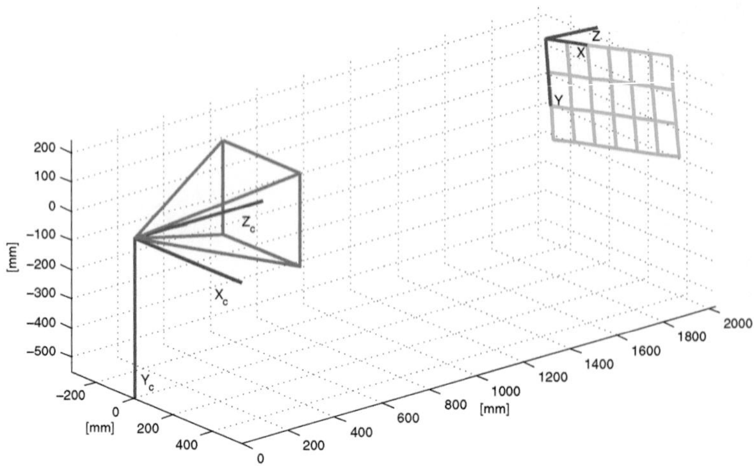

Fig. 4.9 3D position of the sensor plane with respect to the checkerboard

corners P^j, with $j = 1, \ldots, J$, will be taken with respect to the WCS at every acquisition. Due to the planar shape of the sensor and checkerboard, the position of the sensor plane with respect to the WCS, associated with the checkerboard, can be derived by 3D homography, using the correspondences between 3D points P^j and 2D points p_k^j relative to the k-th position of the checkerboard. In the case of a planar target, there is a closed-form solution that allows one to estimate, for each acquisition k, intrinsic parameters $\mathbf{K}$ and extrinsic parameters $\mathbf{R}_k$ and $\mathbf{t}_k$. This solution, however, minimizes an algebraic distance which is not physically meaningful, but it represents an initial guess that can be refined through maximum likelihood inference. In addition, the analytical solution found with this procedure does not account for lens distortion, therefore, distortion parameters $\mathbf{d}_k$ still need to be determined.

The *undistortion* routine is usually implemented by a look-up table, where each pixel of the undistorted image is associated with the coordinates of the pixel in the original image. Since the look-up table is usually derived from equations of the type of (1.14), the coordinates of the original image in general are not integer values, and the intensity value to be associated with each pixel is therefore computed by interpolation of the neighboring pixels. This mapping method is called backward mapping, as opposed to forward mapping techniques, in which pixels in the original image with integer values are mapped to the undistorted image, resulting in floating point locations. Typical interpolation schemes for undistortion include nearest-neighbors, bilinear or more complex interpolation techniques. Higher order techniques are usually more precise, but are usually computationally more expensive. The selection of the interpolation algorithm depends on the application and the input data. A good coverage of feature points in the first step results in a better estimate of such an undistortion map. While the displacement due to radial distortion is on the order

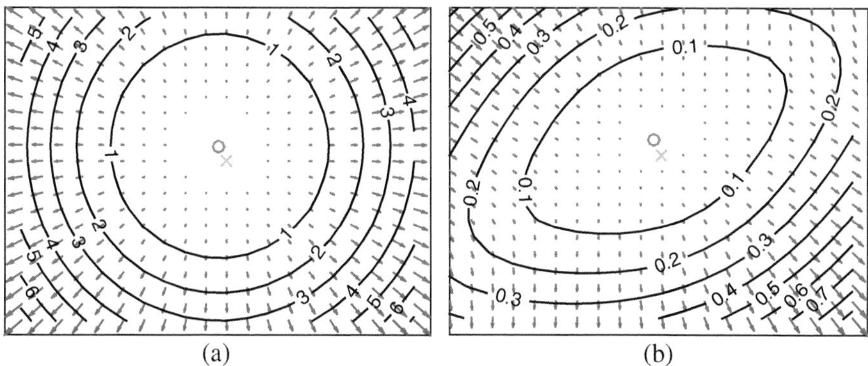

Fig. 4.10 Example of pixel displacement for radial and tangential distortion. (a) Radial component of the distortion model. (b) Tangential component of the distortion model

of pixels, all other sources of a ray's deviation usually cause displacement of less than a pixel from its original position. Figure 4.10 shows the effect of lens distortion divided in radial and tangential terms, by a tool provided by Bouguet [5]. Numbers represent the distance in pixels between corresponding points of the undistorted image and the original distorted image.

Once the 3D position and orientation of the sensor with respect to the checkerboard is known, the 3D coordinates of the checkerboard's corners $\mathbf{P}^j$ can be projected to the sensor plane according to (1.8) followed by distortion (1.14), and compared with the coordinates $\mathbf{p}_k^j$ of the detected 2D points p_k^j. The third refinement step is a maximum likelihood estimate of all the parameters involved, by minimizing the Euclidean distance between the planar positions of the measured and the projected 3D points after anti-distortion, given by

$$\sum_{k=1}^{M} \sum_{j=1}^{J} \|\mathbf{p}_k^j - f(\mathbf{K}, \mathbf{d}, \mathbf{R}_k, \mathbf{t}_k, \mathbf{P}^j)\|_2^2 \tag{4.10}$$

where $f(\mathbf{K}, \mathbf{d}, \mathbf{R}_k, \mathbf{t}_k, \mathbf{P}^j)$ is a function accounting for projection and distortion, such as (4.7). Equation (4.10) extends (1.12) by using also multiple poses, i.e., M poses, besides multiple points, i.e., N points. The minimization of this nonlinear function is usually solved by nonlinear optimization techniques, such as the Levenberg-Marquardt method. The analytical solution found in the previous step can be used as an initial guess for $\mathbf{K}$, $\mathbf{R}_k$ and $\mathbf{t}_k$, while distortion parameters $\mathbf{d}$ can be simply set to 0 if an initial estimation is not available.

The parameters estimated by the previously described procedure therefore are the intrinsic parameters matrix $\mathbf{K}$, the distortion coefficients $\mathbf{d}$, and for each acquisition k, the rotation $\mathbf{R}_k$ and translation $\mathbf{t}_k$ of the calibration target with respect to the camera. The residual error from the minimization of (4.10) gives a quality estimate of the calibration procedure. The geometric calibration routine and many other useful tools are also available in popular open source camera calibration projects, such as the MATLAB Camera Calibration Toolbox [5] and OpenCV [3].

This standard procedure does not provide good results in the case of cameras with wide-angle lenses, fisheye lenses, or in low-cost optics, where the geometry introduces a number of non-idealities and the ideal pin-hole camera model is no longer appropriate. Different solutions, in which a per-pixel map is estimated [11, 67], have been proposed for these systems. Non metric approaches exploiting collinearity, parallelism, and orthogonality of elements in the scene, have recently proved to ensure higher precision in the geometric calibration of such systems [29].

Current photometric calibration methods for standard cameras can be split in two categories: hardware camera register calibration and software image processing calibration. Almost all consumer cameras already come with hardware photometric calibration, providing images which are already equalized. In addition, knowing the optics characteristics in advance allows one to set the camera registers to compensate for optics alteration, like vignetting. It is worth noting that for depth estimation, geometric calibration generally requires much higher accuracy and more precision than photometric calibration. For example, the nominal focal length f value of the camera is usually provided by the manufacturer, but for typical applications its accuracy is not sufficient and users typically perform additional calibration to estimate its correct value. On the contrary, the images provided by these cameras are typically good enough to be directly used, since the associations between intensity values and each camera pixel, i.e. the objective of photometric calibration, is less critical and requires less accuracy for typical applications than the association of the correct optical ray to each pixel, i.e. the objective of geometric calibration. Furthermore, since raw images before their usage typically undergo various pre-processing operations, such as conversion from color to grayscale, gamma correction, or other transformations that depend on the application, there is not a standard photometric calibration procedure. The effects of chromatic aberration can be reduced by separately applying the undistortion procedure (1.14) for each color channel. The per-channel undistortion technique is also of great importance in other applications targeting augmented reality or virtual reality such as [2]. The optical systems adopted in these applications typically frame a display through lenses with a fairly high degree of magnification, associated with significant distortion and chromatic aberration. In order to reduce these effects, most systems apply a pre-distortion correction to the visualized images, individually distorting the red, green, and blue channels and also accounting for chromatic aberration.

4.2.2 Calibration of a Stereo Vision System

Let us recall from Sect. 1.2 that a stereo system is made by two standard cameras framing the same scene. If the two cameras are calibrated and rectified, one can estimate a disparity and associate a depth value for each pixel by (1.26). Therefore, the attribute measured by a stereo system is the depth of the scene points framed by both cameras. A 3D point P can be estimated by triangulation from the 2D coordinates of two conjugate points p_L and p_R via (1.25) and (1.26). Rectification and triangulation procedures depend on the properties of the two standard cameras L and R and on their relative pose. The properties of L and R are modeled as seen in Sect. 4.2.1 by their intrinsic and distortion parameters, while their relative pose is modeled by the relative roto-translation between the reference system L-3D associated with the L camera and the reference system R-3D associated with the R camera.

The calibration of a stereo system consists of estimating the following quantities:

- intrinsic parameters matrices $\mathbf{K}_L$ and $\mathbf{K}_R$
- distortion coefficients $\mathbf{d}_L$ and $\mathbf{d}_R$
- rotation matrix $\mathbf{R}$ and translation vector $\mathbf{t}$ describing the roto-translation between the L-3D and the R-3D reference systems

Let us emphasize that in stereo calibration one seeks a single rotation matrix $\mathbf{R}$ and translation vector $\mathbf{t}$ that relate left and right cameras, different from single camera calibration, where no $\mathbf{R}$ and $\mathbf{t}$ parameters are sought as a final output, even though a list of rotation and translation matrices between the camera and the checkerboard are estimated during the procedure. Once the roto-translation parameters are known, the two images can be rectified[4] so that the epipolar lines are arranged along image rows and the search for conjugate points is simplified.

The most commonly used geometric calibration procedure for stereo cameras is very similar to the one explained in Sect. 4.2.1 for single camera, since it exploits multiple acquisitions of a planar checkerboard to find the calibration parameters. The general procedure can be summarized in three steps:

1. Acquisition of a set of images from the two cameras
2. Independent calibration of the two cameras
3. Refinement of all the parameters through nonlinear optimization, including intrinsic parameters and distortion coefficients of both cameras, pose of the checkerboard for every acquisition, and roto-translation between the two cameras

The first step, requiring the acquisition of a set of images, is identical to the procedure in single camera calibration, therefore, all the previous considerations apply. However, since in this case each acquisition of checkerboard is performed from two cameras with different points of view without perfectly overlapping fields of view, it is important to collect numerous images with the checkerboard visible on both the cameras in this step. There must also be good checkerboard coverage separately on both cameras, in order to estimate a good undistortion map

[4] As already seen in Sect. 1.2, stereo rectification can be also performed upon knowledge of the fundamental matrix $\mathbf{F}$.

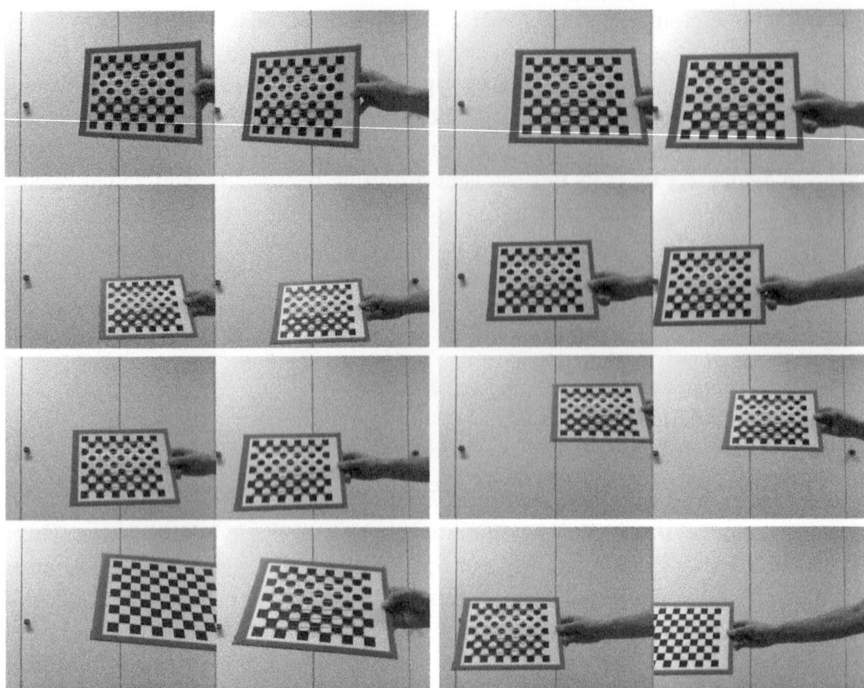

Fig. 4.11 Acquisition of a planar checkerboard for stereo camera calibration. Note that in the last row, the checkerboard is not detected by one of the two cameras

(the undistortion of the images is performed independently on the two cameras). Figure 4.11 shows some acquisitions of the planar pattern with a stereo camera.

Clearly the two intrinsic parameters matrices $\mathbf{K}_L$ and $\mathbf{K}_R$ and the distortion coefficients $\mathbf{d}_L$ and $\mathbf{d}_R$ can be derived independently by single L and R camera calibrations. Once the two standard cameras are calibrated, one can estimate the stereo system extrinsic parameters R and $\mathbf{t}$. Such a two-step operation allows one to reduce the size of the estimated parameters space at each step, simplifying the estimation problem. Better results can be obtained by an overall nonlinear optimization of the type of (4.10), as suggested in [45, Chap. 2]. In this case, the objective function is the reprojection error in the two cameras with additional constraints. If $\mathbf{R}_{Lk}$ and $\mathbf{t}_{Lk}$ are the matrices describing the k-th checkerboard pose with respect to the left camera, and $\mathbf{R}_{Lk}$ and $\mathbf{t}_{Rk}$ with respect to the right camera, given that $\mathbf{R}$ and $\mathbf{t}$ are the rotation and translation matrices relating left and right cameras, the following relationships hold

$$\begin{aligned} \mathbf{R}_{Rk} &= \mathbf{R}_{Lk}\mathbf{R} \\ \mathbf{t}_{Rk} &= \mathbf{R}_{Lk}\mathbf{t} + \mathbf{t}_{Lk}. \end{aligned} \tag{4.11}$$

The functional to minimize, under constraints (4.11), can be written as

$$\sum_{k=1}^{M}\sum_{j=1}^{J}\Big(\delta^j_{Lk}\|\mathbf{p}^j_{Lk}-f(\mathbf{K}_L,\mathbf{d}_L,\mathbf{R}_{Lk},\mathbf{t}_{Lk},\mathbf{P}^j)\|_2^2+$$
$$\delta^j_{Rk}\|\mathbf{p}^j_{Rk}-f(\mathbf{K}_R,\mathbf{d}_R,\mathbf{R}_{Rk},\mathbf{t}_{Lk},\mathbf{P}^j)\|_2^2\Big) \quad (4.12)$$

where $\mathbf{p}^j_{Lk}$ and $\mathbf{p}^j_{Rk}$ are the projections of the point P^j of the k-th checkerboard pose to the left and right camera images respectively, with δ^j_{Lk} (or δ^j_{Rk}) equal to 1 if the point P^j is visible in the k-th image of the left (or right) camera and 0 otherwise. It is worth noting that (4.12) has the same form of (4.10), but concerns two cameras. Combining (4.12) and (4.11), one can see that the number of parameters to estimate is equal to $6N + 6$, while the two-step procedure requires one to estimate $12N$ parameters. With 10 images, this calibration routine requires one to estimate 66 parameters, while the two-step procedure requiring 120 estimated parameters, with a reduction of 54 parameters corresponding almost to 50 %. The single camera calibration is still necessary in order to give an initial guess to the nonlinear optimization, which can be solved using the Levenberg-Marquardt algorithm. A comprehensive treatment of geometric stereo calibration can be found in [6, 69]. Popular open-source stereo calibration algorithms can be found in the MATLAB Camera Calibration Toolbox [5] and in the OpenCV Library [3].

Once the calibration parameters are available, it is possible to undistort and rectify the two images. This section provides some practical considerations about stereo *rectification* introduced in Sect. 1.2. Let us recall that rectification is the process of transforming the two images such that they lie on a common image plane. This process has several degrees of freedom, and for practical reasons, in a configuration of the type of Fig. 1.8, the epipolar lines are constrained to be horizontal. Section 1.2 has already shown the advantages of having row-aligned image planes, for instance, allowing one to restrict the search for corresponding points to one dimension only instead of two. One could choose any direction for epipolar lines and make them for instance parallel and diagonal with respect to the CCS, however, having the epipolar lines horizontal (or vertical if two cameras are on top of one another), allows the search for corresponding points to be more reliable and computationally more tractable, since no interpolation is needed and the data can be reused. There are infinite fronto-parallel planes to choose from, so an additional constraint usually specified is the size of the final images, or the zoom factor to apply to the two images. One would usually like to maximize the common view in the two images and minimize distortion. Common rectification procedures, such as [5, 15, 20], allows one to specify which portion of the images to consider. The output images are usually cropped to the size of the largest common rectangle containing valid pixels, but depending on the application, different options are available. Figure 4.12 shows an example of cropping that considers only valid points in the two images. Although stereo rectification allows one to warp the two images, as they have been acquired by two cameras with row-aligned and co-planar

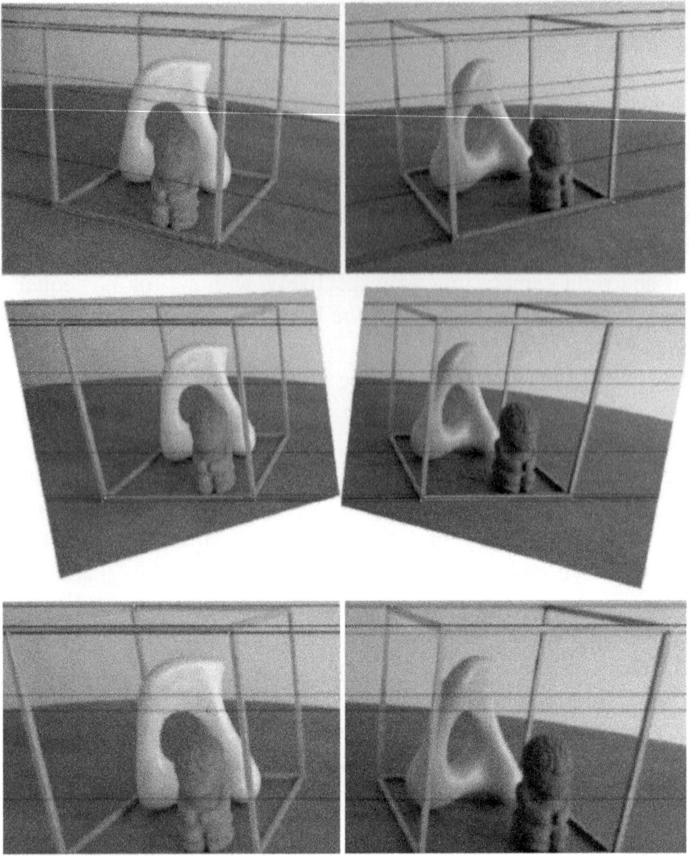

Fig. 4.12 Stereo rectification [42]. From the *top to the bottom*: original images, images rectified, images cropped to the largest rectangle containing valid pixels

image planes, when designing a stereo rig, it is best to arrange the two cameras approximately in a fronto-parallel configuration, as close as possible to a horizontal alignment. This will reduce image distortions in the rectification process. In order to maximize the common field of view, the two cameras are usually displaced in a vergent configuration, as shown in Fig. 1.4.

Similar to undistortion, rectification is also customarily performed by a look-up table and backward mapping. Once the undistortion map and the rectification map are available, the two can be combined in a single map in order to apply undistortion and rectification in a single step. The possibility of combining the two procedures may hide the fact that only the former compensates for systematic errors, but not the latter. Namely, the undistortion of the images acquired by L and R is a compensation for systematic errors, but rectification does not compensate for systematic errors, since it just transforms the acquired images in order to simplify the tasks of stereo vision algorithms.

Photometric calibration requires particular attention for stereo systems. Most stereo matching algorithms rely on the fact that corresponding points in the two images have similar photometric characteristics. Among the most common local matching methods, some assumes that two corresponding pixels have the same color and intensity, for example the sum of absolute differences (SSD). Others, instead, are more robust with respect to different intensities, for example the Normalized Cross Correlation (NCC), invariant to linear illumination change, or census transform [68] invariant to any alteration that preserves the intensity order. Even if there are techniques to compensate for such local intensity differences, it is good practice to reduce the photometric differences before looking for correspondences. Algorithms for single camera photometric calibration can also be applied to the stereo case, with the additional constraint that photometric calibration for stereo images should consider the effects on both images, carefully avoiding the enhancement of photometric differences.

Another problem with stereo systems that does not affect single cameras is the temporal synchronization of left and right cameras. Almost every stereo matching algorithm relies on having two synchronized cameras, since a minimum mismatch in the timing is equivalent to a decalibration of the camera. Dynamic scenes commonly require the time difference between the acquisition of the two cameras to be as small as possible. Best results are obtained by hardware synchronization at the lowest level, but when this is not possible, one needs an accurate software routine to keep the two acquisitions synchronized.

4.2.3 Extension to N-View Systems

In the previous two sections, we introduced the concepts of calibration of a standard camera and a stereo system, i.e., a system with two standard cameras. Of course, one can use more than two cameras to retrieve depth information, so this section introduces the concepts of calibration for a system made of more than two standard cameras. Let us recall that a N-view system is a system made of N cameras as in Fig. 1.10, and that the N cameras can be either a single camera moving to different positions at subsequent times $1, \dots, N$, or different cameras. While the case of a single moving camera leads to structure from motion techniques mentioned in Sect. 1.2.2, the case of N different cameras simultaneously acquiring the scene is also of special interest for depth cameras. In this section we will consider the case of N independent cameras, but all of the reasoning remains valid if a single moving camera is used instead.

The reader will notice that the equations leading to the calibration of a single camera (4.10) and of a stereo camera (4.12) share the same structure, which can be generalized as

$$\min_{\mathbf{K}_n,\mathbf{d}_n,\mathbf{R}_{nk},\mathbf{t}_{nk}} \sum_{n=1}^{N}\sum_{k=1}^{M}\sum_{j=1}^{J} \delta_{nk}^{j} \|\mathbf{p}_{nk}^{j} - f(\mathbf{K}_n, \mathbf{d}_n, \mathbf{R}_{nk}, \mathbf{t}_{nk}, \mathbf{P}^j)\|_2^2 \tag{4.13}$$

where $\mathbf{p}_{nk}^{j}$ is the projection of the 3D feature P^j on the n-th camera at the k-th pose of the checkerboard, δ_{nk}^{j} is 1 if P^j is visible by the n-th camera at the k-th pose and 0 otherwise, and $f(\mathbf{K}_n, \mathbf{d}_n, \mathbf{R}_{nk}, \mathbf{t}_{nk}, \mathbf{P}^j)$ is a function accounting for projection and distortion as in (4.10) and (4.12). Equation (4.13) shows how the calibration routine presented in Sect. 4.2.1 and extended in Sect. 4.2.2 can be further modified to deal with a system made of $N > 2$ cameras. If the minimum in (4.13) is also taken over all the possible P^j, then the problem becomes a N-view reconstruction problem, also known as bundle adjustment.

Recalling supervised and unsupervised calibration methods from Sect. 4.1.3, supervised calibration of single cameras or stereo systems is always feasible, but in the N-camera case there are a number of difficulties. For example, for a 2D calibration plane or a 3D calibration object, the calibration target may not be clearly, simultaneously visible in all views. A possible solution is to use 1D calibration objects, for example [57], with a freely moving bright spot. Another approach is unsupervised calibration, in which structure from motion approaches already seen in Sect. 1.2.2 in connection with 3D reconstruction are used to estimate both the scene structure and the cameras' parameters. In this process, the intrinsic parameters should be known prior [48] or recovered afterwards through self-calibration [61]. Bundle adjustment is the typical tool used to refine the positions of the scene points P^j and the entire set of camera parameters. Usually, single camera calibration is first performed on each camera in the system, in order to reduce the number of unknowns for the bundle adjustment. It is important to constrain the minimization problem when possible, as the number of unknowns grows with the number of cameras in the system. As already seen in Sect. 4.1.3, there exist a great number of methods for structure from motion, also called *self-calibration* when the goal is to find the calibration parameters. The possibility of calibrating a camera from multiple uncalibrated images was first suggested by Faugeras et al. [44] and then extended in many other works considering different constraints like the knowledge of additional information [14, 18, 19, 46, 59, 61]. The reader interested in the theory of self-calibration is referred to [21, Chap. 18].

The case of $N = 3$ cameras, also called a *trinocular* system, is also of great interest for the calibration of consumer depth cameras. The trifocal tensor, which for $N = 3$ corresponds to the fundamental matrix for $N = 2$, has a number of interesting properties (see [21, Chap. 14]). In particular, with three cameras, there are more epipolar geometry constraints than in the two views case, but it is still possible to rectify all three cameras together and have all the images aligned to a common image plane. A point p_1 in camera C_1 has to be in the epipolar line defined by C_1 and C_2 and also in the epipolar line defined by C_1 and C_3. Figure 4.13a shows an example of matching points in the three views, together with the correspondence

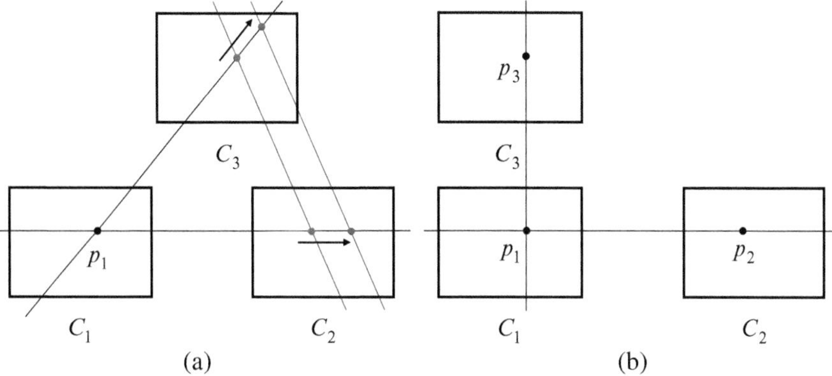

Fig. 4.13 Example of epipolar constraints for a trinocular system: (**a**) Possible matches for the point p_1 are shown in the other two cameras C_1 and C_2; (**b**) Configuration where C_2 and C_3 are horizontally rectified, and C_1 and C_3 are vertically rectified

epipolar lines. This property allows one to make the correspondence problem more robust, since a point must match in three images rather than only two.

The rectification process of the three images is similar to the stereo rectification of Sect. 4.2.2, but in this case there are less degrees of freedom, as the three views impose more constraints. Note that with $N > 3$ it is not possible to rectify all the cameras simultaneously. For stereo rectification, the two optical centers limit the orientation of the common image plane of the two cameras to one degree of freedom, i.e., the rotation around the baseline. In the trinocular case, the common image plane must be parallel to the plane defined by the three optical centers, therefore no adjustment can be made. It is a good practice to place the cameras as close to the ideal configuration as possible, since the trinocular case with respect to the $N = 2$ case has fewer degrees of freedom on epipolar rectification, and the results are more sensitive to the cameras' position [4]. For example, if one wants to rectify the three cameras horizontally and vertically, like the case of Fig. 4.13b, where C_1 and C_2 are horizontally rectified, and C_1 and C_3 are vertically rectified, the cameras must be placed as close as possible to the configuration shown by the figure. Note that this configuration allows one to speed up the matching by reusing data and avoiding interpolation, since the matches can be sought along rows or columns of the images.

Another advantage of trinocular systems over stereo systems is their ability to better detect decalibration. Consider the case of Fig. 4.3a, where a physical alteration of the system causes an horizontal shift of the image. If an additional view is present, e.g., a view above one of the two cameras, an additional check along the epipolar lines of the third camera would detect the error.

The effectiveness of trinocular systems for depth estimation is indirectly supported by their usage within market products. In 2015, Dell released the Dell Venue 8 7840 tablet [1], shown in Fig. 4.14, the first consumer device with three cameras, capable of depth estimation.

Fig. 4.14 Detail of the rear of a Dell Venue 8 7840 tablet with the trinocular camera system. The three cameras are highlighted

Let us finally observe that most state of the art algorithms for N-view system calibration do not consider photometric errors, i.e. color differences between the modeled and the observed scene, but instead minimize a geometric error, e.g., the reprojection error (4.13). Most N-view applications assume a single common color response for all the cameras; however, even if the same type of camera is used, each camera may have different sensor characteristics and responses to ambient conditions. Temperature variations, for instance, can affect the camera's response. Different techniques can be used for photometric calibration of single cameras, as explained in Sect. 4.2.1. However, independent color correction in the case of N-camera systems may lead to notable discontinuities in the texture mapping between neighboring cameras. Better results are achieved by jointly considering the color calibration of all the cameras together with the intrinsic and extrinsic calibration [64].

4.3 Calibration of Depth Cameras

While the geometric working principles of standard cameras involve simple considerations well described by the pin-hole model and perspective projection of Sect. 1.1, depth cameras use additional information to provide a depth estimate of the scene. Structured light depth cameras rely on one or more standard cameras and an additional projector, as seen in Chap. 2, while ToF depth cameras, as shown in Chap. 3, operate on the basis of the RADAR principle, measuring the time difference between the transmitted and received optical signal.

To provide meaningful measurements, depth cameras require a rigorous calibration of each internal component that is not addressed here. In this book we consider structured light and ToF depth cameras as single instruments providing depth information of the scene, although they are actually comprised of different elements, an IR projector and one or more IR cameras for structured light depth cameras, and an array of transmitters and a matrix of receivers for ToF depth cameras.

For practical purposes, additional details on data acquisition for ToF and structured light depth cameras' calibration should be explained. In the case of a ToF depth camera, it is possible to retrieve the amplitude image $\hat{A}_T$ and the depth map $\hat{Z}_T$ in a single step. In the case of a structured light depth camera, however, this requires two steps since the acquisition of the raw IR images requires no pattern projection and the acquisition of the depth maps instead requires the pattern projection. To summarize, the data needed for calibration of depth cameras require a single acquisition in the case of ToF depth cameras and two acquisitions in the case of structured light depth cameras.

Depth cameras are usually pre-calibrated by proprietary algorithms, and the calibration parameters are stored in the device during manufacturing and made accessible to the user only by official drivers. Usually, the manufacturer's calibration does not completely correct depth distortion, and accuracy can be improved by software procedures correcting camera's output data. The correction, however, is based on a specific calibration model whose parameters are identified during the calibration process. This section describes calibration procedures that allow one to fully exploit data provided by depth cameras to reduce systematic errors. The registration of a depth camera with other imaging sensors, like standard cameras, is discussed in the next section.

4.3.1 Calibration of Structured Light Depth Cameras

All structured light depth cameras presented in Chap. 2, share the same key components, a projector and one or two standard cameras. Depth data provided by a structured light depth camera can be obtained only by the camera producer API, since the algorithms used to compute the depth map are confidential.

Some depth cameras also make available the image from the standard camera (usually an IR camera) forming the structured light system, and such an image can be used to improve the depth quality or to calibrate the depth camera with other sensors. The standard camera also represents the viewpoint of the depth map provided by the structured light depth camera, and the availability of the standard camera raw images allows one to use the single camera calibration techniques of Sect. 4.2 to estimate intrinsic parameters. Figure 4.15 shows an example of a

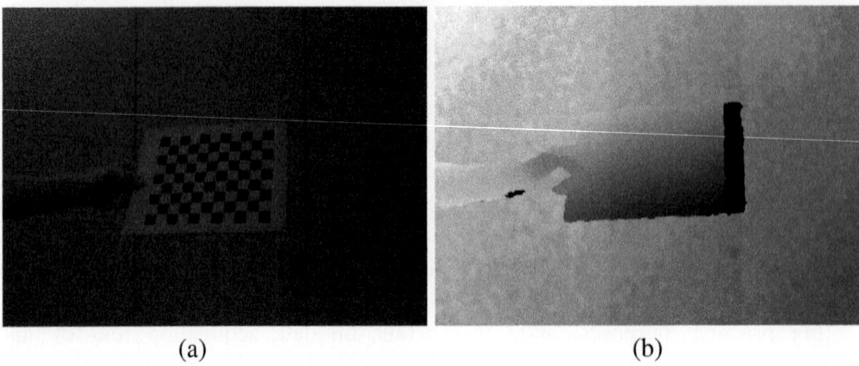

(a) (b)

Fig. 4.15 Example of a checkerboard acquired by a structured light depth camera: (**a**) IR image; (**b**) corresponding depth map

checkerboard acquired by a structured light depth camera. The only difference with the images in Fig. 4.8 is that for structured light depth cameras, the reference camera is an IR camera and therefore no colors are available, only the light intensity at the specific wavelength. Another practical difference with respect to standard color cameras is that to calibrate an IR camera with the procedure described in Sect. 4.2, it is necessary to illuminate the scene by sources emitting light in the IR spectrum, as in the case of sunlight or common incandescent light bulbs. Common fluorescent lamps usually do not emit in the IR bands, therefore are not suited to structured light depth cameras' calibration. This practicality requires particular attention since an accurate calibration requires proper illumination. A non uniform illumination results in darker regions with consequently higher noise making the checkers corners localization less precise.

There are cases in which only the depth map estimated from the camera is available, therefore it is not possible to use the 2D plane calibration with checkerboard pattern. Indeed, the calibration procedure described in Sect. 4.2 requires one to estimate the checkers' corners that are practically not visible in the estimated depth map, as shown from the checkerboard acquisition of Fig. 4.15b. These situations force one to use 3D objects to find the intrinsic parameters of the structured light depth camera. For example, Jin et al. [26] uses a set of cuboids with known sizes as calibration object, and the objective function in the optimization process, instead of considering the reprojection error of 3D points, considers distances and angles between planes, properties that can be robustly measured by the depth map. Figure 4.16 shows an example of the 3D calibration object of Jin et al. [26]. Almost all the approaches based on depth data to find the intrinsic parameters of structured light depth cameras use only planar regions, since corners and edges especially are usually not precise enough. The advantage of these techniques is that depth camera data can be directly used without relying on feature extraction algorithms. The planes in Fig. 4.16b can be manually annotated, without affecting the precision of final results, or automatically detected by a plane fitting technique. The drawback

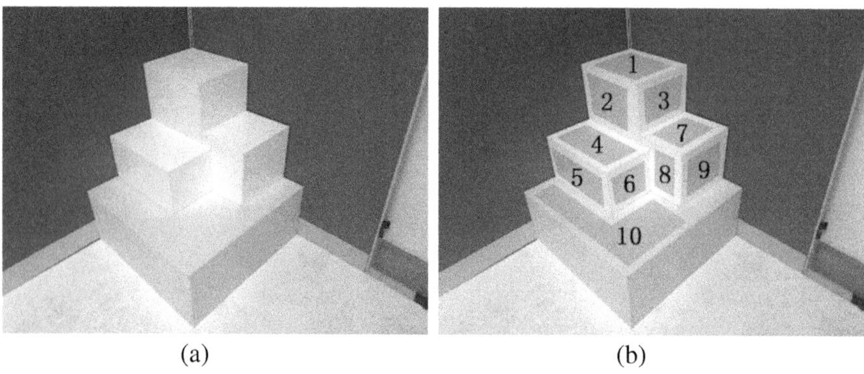

(a) (b)

Fig. 4.16 Example of 3D object used for calibrating a structured light depth camera from its depth data: (**a**) 3D calibration object; (**b**) Annotated planes used by Jin et al. [26]. (Courtesy of the authors of [26])

of these approaches is that the calibration object's precision directly affects the accuracy of the calibration results.

The previous techniques belong to the family of supervised calibration. Different techniques have been proposed to simplify the calibration process without relying on calibration objects of known size. These methods belong to the family of unsupervised calibration as they require little or no knowledge of the framed scene. Among all the methods, the most effective ones rely on simultaneous localization and mapping (SLAM) [34, 60]. The key idea behind methods based on SLAM is that one can first build the scene structure with the most reliable samples in close range and then use the reliable information to calibrate the depth camera, as will be seen in Sect. 7.8 devoted to SLAM with depth cameras.

Photometric calibration for structured light depth cameras is of great importance during the calibration performed by the manufacturer. IR cameras and projectors have to be accurately calibrated, not only for estimating their correct position, but also for guaranteeing reliability of the photometric properties. For the final user, however, the photometric calibration of structured light depth cameras is of limited usefulness, since depth estimation algorithms cannot be changed.

Once the structured light depth cameras are calibrated, systematic errors can be corrected by different techniques. Most of the algorithms refer to the Kinect™ v1, as it reached the market first, but they can also be applied to other products. Many different models have been proposed to compensate for the systematic errors of Kinect™ v1 [7, 23, 26, 60]. The basic corrections are related to the myopic property of structured light depth cameras and consist of a scaled inverse correction and a per-pixel offset.

The scaled inverse correction usually applied to Kinect™ v1 allows one to convert the Kinect disparity unit to actual depth. A commonly used model is

$$\tilde{Z}_D = \frac{1}{\alpha D_D + \beta} \tag{4.14}$$

Fig. 4.17 *Top view* of different acquisitions of a flat wall with a PrimeSense sensor, before and after depth correction as proposed in [60]: (**a**) raw sensor depth data; (**b**) compensated depth data (Courtesy of the authors of [60])

where D_D is the raw disparity map acquired by the sensor, $\tilde{Z}_D$ is the converted depth map, and α and β are two constants to be determined by calibration.

As suggested in [54], a simple per-pixel correction of the type

$$Z_D = \tilde{Z}_D + Z_\delta(u, v) \tag{4.15}$$

where Z_D is the corrected depth map, $\tilde{Z}_D$ the raw depth map from the depth camera, and $Z_\delta(u, v)$ the per-pixel correction map, allows one to improve the depth data accuracy.

The parameters in the previously mentioned depth correction techniques can be obtained by fitting the model to real data. For example, just a few acquisitions of a planar surface from different distances usually suffice to estimate all the parameters.

The example of Fig. 4.17 shows the effects of systematic depth error compensation.

The actual procedures adopted by structured light depth cameras manufacturers in order to calibrate depth cameras and compensate for errors are not publicly available, as they are the intellectual property of the companies. However, this does not penalize the usage of structured light depth cameras, since they are supplied already calibrated with built-in compensation. The calibration procedures described in this section can be used to improve the reliability of the depth camera output and reduce decalibration artifacts, for example due to accidental hits. The most common motivation for geometric calibration of a structured light depth camera is to pair it with a standard color camera, typically present in consumer structured light depth cameras, in order to associate a color to all the points of the depth map that are visible from the color camera's viewpoint. Calibration of a depth camera with a standard color camera is discussed in Sect. 4.4.1 and algorithms to fuse color and depth data are presented in Chap. 5.

4.3.2 Calibration of ToF Depth Cameras

ToF depth cameras could be calibrated like structured light depth cameras and standard cameras in principle, since their image formation can be modeled by perspective projection. The depth measurement of ToF depth cameras is non myopic, as it is derived from time or phase differences between the transmitted and received signal that in principle do not depend on the actual distance. However, a significant number of errors influence the acquired data, due to multiple practical issues presented in Chap. 3. In addition, as reported in [56], lens distortion is high in ToF depth cameras, therefore, proper corrections are necessary to obtain meaningful measurements.

Studies show that ToF depth cameras need a time delay, usually referred to as *pre-heating time*, before providing reliable depth measurements [36, 38] in order to reduce the systematic errors in terms of accuracy. For Kinect^{TM} v2, for example, the accuracy is reduced from 5 [mm] to 1 [mm] after 30 min from the first acquisition. For longer acquisition times the temperature of the camera may increase and affect the measurements, however, passive or active cooling systems usually compensate for such temperature variation.

Averaging multiple acquisitions [35, 36] is another noise reduction strategy requiring a static calibration environment. Interestingly, these works show that averaging more than 50 acquisitions does not increase precision, but gives a smoothing effect especially on corners and far points. When more acquisitions are averaged, the standard deviations of different measurements are reduced but the mean values remain almost constant. This effect can be justified by the nature of IR data's noise. Due to time correlation and not independent laser speckles, once data are averaged, the variance of the mean is not dramatically reduced. Indeed, averaging multiple measurements reduces variance only if the measurements are uncorrelated, as the variance of the mean increases with the average of the correlation.

The geometric calibration of a ToF camera can then be performed by the same techniques used for standard cameras and structured light depth camera calibration, although different methods better suited for ToF data have been proposed. Kahlmann et al. [28], for example, proposes two alternate calibration objects exploiting circular features and IR LEDs. Circular targets have the drawback of occupying many pixels of the acquired images, with a consequent limited resolution and limited number of correspondences. IR LEDs instead, can be easily detected by thresholding the amplitude image but they require an accurate assembly of the electronic components and also assume a non-negligible target size.

The 2D planar checkerboard technique of Sect. 4.2 also remains the most frequently used approach for ToF depth cameras. In this case, the amplitude images $\hat{A}_T$ can be used in place of color images from standard cameras. Another difference with respect to standard cameras is that amplitude images $\hat{A}_T$ usually have lower spatial resolution, making the checkerboard localization less precise. Figure 4.18 shows an example of checkerboard acquired by a ToF depth camera. Scene illumination during the calibration process of ToF depth cameras must comply with the same

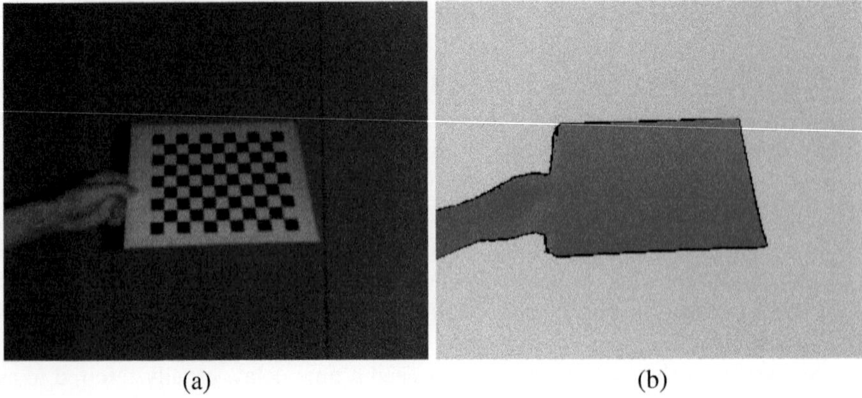

(a) (b)

Fig. 4.18 Example of a checkerboard acquired by a ToF depth camera: (**a**) signal amplitude image; (**b**) corresponding depth map

constraints as seen for structured light depth cameras. Due to the presence of IR filters in the ToF camera receiver, the illumination needs to have enough spectral power in the frequency bands corresponding to that of the receiver, however, the ToF depth camera already includes the illumination system that can be used also for calibration purposes.

The amplitude images can be collected in two different ways, either in the so called *standard mode*, i.e., with the ToF depth camera illuminators active during the acquisition, or in the so called *common mode*, i.e., with ToF camera illuminators off during the acquisition, namely, using the ToF camera as a standard IR camera. The first solution is more direct as it does not require external tools and generally produces better results, but it requires proper integration time setting in order to avoid saturation and reduce noise. The second solution requires an external auxiliary IR illumination system as for structured light depth cameras.

The geometric calibration performed by the previously described technique only relies on amplitude images of the camera and does not consider depth related deformations typical of ToF depth cameras. Since ToF depth cameras measure the time of flight along the optical rays, erroneous measurements are expected to appear along the radial distance as well. The photometric calibration deals with the so called *systematic offsets* of ToF depth camera measurements, due to various factors, such as harmonic distortion and non-zero mean Poisson distributed photon-shot noise (as explained in Chap. 3). It is worth recalling that harmonic distortion, usually referred to as *wiggling error*, depends on the distance from the object to be measured, and that the photon-shot noise depends on the received signal intensity and amplitude. Therefore the systematic offsets ultimately depend on the measured distance, amplitude and intensity.

The most crucial error source is due to the systematic wiggling error, altering the depth measurement toward or away from the camera depending on the true distance. As explained in Chap. 3, this error is due to the non ideal modulation and sampling of the signal that lead to periodic errors, causing the measured depth to oscillate around the real value. The actual form of these oscillations depends on the modulation and demodulation techniques [37]. The calibration of this systematic offset can be performed by measuring targets characterized by different known reflectivity placed at preassigned distances and by comparing the measured distances against the actual ones. A comprehensive description and analysis of such procedures can be found in [27, 52, 73]. Such measurements can be used both in a model-based approach as in [52] where the model is a polynomial function, or in a model-less approach as in [27, 73] where a look-up table describes the relationship between measured and actual radial distances. More accurate schemes, such as [40], rely on B-splines interpolation to model distance deviation.

Systematic errors concerning both aspects of geometric calibration correction such as undistortion and radial distance measurement correction can be independently compensated. Undistortion for the ToF depth cameras can be implemented as for the case of standard cameras. Systematic radial distance measurement errors in the case of model-based calibration can be compensated by inverting the fitted functional-model, and in the case of model-less calibration, by a look-up-table correction. Figure 3.14 shows the reduction of the wiggling effect after this correction. More refined corrections, such as [31], involve both a depth correction to compensate errors along the optical rays and an optical rays direction alignment to compensate for angular misalignments. Another method accounting for both geometric calibration and depth measurement correction requiring less input data is presented in [40], where multiple error sources are combined in the same model accounting for wiggling adjustment and intensity related artifacts.

Another photometric aspect that requires particular attention is the depth dependence from the reflected intensity. Lindner et al. [39] proposed an intensity based depth calibration to compensate for the depth measurements alteration due to the total amount of light received by the sensor. This effect is clearly visible in Fig. 4.19 where dark and bright checkers do not appear in the same plane. When depth data are used to acquire the checkerboard used for calibration, it is advisable to avoid using the raw depth map directly, instead, first fitting a plane with all the points of the depth map inside the checkerboard and then projecting the feature points to that plane, in order to obtain more accurate measurements. Due to the presence of outliers in ToF depth measurements, it is customary to compute the plane parameters by robust estimation methods, e.g., RANSAC-based approaches as described in [17]. Amplitude A and intensity B images can eventually be used to remove noisy points associated with low SNR according to the model (3.47) derived in earlier works [9, 37] and extended in [47]. Equation (3.47) shows that the noise variance is inversely proportional to the signal-to-noise ratio $A\sqrt{2}/\sqrt{B}$.

There are many other sources of error, as shown in Chap. 3, that cannot be directly addressed by calibration and for which different solutions have been proposed. These additional errors are usually scene dependent or random errors and post processing of the acquired data is needed in order to increase the reliability of ToF depth measurements. Better calibration results can be obtained if a color camera is paired with the ToF depth camera (as discussed in Sect. 4.4.1), with

Fig. 4.19 Intensity related depth artifacts on a ToF depth camera. Dark checkers reflect less light and appear as if they are farther from the camera

larger improvements compared to the structured light case, since ToF depth cameras typically have low resolution and can benefit from the higher spatial resolution of the color camera. In addition, unsupervised calibration methods based on SLAM, mentioned in the previous section, can be applied to ToF depth cameras as well.

4.4 Calibration of Heterogeneous Imaging Systems

The previous sections described the calibration of standard cameras and depth cameras as single entities, however, different cameras may be jointly used in order to provide additional information or deliver more accurate depth maps, as will be seen in Chap. 5. In this book we refer to these systems made by different imaging devices as *heterogeneous imaging systems*.

A heterogeneous imaging system is a setup made by multiple imaging devices of different natures measuring various attributes of the same scene, such as color and geometry. The attributes measured by a specific imaging device depend on its spatial position, hence the measurements of different imaging systems can only be related based on the knowledge of their relative positions. The calibration of an heterogeneous imaging system consists of the calibration of each imaging device forming the system and the estimation of the relative positions between such imaging devices. The mapping of data from different cameras to a common reference system requires one to deal with many issues, such as occlusions and different resolutions. Data fusion techniques addressing these problems are presented in Chap. 5.

The remainder of this section considers the calibration of systems made by a depth camera and color camera to provide combined depth and color information, for example, and systems made by a depth camera and a stereo vision system, usually deployed to provide more reliable depth maps.

4.4.1 Calibration of a Depth Camera and a Standard Camera

Most consumer depth camera products presented in Chaps. 2 and 3 feature a color camera in addition to the depth sensor. This family of devices are usually called RGB-D cameras and their usage has recently become more popular for applications like virtual reality, 3D scanning and gesture recognition, some of which are presented in the next chapters. The setup of a depth camera and a color camera is easily reproducible by pairing the two independent devices in a rigid system. Consumer depth cameras come already calibrated but systems made by different cameras, although independently calibrated, need to be jointly calibrated before use.

The coupling of color cameras with depth cameras provides substantial improvements since it combines the advantages of color and depth data. High resolution color cameras make the pose estimation of the calibration object more stable and accurate, while depth cameras already deliver the 3D scene representation without additional computation.

In principle, as long as the cameras provide a standard image (a color image for standard cameras and an IR or amplitude image for depth cameras) the RGB-D system can be calibrated by the method presented in Sect. 4.2.3 for standard cameras. The first step is the independent calibration of the depth camera and the standard camera. It is therefore necessary to calibrate and compensate the measurements of the standard camera and those of the depth camera as discussed in the previous sections. The second step is to estimate the relative position between the depth camera and the standard camera, i.e., the estimate of the relative roto-translation between the reference systems associated with the two imaging systems. This can be accomplished by the recognition of saliency points of suitable calibration objects (e.g., checkerboards) from the data acquired by the different cameras. A final nonlinear minimization (4.13) is usually performed to improve the results.

Other different techniques have been proposed for the particular case of RGB-D camera calibration, especially for the Kinect^{TM} v1 [7, 23, 54, 71], but all share the same basic principles. In order to account for depth related artifacts of the measurements, one has to consider depth data which the standard procedure described before does not account for in the calibration process. Checkers' corners are good features to be detected in color or IR images, however, as previously seen, they are not easily detectable in depth maps. Depth discontinuities, usually associated with color discontinuities, are possible features, however, due to occlusions and other problems, one cannot assume that color and depth discontinuities coincide and are visible from both imaging systems. The most reliable features for depth data are

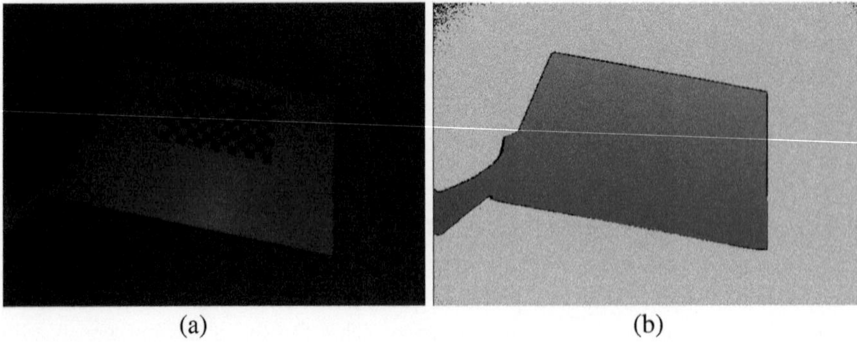

Fig. 4.20 Checkerboard attached to a plane used to calibrate depth cameras (a ToF depth camera in this example): (**a**) amplitude image; (**b**) depth map. Note that the calibration checkers are not visible in the depth map

planar surfaces that can be easily combined with the 2D checkerboard as shown in Fig. 4.20. Equation (4.13) can be further extended to consider the difference between the depth measured by the depth camera and the one computed from a model. The samples to consider for this term can be ones belonging to the planar surface, which can be easily modeled and the feature points easily detected either by manual or automatic procedures.

4.4.2 *Calibration of a Depth Camera and a Stereo Vision System*

The calibration of an heterogeneous system made of a depth camera and a stereo system requires careful attention. In principle, one could consider this system as a N-view system and apply a calibration procedure of the type described in Sect. 4.2.3. However, the two cameras L and R of the stereo system S may not be separately accessible and it may not be possible to change the depth computation algorithm. In addition, the joint calibration of all three single cameras of the heterogeneous system, i.e., depth camera D, left camera L and right camera R, results in worse depth estimates by the stereo system S if all cameras are rectified together. This effect is due to the rectification process described in Sect. 4.2.1 and the effect of minimization (4.13), which tries to optimize the reprojection error on all of the cameras without considering that L and R are part of the same stereo system S. In this type of heterogeneous system therefore, it is customary to rectify only the two cameras of the stereo system and to apply only undistortion correction to the depth camera. In this way, the stereo system performance is not affected by the presence of the additional depth camera. For these reasons, the calibration of heterogeneous systems consisting of a depth camera and a stereo vision system require techniques different from the previous approaches.

Unlike standard cameras, depth cameras and stereo systems can both produce depth measurements. Given a calibration object such as the standard checkerboard with associated J features (i.e., the checkerboard corners), one can obtain the representation of the 3D features P^j, $j = 1, \ldots, J$, with respect to the depth camera D-CCS and with respect to stereo system S-CCS, obtaining two sets of point coordinates $\mathbf{P}_D^j$ and $\mathbf{P}_S^j$ respectively (let us recall from Chap. 1 that the stereo camera reference system is the reference system of one of the two cameras, typically the left one). In other words, the depth camera D and the stereo system S provide the same cloud of points expressed with respect to two different reference systems. The computation of the roto-translation between the same set of points in two different reference systems is known as the *absolute orientation problem* and has a closed-form solution given by the Horn algorithm [24], which estimates the rotation matrix $\hat{\mathbf{R}}$ and the translation vector $\hat{\mathbf{t}}$ minimizing the distance between the 3D positions of P^j measured by the two imaging systems D and S, namely

$$\left[\hat{\mathbf{R}}, \hat{\mathbf{t}}\right] = \underset{\mathbf{R},\mathbf{t}}{\operatorname{argmin}} \frac{1}{J} \sum_{j=1}^{J} \left\| \mathbf{P}_D^i - [\mathbf{R}\mathbf{P}_S^j + \mathbf{t}] \right\|_2. \tag{4.16}$$

An alternate way of computing $\hat{\mathbf{R}}$ and $\hat{\mathbf{t}}$ is to minimize the reprojection error once again by (4.13). Assuming the stereo pair is already calibrated and rectified, it suffices to perform the minimization with respect to the rotation matrix $\mathbf{R}$ and the translation vector $\mathbf{t}$ describing the roto-translation of the left camera with respect to the depth camera. Equation (4.13) under these assumptions becomes

$$\left[\hat{\mathbf{R}}, \hat{\mathbf{t}}\right] = \underset{\mathbf{R},\mathbf{t}}{\operatorname{argmin}} \frac{1}{2J} \sum_{j=1}^{J} \left[\left\| \begin{bmatrix} \mathbf{p}_L^j \\ 1 \end{bmatrix} - \mathbf{K}_S \left[\mathbf{R}\mathbf{P}_D^j + \mathbf{t}\right] \right\|_2 + \left\| \begin{bmatrix} \mathbf{p}_R^j \\ 1 \end{bmatrix} - \mathbf{K}_S \left[\mathbf{R}\mathbf{P}_D^j + \mathbf{t} - \mathbf{b}\right] \right\|_2 \right] \tag{4.17}$$

where $\mathbf{K}_S$ is the intrinsic parameters matrix of both the L and R cameras of the rectified stereo vision system with baseline b, where $\mathbf{b} = [-b, 0, 0]^T$ is the vector with the coordinates of the origin of the R-3D reference system with respect to the L-3D reference system. We are using the left camera as the reference for the stereo system, i.e., S-3D $=$ L-3D within the assumptions and conventions of Chap. 1. Practical implementation details about the two approaches can be found in [12, 73].

Summarizing, there are two main algorithmic choices in the calibration of a system made by a depth camera D and a stereo system S:

- The first choice is the selection between the homography-based approach and the direct depth measurement approach for obtaining the 3D coordinates of the checkerboard corners acquired by D and the reference camera L of the stereo system. The main advantage of the former approach is that the calibration precision and accuracy are not affected by the precision and the accuracy of the depth measurements of D and S. The main advantage of the latter approach, instead, is that the calibration accounts for the depth nature of the measurements performed by the depth camera and stereo system, leading therefore to a better synergy between the data acquired by the two imaging systems.

- The second choice concerns the selection of the optimization criterion for the estimation of **R** and **t**. The two options are the *absolute-orientation* approach and the *reprojection-error minimization* approach. The former generally implies an easier task and an exact solution, but the latter leads to better performance for most applications. The choice between absolute orientation or reprojection error minimization depends on the specific application. For instance, if the application calls for the fusion of the 3D scene geometry estimates from the stereo vision system with those from the depth camera, it is worth optimizing absolute orientation, since in this case a good synergy between the data acquired by the two depth imaging systems is important. If the application instead reprojects the 3D geometry estimates performed by D on the images acquired by L and R for further processing, it is worth minimizing the reprojection error.

4.4.3 Calibration of Multiple Depth Cameras

Sections 4.3.1 and 4.3.2 showed that most of the time, structured light and ToF depth camera calibration can be performed by the methods introduced in Sect. 4.2 for standard camera calibration. If the depth cameras to be calibrated provide an image from the depth camera viewpoint, then the problem of multiple depth camera calibration corresponds to the N-view system calibration of Sect. 4.2.3. The required images can be, for example, the IR reference camera for structured light depth cameras, or signal amplitude or intensity for ToF depth cameras. Intrinsic parameters can be estimated by the methods of Sects. 4.3.1 and 4.3.2, and used in the non-linear minimization (4.13) to refine the initial estimate and determine the roto-translation matrices for each depth camera with respect to a common reference system.

Structured light and ToF depth cameras, with respect to standard cameras, are active light imaging systems and their accuracy is strictly related to the photometric properties of the acquired signal. The combined usage of multiple depth cameras may cause signal interference. Different techniques have been proposed to reduce the interference with the addition of external devices, such as in [8, 43] where artificial and controlled motion blur is introduced in the system in order to resolve the pattern of each structured light depth camera. Interference of multiple ToF depth cameras can be mitigated by altering the transmitted signal shape, e.g, by setting different modulation frequencies on each ToF depth camera [32]. Some camera manufacturers allow one to change the modulation in the camera settings. The interference effect in this case is reduced, since the emitted signal modulation of one camera does not match that of the other cameras, and it just contributes as additional noise.

Often, multiple depth cameras are used together to increase the field of view of the estimated depth maps, or to assign a depth value to regions occluded from one camera, without having to move any depth camera or rely on 3D reconstruction algorithms. In these cases, the depth cameras are placed so that there is minimal or no overlap between their fields of view, thus reducing possible interference but making calibration more cumbersome. After calibration it is possible to stitch together depth maps from different cameras in order to obtain a wider depth map with a larger field of view and possibly fewer occluded regions. Another possibility is to project the points of each depth map to a common point cloud. In this last case, it is also possible to refine the extrinsic calibration parameters by the application of the pairwise registration and global refinement techniques presented in Chap. 7.

4.5 Conclusions and Further Readings

Calibration of imaging devices is a fundamental requirement in 3D structure estimation for which both photometric and geometric calibration must be considered, as seen in this chapter. In general, the correspondence association of camera pixels and optical rays, performed by geometric calibration, is required by every standard or depth camera to provide useful data. Photometric calibration instead is usually required to improve data provided by the imaging system and reduce potential errors and possible systematic offset, as is shown for ToF depth cameras.

The calibration of a standard camera is of fundamental relevance for the calibration of all other imaging systems, both from a conceptual and operational point of view. The reader interested in more details about camera calibration can find image processing and corner detection techniques in [16, 53], projective geometry in [21, 63] and numerical methods in [49]. Classical camera calibration approaches are described in [11, 22, 69] and some theoretical and practical hints can also be found in [6], [45, Chap. 2] and in open source implementations like MATLAB Camera Calibration Toolbox [5] and calibration routines of OpenCV [3].

ToF depth camera calibration theory and practice is presented in [27, 38, 52, 73] and basic fundamental readings for the calibration of structured light depth cameras are [62, 72]. Depth cameras alone usually do not provide color information associated with the 3D geometry, so a more interesting setup considers a depth camera and one or more color cameras together. The joint calibration of color and depth cameras has become more popular since the recent introduction of consumer depth cameras featuring both color and depth cameras together as seen in Chaps. 2 and 3.

In a system made by multiple imaging devices, calibration allows them to work together in order to improve depth estimates with respect to those of each single imaging device alone. Chapter 5 will describe how to use multiple imaging devices to fuse data from depth and standard cameras, and Chap. 6 will describe how to combine color and depth data to improve scene segmentation, a task traditionally approached by standard color cameras only.

References

1. Dell Venue 8 7840 (2016), http://www.dell.com/us/p/dell-venue-8-7840-tablet/pd
2. Oculus Rift (2016), https://www.oculus.com
3. OpenCV (2016), http://opencv.org
4. L. An, Y. Jia, J. Wang, X. Zhang, M. Li, An efficient rectification method for trinocular stereovision, in *Proceedings of International Conference on Pattern Recognition* (2004), pp. 56–59
5. J.Y. Bouguet, Camera calibration toolbox for matlab, www.vision.caltech.edu/bouguetj/calib_doc/
6. G. Bradski, A. Kaehler, *Learning OpenCV: Computer Vision with the OpenCV Library* (O'Reilly, Sebastopol, CA, 2008)
7. N. Burrus, Kinect Calibration (2016), http://nicolas.burrus.name/index.php/Research/KinectCalibration
8. D.A. Butler, S. Izadi, O. Hilliges, D. Molyneaux, S. Hodges, D. Kim, Shake'n'sense: reducing interference for overlapping structured light depth cameras, in *Proceedings of ACM Annual Conference on Human Factors in Computing Systems* (New York, 2012), pp. 1933–1936
9. B. Buttgen, P. Seitz, Robust optical time-of-flight range imaging based on smart pixel structures. IEEE Trans. Circuits Syst. I: Reg. Pap. **55**(6), 1512–1525 (2008)
10. J. Chang, H. Kang, M.G. Kang, Correction of axial and lateral chromatic aberration with false color filtering. IEEE Trans. Image Process. **22**(3), 1186–1198 (2013)
11. D. Claus, A.W. Fitzgibbon, A rational function lens distortion model for general cameras, in *Proceedings of IEEE Conference on Computer Vision and Pattern Recognition* (2005)
12. C. Dal Mutto, P. Zanuttigh, G.M. Cortelazzo, A probabilistic approach to tof and stereo data fusion, in *Proceedings of 3D Data Processing, Visualization and Transmission* (Paris, 2010)
13. O. Faugeras, *Three-Dimensional Computer Vision: A Geometric Viewpoint* (MIT Press, Cambridge, MA, 1993)
14. Y. Furukawa, J. Ponce, Accurate camera calibration from multi-view stereo and bundle adjustment, in *Proceedings of IEEE Conference on Computer Vision and Pattern Recognition* (2008), pp. 1–8
15. A. Fusiello, E. Trucco, A. Verri, A compact algorithm for rectification of stereo pairs. Mach. Vis. Appl. **12**, 16–22 (2000)
16. R.C. Gonzalez, R.E. Woods, *Digital Image Processing* (Addison-Wesley Longman Publishing Co., Inc., Reading, Ma, 2001)
17. M. Hansard, R. Horaud, M. Amat, S. Lee, Projective alignment of range and parallax data, in *Proceedings of IEEE Conference on Computer Vision and Pattern Recognition* (2011), pp. 3089–3096
18. R.I. Hartley, Estimation of relative camera positions for uncalibrated cameras, in *Proceedings of European Conference on Computer Vision* (Springer-Verlag, Berlin, 1992), pp. 579–587
19. R.I. Hartley, Euclidean reconstruction from uncalibrated views, in *Proceedings of Joint European - US Workshop on Applications of Invariance in Computer Vision* (Springer, London, 1994), pp. 237–256
20. R.I. Hartley, Theory and practice of projective rectification. Int. J. Comput. Vis. **35**(2), 115–127 (1999)
21. R.I. Hartley, A. Zisserman, *Multiple View Geometry in Computer Vision* (Cambridge University Press, Cambridge, 2004)
22. J. Heikkila, O. Silven, A four-step camera calibration procedure with implicit image correction, in *Proceedings of IEEE Conference on Computer Vision and Pattern Recognition* (1997)
23. C.D. Herrera, J. Kannala, J. Heikkilä, Joint depth and color camera calibration with distortion correction. IEEE Trans. Pattern Anal. Mach. Intell. **34**(10), 2058–2064 (2012)
24. B.K.P. Horn, Closed-form solution of absolute orientation using unit quaternions. J. Opt. Soc. Am. **4**, 629–642 (1987)

25. B.K.P. Horn, Shape from shading, in *Obtaining Shape from Shading Information* (MIT Press, Cambridge, MA, 1989), pp. 123–171
26. B. Jin, H. Lei, W. Geng, Accurate intrinsic calibration of depth camera with cuboids, in *Proceedings of European Conference on Computer Vision (ECCV)*. Lecture Notes in Computer Science, vol. 8693 (Springer, Heidelberg, 2014), pp. 788–803
27. T. Kahlmann, H. Ingensand, Calibration and development for increased accuracy of 3d range imaging cameras. J. Appl. Geod. **2**, 1–11 (2008)
28. T. Kahlmann, F. Remondino, H. Ingensand, Calibration for increased accuracy of the range imaging camera SwissRangerTM, in *Proceedings of ISPRS Com. V Symposium* (Dresden, 2006), vol. 36, pp. 136–141.
29. K. Kanatani, Calibration of ultrawide fisheye lens cameras by eigenvalue minimization. IEEE Trans. Pattern Anal. Mach. Intell. **35**(4), 813–822 (2013)
30. J. Kannala, S.S. Brandt, A generic camera model and calibration method for conventional, wide-angle, and fish-eye lenses. IEEE Trans. Pattern Anal. Mach. Intell. **28**(8), 1335–1340 (2006)
31. Y.M. Kim, D. Chan, C. Theobald, S. Thrun, Design and calibration of a multi-view tof sensor fusion system, in *Proceedings of IEEE Conference on Computer Vision and Pattern Recognition* (2008)
32. Y.M. Kim, C. Theobald, J. Diebel, J. Kosecka, B. Miscusik, S. Thrun, Multi-view image and tof sensor fusion for dense 3d reconstruction, in *Proceedings of 3DIM Conference* (2009)
33. H. Komagata, I. Ishii, A. Takahashi, D. Wakatsuki, H. Imai, A geometric method for calibration of internal camera parameters of fish-eye lenses. Syst. Comput. Jpn. **38**(12), 55–65 (2007)
34. R. Kummerle, G. Grisetti, W. Burgard, Simultaneous calibration, localization, and mapping, in *Proceedings of IEEE/RSJ International Conference on Intelligent Robots and Systems* (2011), pp. 3716–3721
35. E. Lachat, H. Macher, T. Landes, P. Grussenmeyer, Assessment and calibration of a RGB-D camera (Kinect v2 Sensor) towards a potential use for close-range 3D modeling. Remote Sens. **7**, 13070–13097 (2015)
36. E. Lachat, H. Macher, M.-A. Mittet, T. Landes, P. Grussenmeyer, First experiences with Kinect v2 sensor for close range 3D modelling, in *ISPRS - International Archives of the Photogrammetry, Remote Sensing and Spatial Information Sciences* (2015), pp. 93–100
37. R. Lange, 3D Time-Of-Flight distance measurement with custom solid-state image sensors in CMOS/CCD-technology. Ph.D. thesis, University of Siegen, (2000)
38. D. Lefloch, R. Nair, F. Lenzen, H. Schäfer, L. Streeter, M.J. Cree, R. Koch, A. Kolb, Technical foundation and calibration methods for time-of-flight cameras, in *Time-of-Flight and Depth Imaging. Sensors, Algorithms, and Applications*, ed. by M. Grzegorzek, C. Theobalt, R. Koch, A. Kolb. Lecture Notes in Computer Science, vol. 8200 (Springer, Berlin/Heidelberg, 2013), pp. 3–24
39. M. Lindner, A. Kolb, Calibration of the intensity-related distance error of the pmd tof-camera, in *Proceedings of SPIE: Intelligent Robots and Computer Vision XXV*, vol. 6764 (2007), pp. 6764–6735
40. M. Lindner, I. Schiller, A. Kolb, R. Koch, Time-of-flight sensor calibration for accurate range sensing. Comput. Vis. Image Underst. **114**(12), 1318–1328 (2010)
41. A.L. Lluis-Gomez, E.A. Edirisinghe, Chromatic aberration correction in raw domain for image quality enhancement in image sensor processors, in *Proceedings of IEEE International Conference on Intelligent Computer Communication and Processing* (2012), pp. 241–244
42. C. Loop, Z. Zhang, Computing rectifying homographies for stereo vision, in *Proceedings of IEEE Conference on Computer Vision and Pattern Recognition* (1999), p. 131
43. A. Maimone, H. Fuchs, Reducing interference between multiple structured light depth sensors using motion, in *Proceedings of IEEE Virtual Reality* (Washington, DC, 2012), pp. 51–54
44. S.J. Maybank, O.D. Faugeras, A theory of self-calibration of a moving camera. Int. J. Comput. Vis. **8**(2), 123–151 (1992)

45. G. Medioni, S.B. Kang, *Emerging Topics in Computer Vision* (Prentice Hall PTR, Upper Saddle River, NJ, 2004)
46. P.R.S. Mendonca, R. Cipolla, A simple technique for self-calibration, in *Proceedings of IEEE Conference on Computer Vision and Pattern Recognition* (1999), p. 505
47. F. Muft, R. Mahony, Statistical analysis of measurement processes for time-of flight cameras, in *Proceedings of SPIE* (2009)
48. D. Nister, An efficient solution to the five-point relative pose problem. IEEE Trans. Pattern Anal. Mach. Intell. **26**(6), 756–770 (2004)
49. J. Nocedal, S. J. Wright, *Numerical Optimization* (Springer, New York, 2000)
50. M. Pollefeys, R. Koch, L. Van Gool, Self-calibration and metric reconstruction in spite of varying and unknown internal camera parameters, in *Proceedings of International Conference on Computer Vision* (1998), pp. 90–95
51. S.F. Ray, *Applied Photographic Optics*, 3rd edn. (Focal Press, Oxford, 2002)
52. I. Schiller, C. Beder, R. Koch, Calibration of a pmd-camera using a planar calibration pattern together with a multi-camera setup, in *Proceedings of ISPRS Conference* (2008)
53. J. Shi, C. Tomasi, Good features to track, in *Proceedings of IEEE Conference on Computer Vision and Pattern Recognition* (1994)
54. J. Smisek, M. Jancosek, T. Pajdla, 3d with kinect, in *Proceedings of IEEE Workshop on Consumer Depth Cameras for Computer Vision* (2011)
55. C. Steger, M. Ulrich, C. Wiedemann, *Machine Vision Algorithms and Applications* (Wiley-VCH, New York, 2007)
56. D. Stoppa, F. Remondino (eds.), *TOF Range-Imaging Cameras* (Springer, Berlin, 2012)
57. T. Svoboda, D. Martinec, T. Pajdla, A convenient multi-camera self-calibration for virtual environments. PRESENCE: Teleoperators Virtual Environ. **14**(4), 407–422 (2005)
58. R. Szeliski, *Computer Vision: Algorithms and Applications* (Springer, New York, 2010)
59. T. Taketomi, J. Heikkila, Zoom factor compensation for monocular slam, in *Proceedings of IEEE Virtual Reality* (2015), pp. 293–294
60. A. Teichman, S. Miller, S. Thrun, Unsupervised intrinsic calibration of depth sensors via SLAM, in *Proceedings of Robotics: Science and Systems* (2013)
61. B. Triggs, Autocalibration and the absolute quadric, in *Proceedings of IEEE Conference on Computer Vision and Pattern Recognition* (1997), pp. 609–614
62. M. Trobina, Error model of a coded-light range sensor. Technical report, Communication Technology Laboratory Image Science Group, ETH-Zentrum (1995)
63. E. Trucco, A. Verri, *Introductory Techniques for 3-D Computer Vision* (Prentice Hall PTR, Upper Saddle River, 1998)
64. G. Unal, A. Yezzi, S. Soatto, G. Slabaugh, A variational approach to problems in calibration of multiple cameras. IEEE Trans. Pattern Anal. Mach. Intell. **29**(8), 1322–1338 (2007)
65. J. Weng, P. Cohen, M. Herniou, Camera calibration with distortion models and accuracy evaluation. IEEE Trans. Pattern Anal. Mach. Intell. **14**(10), 965–980 (1992)
66. R.J. Woodham, Photometric method for determining surface orientation from multiple images. Opt. Eng. **19**(1), 191139–191139 (1980)
67. R. Wu, L. Wang, X. Cheng, A piecewise correction method for high-resolution image of wide-angle lens, in *Proceedings of International Congress on Image and Signal Processing* (2013), pp. 331–335
68. R. Zabih, J. Woodfill, Non-parametric local transforms for computing visual correspondence. Lecture Notes in Computer Science (Springer, Berlin, 1994)
69. Z. Zhang, A flexible new technique for camera calibration. IEEE Trans. Pattern Anal. Mach. Intell. **22**, 1330–1334 (1998)
70. Z. Zhang, Camera calibration with one-dimensional objects, IEEE Trans. Pattern Anal. Mach. Intell. **26**(7), 892–899 (2004)
71. C. Zhang, Z. Zhang, Calibration between depth and color sensors for commodity depth cameras, in *Proceedings of IEEE International Conference on Multimedia and Expo* (2011), pp. 1–6

72. L. Zhang, B. Curless, S.M. Seitz, Rapid shape acquisition using color structured light and multi-pass dynamic programming, in *Proceedings of IEEE International Symposium on 3D Data Processing, Visualization, and Transmission* (2002), pp. 24–36
73. J. Zhu, L. Wang, R. Yang, J. Davis, Fusion of time-of-flight depth and stereo for high accuracy depth maps, in *Proceedings of IEEE Conference on Computer Vision and Pattern Recognition* (2008)

Chapter 5
Data Fusion from Depth and Standard Cameras

As discussed in Chaps. 2 and 3, data provided by depth cameras have several limitations. In particular, data from depth cameras are usually noisier and at a much lower resolution than data from standard cameras, because depth camera technology is still far from the maturity of standard camera technology. This fact suggests that combining depth cameras with standard cameras may lead to more accurate 3D representations than those provided by depth cameras alone, and that the higher resolution of standard cameras may be exploited to obtain higher resolution depth maps. Furthermore, depth cameras can only provide scene geometry information, while many applications, e.g., 3D reconstruction, scene segmentation, and matting, also need the scene's color information.

ToF depth cameras are generally characterized by low spatial resolution, as seen in Chap. 3, and structured light depth cameras by poor edge localization, as discussed in Chap. 2. Therefore, a depth camera alone is not well suited for high-resolution and precise 3D geometry estimation, especially near depth discontinuities. If such information is desired, as is usually the case, it is worth coupling a depth camera with a standard camera. For instance, hand or body gesture recognition and scene matting applications might benefit from this kind of heterogeneous acquisition system. In addition, it is possible to consider an acquisition system made by a depth camera and a stereo system where both sub-systems are able to provide depth information and take advantage of the depth measurements' redundancy. This solution can also reduce occlusion artifacts between color and 3D geometry information and it may also be beneficial, for example, in 3D video production and 3D reconstruction.

In synthesis, the quality of acquired depth data can be improved by combining high resolution color data, particularly in critical situations typical of each family of depth cameras.

This chapter analyzes issues and benefits related to the combination of depth cameras with standard cameras. For a synergic combination, all data fusion approaches assume a calibration step where depth maps and standard camera images

P. Zanuttigh et al., *Time-of-Flight and Structured Light Depth Cameras*,
DOI 10.1007/978-3-319-30973-6_5

are registered, as discussed in Sect. 4.4 and briefly recalled in the next section for user convenience. The several ways of combining standard cameras and depth data can be categorized according to the following two families:

- Approaches exploiting a single color camera, discussed in Sect. 5.2, in which there is a single source of depth information and the main goal is to improve its quality and resolution by using filtering, interpolation, or enhancement techniques guided by the color data. As shown in Fig. 5.1a, the two devices are first calibrated, depth maps are possibly pre-processed and reprojected on the color data, and finally, color and depth data are processed together.
- Approaches exploiting two or more color cameras, considered in Sect. 5.3, in which two depth sources, the depth camera and the stereo vision system, are combined together. In the general pipeline outlined in Fig. 5.1b, after calibration, the two different depth fields are computed and reprojected on a common reference system and finally the two depth fields are fused into the final depth map, typically also using some reliability information at each location. This family of approaches is called *late fusion* since the fusion process directly combines the depth maps from the two systems.

  An alternate solution, shown in Fig. 5.1c, is to directly compute the final depth map by using both the information from the color cameras and from the depth sensor. This is typically achieved by modified stereo vision algorithms where the correspondence problem is constrained using depth data. Since this family of approaches uses raw data from the sensors, before the depth map computation in the case of the stereo system, it is usually referred to as *early fusion*.

The increase of spatial resolution obtainable by pairing a depth camera with a standard camera is often referred to as spatial *super-resolution*. It can be achieved either by local approaches, independently working on each spatial location, or by global approaches, typically exploiting probabilistic models defined on the whole depth map. Each approach has advantages and disadvantages, as described further in detail in Sect. 5.2. Also, in the fusion of depth data from a depth camera with depth data from a stereo system, both local and global approaches can be used, as will be shown in Sect. 5.3.

5.1 Acquisition Setup with Multiple Sensors

Before introducing the various super-resolution and data fusion algorithms, this section presents different acquisition setups typical of standard and depth camera data fusion. The different possible camera arrangements will be presented first and registration issues will be discussed next.

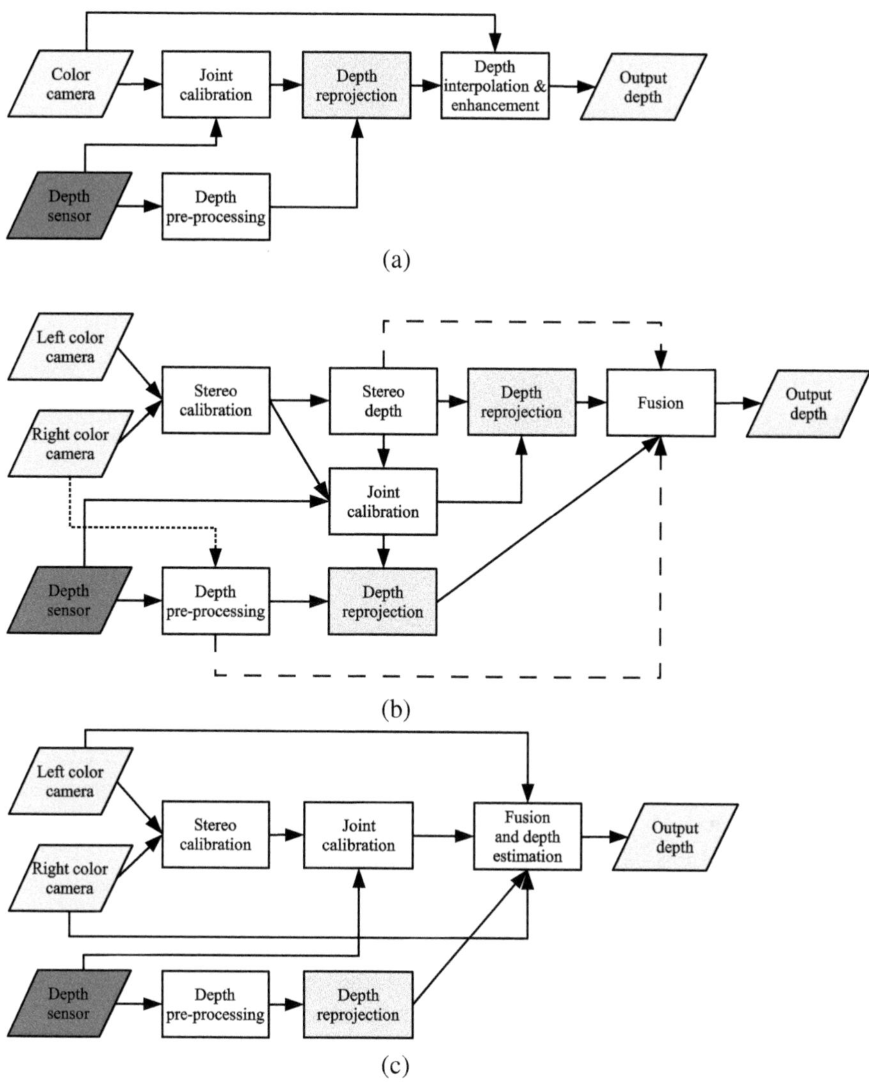

Fig. 5.1 Basic pipelines: (**a**) Depth camera and a single color camera; (**b**) Depth camera and two color cameras (late fusion); (**c**) Depth camera and two color cameras (early fusion). The optional data paths are shown as *dashed lines*

5.1.1 *Example of Acquisition Setups*

In order to jointly acquire color and depth information, one can use different acquisition setups. The simplest solution is to use a single color camera C together with the depth camera D, as in Fig. 5.2a. In this case, depth information is only provided by the depth camera. Depth and color data can be reprojected to a common reference viewpoint in order to align color and depth images. Notice that the color

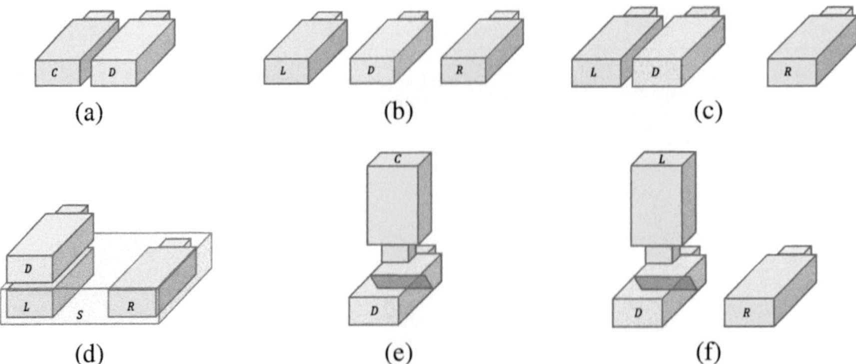

Fig. 5.2 Acquisition setup made by: (**a**) Depth camera and single color camera; (**b**) Depth camera and two color cameras (symmetric); (**c**) Depth camera and two color cameras (asymmetric); (**d**) Depth camera and stereo system (vertical arrangement); (**e**) Depth camera and single color camera (exploiting a mirror to have the same viewpoint for depth and color camera); (**f**) Depth camera and two color cameras (using a mirror to achieve the same viewpoint for depth and reference camera)

camera is typically placed as close as possible to the depth camera in order to minimize occlusions in the reprojected data. The setups with two color cameras *L* and *R* (see Fig. 5.2) are more interesting, since the two color cameras can be used as a stereo vision system, thus providing a separate source of depth information that can be combined with the depth camera data. Furthermore, two color cameras provide a representation of the scene from two different viewpoints, reducing the number of occluded points without a valid color value. The simplest solution for this setup is to place the depth camera *D* in the middle between the two color cameras *L* and *R* of a stereo system *S*, as in Fig. 5.2b. However, in order to further reduce the number of occluded points it is common to place the depth camera *D* closer to one of the two color cameras, e.g. *L*, like in Fig. 5.2c. Having the two viewpoints very close to each other is also useful in reducing reprojection errors due to the limited accuracy of the calibration or depth data noise. Another possibility, shown in Fig. 5.2d, is to place the depth camera *D* over one of the color cameras, e.g. *L*. This is a good solution if the stereo system *S* is a closed structure and the region between the two cameras is not accessible, or when the size of the cameras prevents them from being placed close to one another in the horizontal arrangement. Notice that the ideal setup of Fig. 5.2e, in which the color and depth cameras have exactly the same viewpoint, does not require data reprojection. This configuration can be built by mirroring systems combining both color and depth camera rays. This principle can be extended to the setup with a depth camera *D* and two standard cameras *L* and *R*, as shown in Fig. 5.2f. This ideal configuration has been studied in research projects, e.g. [1] and in a prototype camera (Fig. 5.3c) developed by Arri for the EU-funded SCENE project [17]. Despite the interest, this setup is uncommon since it is complex to build and maintain because of calibration issues associated with the presence of the mirroring system.

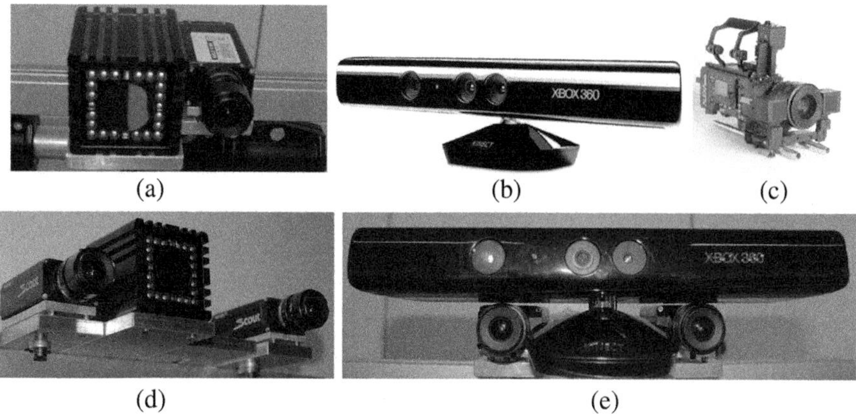

Fig. 5.3 Examples of real acquisition setups: (**a**) MESA Imaging SR4000 and a color camera; (**b**) Kinect™ v1 depth and color camera; (**c**) Arri prototype color camera with a ToF depth camera and a mirror system (**d**) MESA Imaging SR4000 and two color cameras forming a stereo vision system; (**e**) Kinect™ v1 and two color cameras forming a stereo vision system

Figure 5.3 shows examples of real acquisition systems combining color and depth cameras. Consider also that most consumer depth sensors typically include both a depth camera and standard color camera in order to provide both depth and color data. Most consumer depth camera products can be modeled as the acquisition setup of Fig. 5.2a. The first row of Fig. 5.3 shows three setups made by a depth camera and a single standard camera, while the second row shows two setups made by a depth camera and a few standard cameras forming a stereo system.

5.1.2 Data Registration

Let us consider first the setup of Fig. 5.2a made by a depth camera D and a standard color camera C. Note that if C and D are in different positions, as is usually the case, color and depth data refer to two different viewpoints. This can only be avoided by using an optical splitter on both devices similar to the setup of Fig. 5.2e proposed in [1]. This approach, however, introduces a number of undesirable effects. Notably, splitters affect the optical power and thus the distances measured by ToF depth cameras [28] and the IR emitters must be moved out of the ToF depth camera, increasing system costs.

For all the reasons above, in order to reduce the artifacts introduced by different viewpoints, it is simpler to place depth and standard cameras on a rig as close to one another as possible. Notice that many consumer depth cameras, including both versions of the Kinect™, embed both depth and color cameras in the same device, usually placing them in very close proximity.

The top row of Fig. 5.4 shows an example of data acquired by a high resolution color camera and a ToF depth camera. The second row of Fig. 5.4 shows data from the same scene acquired by a Kinect™ v1.

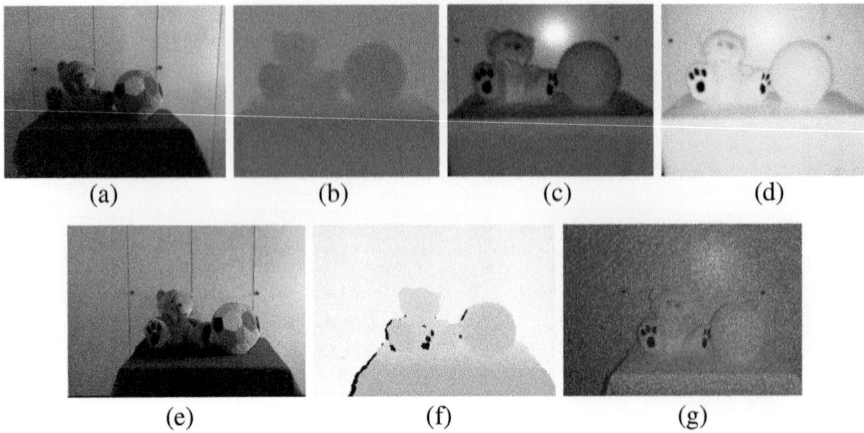

Fig. 5.4 Data acquired by depth and color cameras. *Top row* shows a setup made by a ToF depth camera and a standard camera: (**a**) color image acquired by the standard camera; (**b**) depth map acquired by the ToF depth camera; (**c**) amplitude image acquired by the ToF depth camera; (**d**) confidence map of ToF data. *Second row* shows data from Kinect^{TM} v1: (**e**) color image acquired by the Kinect^{TM} v1; (**f**) depth map acquired by the Kinect^{TM} v1; (**g**) IR image acquired by the Kinect^{TM} v1

Let us recall that it is necessary to calibrate the acquisition setup, as seen in Chap. 4, to obtain a common reference system for the color and depth data streams provided by the two imaging systems. Once a common reference system is defined, one can obtain the following data:

- color image and depth map from the color camera viewpoint;
- color image and depth map from the depth camera viewpoint;
- colored 3D point cloud representing the scene;
- textured 3D mesh representing the scene.

While images and depth maps are typically more suited to 3D video applications, 3D textured meshes or colored point clouds are more suited to 3D scene reconstruction or navigation.

In order to obtain an image I_C and a depth map Z_C from the reference viewpoint of the color camera C defined on a lattice Λ_C, each sample $p_D \in Z_C$ acquired by the depth camera D, associated with a 3D point P with coordinates $\mathbf{P}_D = [x_D, y_D, Z_D]^T$, must be reprojected to pixel $p_C \in I_C$ with coordinates $\mathbf{p}_C = [u_C, v_C]^T$. Considering both the systems already calibrated and undistorted as shown in Chap. 4, given the intrinsic matrices $\mathbf{K}_C$ and $\mathbf{K}_D$ of color and depth cameras, and extrinsic parameters $\mathbf{R}$ and $\mathbf{t}$ providing the roto-translation of the D-CCS with respect to the C-CCS, the reprojection can be performed according to

$$\begin{bmatrix} \mathbf{p}_C \\ 1 \end{bmatrix} = \frac{1}{z_C} \mathbf{K}_C \left[\mathbf{R} \mid \mathbf{t} \right] \begin{bmatrix} \mathbf{P}_D \\ 1 \end{bmatrix} \quad (5.1)$$

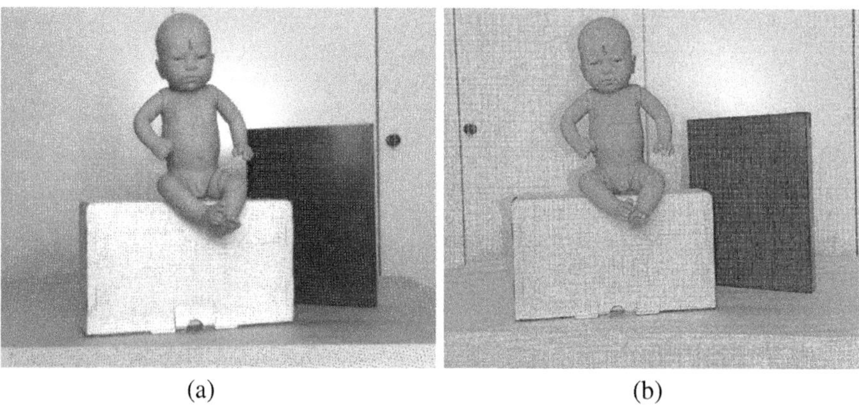

(a) (b)

Fig. 5.5 Reprojection of depth samples from a depth camera on the image from a color camera: (**a**) data acquired by a ToF depth camera; (**b**) data acquired by a structured light depth camera

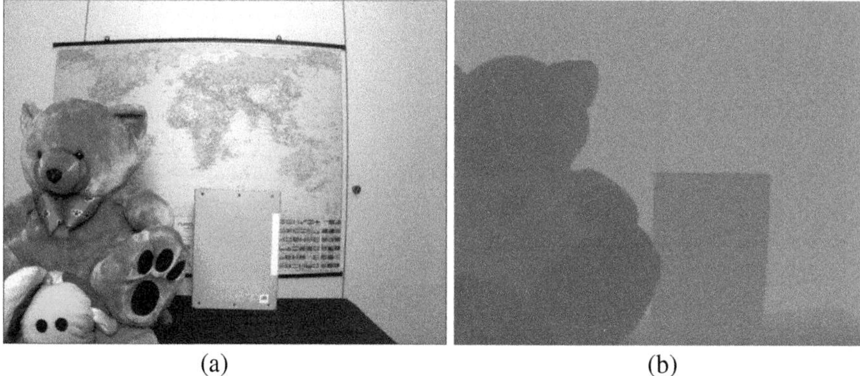

(a) (b)

Fig. 5.6 Example of occlusion issues: (**a**) color camera image; (**b**) ToF depth map. Note how the map in the region between the teddy bear and the collector is visible from the viewpoint of the color camera but not from the viewpoint of the ToF depth camera

where $z_C = \mathbf{r_3}^T\mathbf{P}_D + t_3$ is the depth associated with the point P in the C-CCS. The reprojection of depth data samples $p_D \in Z_C$ on the color image I_C produces a low resolution depth map Z_C^{low} defined on a subset of Λ_C, since the resolution of I_C is typically higher than that of Z_D. This is clearly shown by the examples in Fig. 5.5.

The reprojected depth samples need to be interpolated in order to associate a depth value to each pixel of Λ_C and obtain from Z_C^{low} a depth map Z_C with spatial resolution equal to that of I_C. Interpolation strategies for this task will be discussed in Sect. 5.2.1. The artifacts introduced by depth interpolation, however, may cause errors in the occluded regions, as some samples may be reprojected on wrong locations occluded from the C viewpoint. Figure 5.6 shows an example of occlusions caused by the different viewpoint of the two cameras. Some regions visible from the depth camera viewpoint are not visible from the color camera viewpoint and vice-versa.

The occluded samples must be removed, as they are artifacts of the reprojection and their presence may be critical for many applications. Removal can be achieved by building a 3D mesh from the depth map Z_D and then rendering it from the color camera viewpoint. The Z-buffer associated with the rendering can then be used in order to recognize the occluded samples. An example of Z-buffer based algorithms is presented in [27]. Another possibility [37] is to use the 3D geometry acquired from the depth camera D in order to convert I_C to an orthographic image where all the optical rays are parallel. Such a conversion makes the fusion of depth and color data trivial.

A different approach is needed to represent depth and color data from the D viewpoint. The color camera provides only 2D information and the acquired pixels cannot be directly reprojected to the D viewpoint like in (5.1). In order to get a color image from the D viewpoint it is necessary to build a 3D mesh from the depth map Z_D acquired by D, reproject the 3D vertices of the mesh visible from the C viewpoint to I_C, and associate a color to the corresponding pixels on Z_D.

From the D viewpoint one obtains a low resolution color image I_D defined on the 2D lattice Λ_D associated with the D-2D reference system, and a low resolution depth map Z_D defined on the same lattice. Since the final resolution is equal to the resolution of depth data, usually lower than the one of the color camera, no special interpolation strategies are needed. Interpolation is only needed to associate a color to the pixels of Z_D occluded from the C viewpoint.

As previously mentioned, representations in terms of images and depth maps are typical of 3D video applications, but in other fields, such as 3D reconstruction, colored point-clouds or textured meshes are more common scene descriptions. In this case, the geometric description, i.e., the point cloud or the mesh, can be obtained from the depth measurements and calibration information. The color of each acquired point or mesh triangle can then be obtained by projecting the point cloud points or the triangle vertices by (5.1) on the color camera image. After computing the point coordinates on the camera image, the actual color value can be computed from the closest image samples by image interpolation (e.g., bilinear, bicubic, spline, etc.). A similar procedure can be used for the texture coordinates of each 3D triangle of the 3D mesh.

If the setup features a stereo system S made by two color cameras and a depth camera D, multiple color and depth data streams are available, allowing for two major configurations. The first is to independently apply the previously described methods to each of the two cameras in order to reproject the acquired data on the D viewpoint or the viewpoint of one of the two cameras. The second possibility is to use stereo vision techniques [43] applied to the data acquired from the stereo system S in order to compute the 3D positions of the acquired points. The 3D points obtained from the stereo can then be combined with the 3D data obtained from the depth camera D with respect to a 3D reference system (either the D-3D or the S-3D reference systems) as described in Sect. 5.3. In this particular case, the adopted approach generally depends on the data fusion algorithm.

5.2 Fusion of a Depth Camera with a Single Color Camera

As seen in the previous section, calibration and reprojection make available an image I_C (or I_D) and a depth map Z_C (or Z_D) referring to the same viewpoint. Depth data typically have lower resolution, higher noise levels, and less accurate localization of the edges. For these reasons, exploiting color data to improve the depth representation, in particular using the better localization of edges in color images, is a very interesting option called depth data *super-resolution*. It is possible to divide the proposed methods into two main families. In the first family, local filtering and interpolation algorithms guided by color data are applied to depth data. The second one instead uses suitable global energy functions containing terms depending on both depth and color data that are optimized in order to compute the output depth map. These two families capture most of the approaches present in the literature and will be described in detail over the next two sections. Other approaches not belonging to these two main families exist as well. For example, the authors of [30] notice that color and depth patches lie in a low-dimensional subspace and perform the depth enhancement by assembling the patches in a matrix enforcing the low-rank constraint.

5.2.1 Local Filtering and Interpolation Techniques

The basic idea of this family of methods is to use color information to assist and improve filtering and interpolation of depth data. Color data are typically employed to properly locate the different scene surfaces and the edges between them. Some methods directly use color data while some use the output of edge detection or segmentation algorithms applied to color data.

Among the various possible approaches, range domain filtering techniques guided by color data to interpolate and filter depth information represent the most common solution. There are several methods based on this idea, e.g. [24, 46], while some other recent approaches exploit it as a pre-processing step before fusing color and depth data [8, 10]. The bilateral filter [45] is one of the most effective and commonly used edge-preserving filtering schemes. This filter, as with most standard filters used in image processing, computes the output value $I_f(p^i)$ at pixel p^i as the weighted average of the pixels $p^n, n = 1, \ldots, N$ in a window W^i surrounding p^i. The key difference with respect to standard approaches is that the weights do not depend only on the spatial distance between p^i and each p^n in W^i but also on the difference between the color value of p^i and p^n. More precisely $I_f(p^i)$ is computed as

$$I_f(p^i) = \frac{1}{n_f^i} \sum_{p^n \in W^i} G_s(p^i, p^n) G_r(I(p^i), I(p^n)) I(p^n) \tag{5.2}$$

where the normalization factor n_f^i is given by

$$n_f^i = \sum_{p^n \in W^i} G_s(p^i, p^n) G_r(I(p^i), I(p^n)). \tag{5.3}$$

Expression (5.2) shows that the filter output at p^i is the weighted average of the image samples in W^i. The weights are given by the product of a function G_s depending on the spatial distance between p^i and p^n with a function G_r depending on the color difference between $I(p^i)$ and $I(p^n)$. The spatial weighting function G_s is the standard Gaussian function acting as a spatial low-pass filter

$$G_s(p^i, p^n) = \frac{1}{2\pi\sigma_s^2} e^{-\frac{(u^i - u^n)^2 + (v^i - v^n)^2}{2\sigma_s^2}}. \tag{5.4}$$

In [45] the color weighting term G_r is also a Gaussian function applied to the difference between the colors of the two samples

$$G_r(p^i, p^n) = \frac{1}{2\pi\sigma_r^2} e^{-\frac{\Delta(I(p^i), I(p^n))^2}{2\sigma_r^2}} \tag{5.5}$$

where $\Delta(I(p^i), I(p^n))$ is a suitable measurement of the difference between the colors of the two samples p^i and p^n, such as their Euclidean distance in the CIELab color space [36] or other color metrics.

In this setup, where the available data are a low resolution depth map Z_D and a high resolution color image I_C, a modified version of the bilateral filter allows one to obtain a high resolution depth map defined on lattice Λ_C. As noted in Sect. 5.1, the reprojection of Z_D on Λ_C gives a low resolution sparse depth map Z_C^{low} defined on a subset $\Gamma_L \triangleq \{p^l, l = 1, \ldots, L\}$ of Λ_C. The high resolution color image I_C associates a color value $I_C(p^i)$ to each sample $p^i \in \Lambda_C$. Two different approaches can be considered in order to apply bilateral filtering on the considered representation.

The first one computes the filter output $I_f(p^i)$ for each sample of the high resolution depth map using only the samples $I_f(p^l)$ that are inside the window W^i and have an associated depth value, i.e., (5.2) can be rewritten as

$$Z_{Cf}(p^i) = \frac{1}{\overline{n}_f^i} \sum_{p^l \in \overline{W}^i} G_s(p^i, p^l) G_r(I_C(p^i), I_C(p^l)) Z_C^{low}(p^l) \tag{5.6}$$

where $\overline{W}^i = W^i \cap \Gamma_L$ is the set of the low resolution depth map points falling inside window W^i, and the normalization factor $\overline{n}_f^i$ is computed as

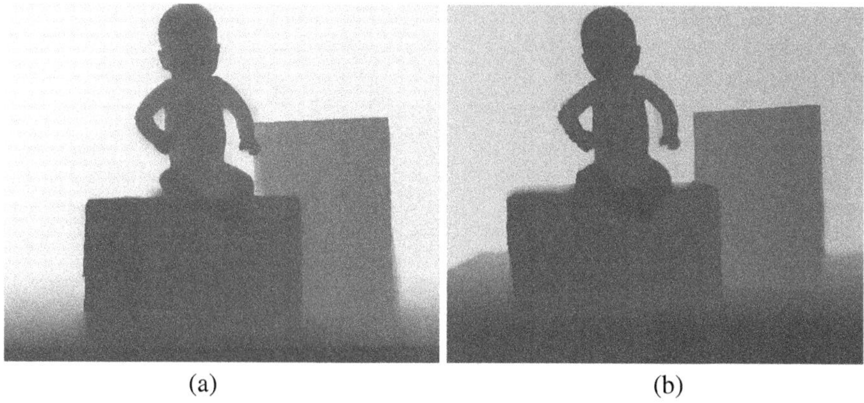

(a) (b)

Fig. 5.7 Bilateral upsampling of depth data: (**a**) from ToF depth camera; (**b**) from structured light depth camera

$$\overline{n}_f^i = \sum_{p^l \in \overline{W}^i} G_s(p^i, p^l) G_r(I_C(p^i), I_C(p^l)). \tag{5.7}$$

It is important to note that in this case, the range weighting factor G_r is not computed on the depth values but as the difference between the color of p^i and of p^l on the corresponding color image I_C acquired by the color camera. This approach is commonly called *cross bilateral filtering* to underline that the range weights come from a domain different from the one of the filtering operation, i.e. the filtering concerns depth data but the weights corresponding to the G_r component are computed from associated color data. Figure 5.7 shows an example of results given by this approach. Note how the use of the high resolution color information throughout the range weighting factor G_r allows one to correctly locate and preserve edges.

The second possibility is given by the following two-step procedure:

a) interpolate the depth map Z_C^{low} by a standard image interpolation algorithm (e.g., spline interpolation) in order to obtain a high resolution depth map $\tilde{Z}_C$ defined on the high resolution lattice Λ_C;
b) obtain the final high resolution depth map Z_C by applying the bilateral filter to $\tilde{Z}_C$ with the range weighting factor G_r computed as before, i.e., as the difference between the color of p^i and of p^n on the corresponding color image I_C, i.e.

$$Z_C(p^i) = \frac{1}{n_f^i} \sum_{p^n \in W^i} G_s(p^i, p^n) G_r(I_C(p^i), I_C(p^n)) \tilde{Z}_C(p^n) \tag{5.8}$$

with normalization factor

$$n_f^i = \sum_{p^n \in W^i} G_s(p^i, p^n) G_r(I_C(p^i), I_C(p^n)). \tag{5.9}$$

Note that after the interpolation of step (a), both the depth map and the color image are defined on the same lattice Λ_C. This allows one to use all the samples of window W^i for the interpolation. A similar approach has been employed in [24] with the non-local means filter [7] in place of the bilateral filter. The main difference between a non-local means and bilateral filter is that the range weighting factor of the former is computed from two windows surrounding the two samples p^i and p^n instead of just from the two samples p^i and p^n. Furthermore, this approach explicitly handles depth data outliers, which are quite common in depth maps acquired by current depth cameras.

Bilateral filtering also suffers the drawback that since the range distance is just given by the difference between the p^i and p^n samples, without considering the pixels between them, filtering operations could cross thin structures. For example, two regions of similar color at different depths with a small thin structure between them could be joined together by this approach. In order to overcome this limitation, one may, for example, use geodesic distances over the color frame for the selection of the samples to be used in the interpolation, as proposed in [29].

An alternative to bilateral filtering in order to perform a color guided interpolation is to use a modified median filtering scheme. The upsampling algorithm proposed in [12] computes the high resolution depth samples $Z_C(p^i)$ as the median of the low resolution depth samples inside the window W^i that have a corresponding color value not too different from the one of the considered sample, i.e.

$$Z_C(p^i) = \text{median}\left(Z_C(p^n) \mid p^n \in W^i \wedge \left|I_C(p^i) - I_C(p^n)\right| < \epsilon\right) \tag{5.10}$$

where the second constraint on p^n restricts the pixel selection, thus preserving depth discontinuities and filtering out depth data noise. This approach has a behavior similar to cross bilateral filtering except that a hard decision on which samples to use is made in place of the soft decision given by the range term of the bilateral filter. Furthermore, the spatial filtering component is a median filter instead of the Gaussian filter used by the bilateral filter. As discussed in Chap. 3, median filters are well-suited for removing noise from ToF depth maps.

Another alternative to bilateral filtering is the extraction of explicit edge information from the color data by edge detection or segmentation techniques, using this information to assist the interpolation process. The segmentation process divides the color image in a set of regions (*segments*) ideally corresponding to the different objects in the scene. It is reasonable to assume that inside each segmented patch the depth varies smoothly and that sharp depth transitions between different scene objects occur at the boundaries between different segments. Details about segmentation algorithms using color and depth data will be introduced in Chap. 6.

Figure 5.8 shows an example of segments containing the reprojected depth samples Z_C^{low} from a depth camera. The spatial resolution of depth map Z_C^{low} may be increased to that of Z_C by interpolation schemes computing the missing depth values only from the neighboring depth values inside each segment, i.e., each interpolated depth sample is only a function of the samples within the same segment. This is

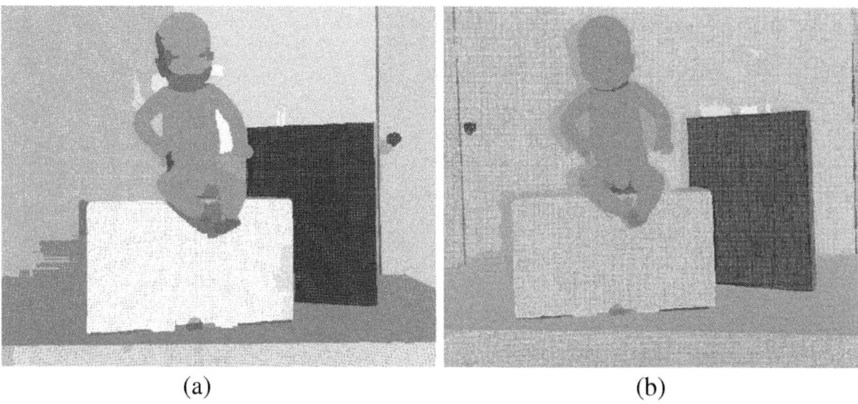

(a) (b)

Fig. 5.8 Reprojection of depth samples from a depth camera on the segmented image of a color camera: (**a**) data acquired by a ToF camera; (**b**) data acquired by a structured light depth camera

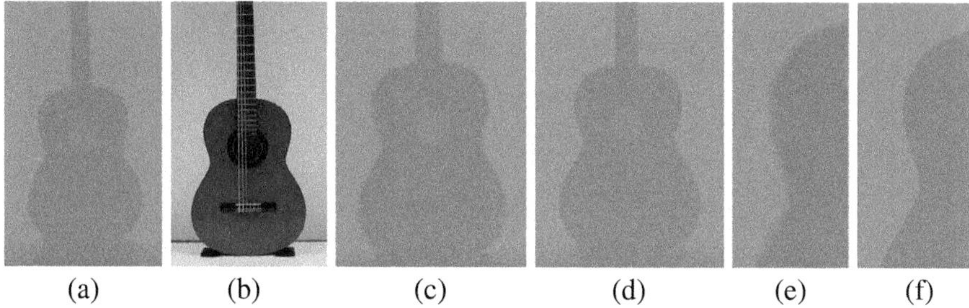

(a) (b) (c) (d) (e) (f)

Fig. 5.9 Super-resolution assisted by color segmentation [14]: (**a**) low resolution depth map acquired by a ToF camera; (**b**) image acquired by a color camera; (**c**) high resolution depth map obtained by Lanczos interpolation; (**d**) high resolution depth map interpolated with the aid of color segmentation; (**e**) detail of the high resolution depth map of (**c**); (**f**) detail of the high resolution depth map of (**d**)

equivalent to confining the low-pass action of the interpolation within each segment and preserving sharp transitions between different segments.

This concept inspired the methods of [14, 15], where the depth values are computed by bilinear interpolation of the reprojected depth samples inside each segment. In particular, these approaches use some heuristics on the 3D surface shape to compute the position that the depth samples reprojected outside the current segment would have if they lay on an extension of the surface corresponding to the considered segment. The samples predicted in this way are then used to improve interpolation accuracy at edge locations. These approaches can produce depth maps with sharp and precisely located edges as shown in Figs. 5.9 and 5.10. The biggest drawback is that they require very accurate alignment between the segmented image and the depth data, and their performance are thus limited by segmentation errors and calibration accuracy.

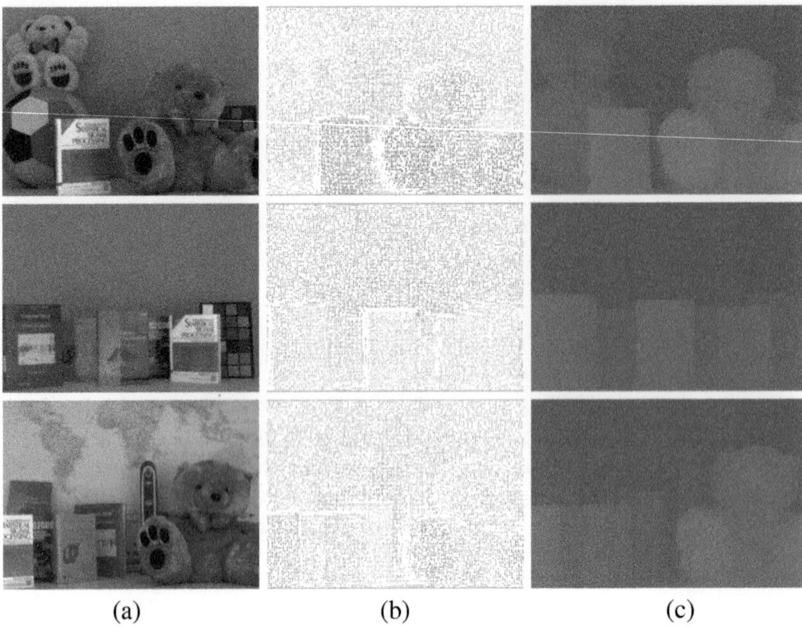

(a) (b) (c)

Fig. 5.10 Interpolation of three complex scenes with the approach of [15]: (**a**) color image; (**b**) sparse depth data acquired by the ToF depth camera and reprojected on the camera lattice; (**c**) interpolated depth map

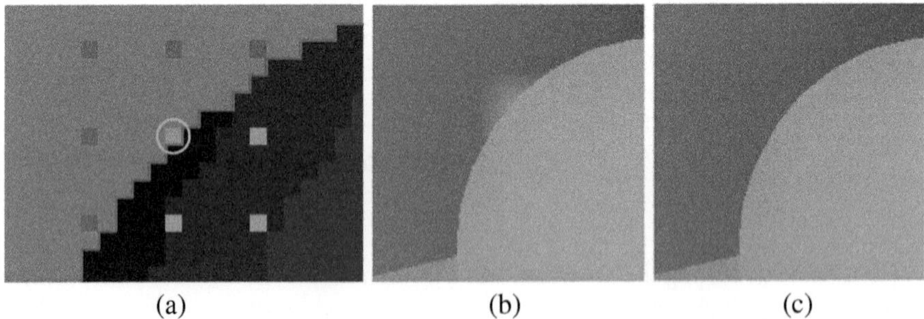

(a) (b) (c)

Fig. 5.11 Artifacts due to segmentation and calibration inaccuracies: (**a**) a reprojected depth value (*circled in green*) has been assigned to the background but it belongs to the foreground; (**b**) high resolution depth map obtained by interpolating the depth data inside the different segments; (**c**) high resolution depth map obtained by the method of [14] which also takes into account data reliability

Segmentation is a very challenging task and, despite the significant research activity in this field, there are currently no completely reliable procedures for generic scenes. As expected, segmentation errors or inaccuracies can cause wrong assignments of depth samples, leading to artifacts in the estimated depth maps of the type shown in Fig. 5.11. For example, the reprojected depth sample inside the green circle of Fig. 5.11a belongs to the foreground but it has been incorrectly assigned to the background because of calibration and segmentation inaccuracies. Figure 5.11b shows how the incorrectly assigned depth sample is propagated inside the background by the interpolation algorithm. Inaccuracies in the calibration process or the acquired depth samples can similarly bring the reprojected depth samples close to the boundaries between segments to "cross" them and be assigned to wrong regions.

A possible solution to segmentation issues is to replace segmentation with edge detection, which is a simpler and more reliable operation. However, cracks of the edges may allow the reprojected depth samples to be used for interpolation out of the corresponding regions, with consequent artifacts in the interpolated depth. False or double edges can affect the interpolation process as well. Artifacts due to the second issue may be reduced by more accurate calibration procedures. Another possibility is to either exclude or underweight reprojected depth values too close to the edges [14] in order to eliminate unreliable data from the interpolation process. Figure 5.11c shows how this provision eliminates some interpolation artifacts.

Finally it is also possible to combine this family of approaches with the range filtering approach, for example, the works of [8, 10] exploit an extended bilateral filtering scheme where the weights not only depend on the spatial and range distance but also on the segmentation information. More precisely, in these approaches the filter acts inside each segment as a standard low pass interpolation filter, while when considering samples in different segments the filter activates the range term of the bilateral filter.

5.2.2 Global Optimization Based Approaches

Another solution to depth and color data fusion is given by the approaches based on the optimization, usually by probabilistic methods, of a global energy function accounting for both depth and color clues.

Before describing this family of methods, we introduce a few basic concepts and notation. A depth map Z can be considered as the realization of a random field $\mathcal{Z}$ defined over a 2D lattice Λ_Z. The random field $\mathcal{Z}$ can be regarded as the juxtaposition of the random variables $Z^i \triangleq \mathcal{Z}(p^i)$, with $p^i \in \Lambda_Z$. The neighbors of p^i are denoted as $p^{i,n} \in N(p^i)$, where $N(p^i)$ is a neighborhood of p^i. Neighborhood $N(p^i)$ can either be the 4-neighborhood, the 8-neighborhood, or even a generic window W^i centered at p^i. The random variable Z^i assumes values in the discrete alphabet $z^{i,j}, j = 1, \ldots, J_i$ in a given range $\mathcal{R}^i$. A specific realization of Z^i is denoted as z^i.

The high resolution version of the depth distribution can be seen as a Maximum-a-Posteriori (MAP) estimate of the most probable depth distribution $\hat{Z}$ given the acquired measurements M, i.e.

$$\hat{Z} = \underset{Z \in \mathcal{Z}}{\operatorname{argmax}}\, P(Z|M) \tag{5.11}$$

where $\mathscr{Z}$ is the random field of the framed scene depth distribution and Z is a specific realization of $\mathscr{Z}$. The random field $\mathscr{Z}$ is defined on a lattice Λ_Z that can be the depth camera lattice Λ_D, the color camera lattice Λ_C, or an arbitrary lattice other than the source data lattices (e.g., a lattice associated with a virtual camera placed between D and C). The choice of this lattice is the first major design decision. In this section, we assume the use of a suitable high resolution lattice (e.g., Λ_C) where depth and color data have been reprojected and interpolated as described in Sect. 5.2.1. The random field $\mathscr{Z}$ is therefore made by a set of random variables $Z^i \triangleq \mathscr{Z}(p^i), p^i \in \Lambda_Z$.

From Bayes' rule, (5.11) can be rewritten as

$$\hat{Z} = \underset{Z \in \mathscr{Z}}{\operatorname{argmax}} \frac{P(M|Z)P(Z)}{P(M)} = \underset{Z \in \mathscr{Z}}{\operatorname{argmax}} P(M|Z)P(Z) \tag{5.12}$$

where $P(M|Z)$ is the measurement's likelihood given scene depth distribution Z, and $P(Z)$ is the prior probability of the scene depth distribution. The probability of the measurement $P(M)$ does not depend on Z and can be eliminated from the formulation. In this section, the set of measurements M corresponds to Z_D and the color image I_C, while in Sect. 5.3 the set of measurements will correspond to depth data from the depth camera D and the stereo vision system S. The field $\mathscr{Z}$ can be modeled as a Markov Random Field (MRF) defined on the considered lattice. Markov Random Fields extend the concept of Markov chains to regular 2D fields such as images or depth maps. A detailed presentation of the theory behind these representations is out of the scope of this book but can be found in other resources, e.g. [26]. MRFs are suited to model relationships between neighboring pixels in the high resolution lattice, i.e., the typical structure of depth information made by smooth regions separated by sharp edges.

Let us recall that in a MRF $\mathscr{Z}$

$$\begin{aligned} P(Z^i|Z^n : Z^n \triangleq \mathscr{Z}(p^n), \forall p^n \in \Lambda_Z \setminus \{p^i\}) = \\ = P(Z^i|Z^n : Z^n \triangleq \mathscr{Z}(p^n), \forall p^n \in N(p^i)), \quad \forall p^i \in \Lambda_Z \end{aligned} \tag{5.13}$$

where $N(p^i)$ is a suitable neighborhood of p^i. It is possible to demonstrate that $\mathscr{Z}$ is characterized by a Gibbs distribution [26]. Therefore, the Maximum-a-Posteriori (MAP) problem of (5.12) after some manipulation can be expressed as the minimization of an energy function

$$\begin{aligned} U(Z) &= U_{data}(Z) + U_{smooth}(Z) = \\ &= \sum_{p^i \in \Lambda_Z} V_{data}(p^i) + \sum_{p^i \in \Lambda_Z} \sum_{p^n \in N(p^i)} V_{smooth}(p^i, p^n). \end{aligned} \tag{5.14}$$

Energy $U(Z)$ is the sum of two energy terms: a *data* term U_{data} modeling the probability that the depth assumes a certain value at p^i, and a *smoothness* term

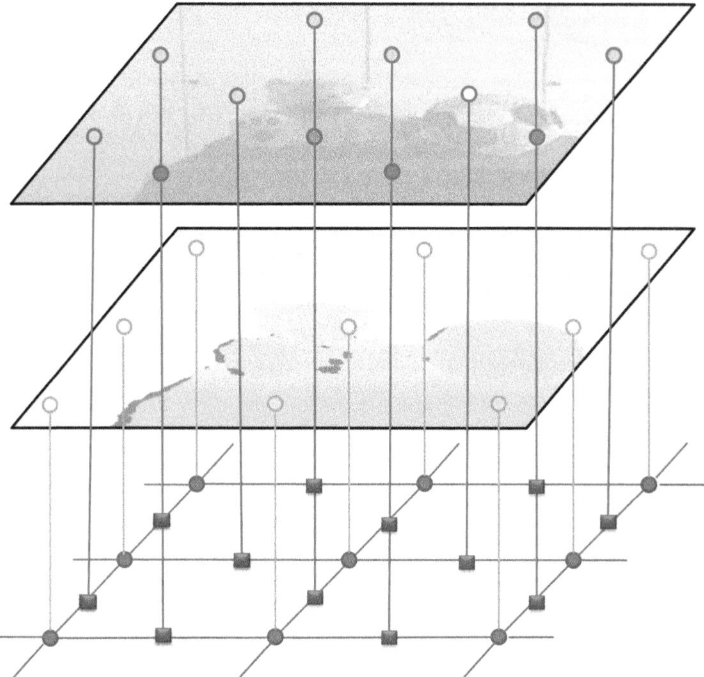

Fig. 5.12 Pictorial illustration of the energy functions used in depth super-resolution: the V_{data} term (*shown in blue*) depends on the depth camera measurements, while the V_{smooth} term affects the strength of the links (*in red*) representing the dependency of each depth sample from its neighbors

U_{smooth} modeling the dependency of the depth value at p^i from its neighbors $p^n \in N(p^i)$. The first term typically accounts for the depth camera measurements, while the second term models the scene geometry, enforcing smooth depth regions separated by sharp edges. A simple possibility would be to use U_{smooth} to enforce the smoothness of the depth map, however, using this method the available high resolution color information would not be exploited. A more interesting option used in several approaches [11, 23, 34] is to take color information into account in the construction of the MRF associated with depth data. As shown in Fig. 5.12, the dependency of Z^i on its neighbors $Z^n \triangleq \{\mathscr{Z}(p^n), p^n \in N(p^i)\}$ is stronger when samples have similar colors and weaker when samples have different colors. This assumption models that samples with similar colors are more likely to belong to the same object and thus have similar depth values, while depth information edges are probably aligned with color data edges.

Typically the data term U_{data} is the sum of a set of potentials $V_{data}(p^i)$, one for each sample of the lattice (the blue circles in Fig. 5.12). The potential $V_{data}(p^i)$ based on the depth camera measurements can be expressed as a function of the difference between the considered value Z^i and value $Z_D(p^i)$ actually measured by the depth camera. Many proposed approaches use an expression of the form

$$V_{data}(p^i) = k_1[z^i - Z_D(p^i)]^2 \tag{5.15}$$

where k_1 is a constant weighting factor used to balance the data term with respect to the smoothness term U_{smooth}.

The smoothness term U_{smooth} modeling the dependency of each sample on its neighbors (the red squares in Fig. 5.12) can be expressed as the sum of a set of potentials $V_{smooth}(p^i, p^n)$ defined on every couple of neighboring points on the considered lattice. The potentials $V_{smooth}(p^i, p^n)$ can be defined in different ways, but the most common solution is to model them as a function of the difference between the considered value z^i and its neighbors z^n, i.e.

$$V_{smooth}(p^i, p^n) = V_{smooth}(z^i, z^n) = w_{i,n}[z^i - z^n]^2 \tag{5.16}$$

where $w_{i,n}$ are suitable weights modeling the strength of the relationship between close points. The weights are the key elements for this family of approaches: weights $w_{i,n}$ not only model the depth map smoothness but also provide the link between the depth data and the associated color image acquired by the standard camera. If z^i and z^n are associated with samples of similar color on the camera image they probably belong to the same region and thus there should be a strong link between them. If they correspond to samples of a different color, the smoothness constraint should be weaker. The simplest way to account for this rationale is to compute the weight $w_{i,n}$ linking z^i to z^n as a function of the color difference between the associated color samples $I_C(p^i)$ and $I_C(p^n)$, for instance by an exponential function [11, 23]

$$w_{i,n} = e^{-k_2 \Delta(I_C(p^i), I_C(p^n))^2} \tag{5.17}$$

where $\Delta(I_C(p^i), I_C(p^n))$ is a measure of the color difference between the two samples at locations p^i and p^n (e.g., in the simplest case, the absolute value of the difference between their intensity values), and k_2 is a tuning parameter. Needless to say, this simple model can be easily extended in order to feed the probabilistic model with many other clues, such as the segmentation of the color image, the output of an edge detection algorithm, or other characteristics of the image [34].

Depth distribution $\hat{Z}$ is finally computed by finding the depth values that maximize the MAP probability (5.14) modeled by the MRF. This problem cannot be subdivided into smaller optimization problems and needs to be solved by complex global optimization algorithms, such as loopy belief propagation (LBP) [42] or graph cuts (GC) [5]. The choice of the specific optimization algorithm is another important element that affects the quality of the final interpolated depth. The theory behind these powerful optimization methods can be found in [4, 26, 43, 44].

5.3 Fusion of a Depth Camera with a Stereo System

A more interesting method to overcome the limitations of current depth cameras and provide both color and 3D geometry information is to use an acquisition setup made by a depth camera D and a few color cameras L and R used as an autonomous stereo system S. In this case, both the stereo system S and the depth camera D can independently estimate the 3D scene geometry. Moreover, each camera in the stereo vision system provides additional color information. The rationale behind the fusion of data from a depth camera and a stereo vision system is to provide 3D geometry estimates with accuracy, precision, depth resolution, spatial resolution, and robustness higher than those of the original data from the depth camera or stereo system alone.

The components of such an acquisition system are:

- The color images pair $I_S = \{I_L, I_R\}$ acquired by the stereo system S, where I_L is the image acquired by the left camera L and I_R is the image acquired by the right camera R. The image I_L is defined on lattice Λ_L associated with the L-2D reference system, and the image I_R is defined on lattice Λ_R associated with the R-2D reference system.
- The depth map Z_S estimated by the stereo vision system from the color images I_L and I_R, conventionally defined with respect to the L-2D on lattice Λ_L. Some approaches also compute a confidence map C_S that represents the reliability of stereo data at each location.
- The depth map Z_D and additional information that depends on the selected depth camera, like amplitude, intensity, or confidence data (for example, many ToF cameras provide depth and amplitude information). The set of possible data available for the depth camera is $F_D = \{Z_D, A_D, B_D, C_D\}$, where A_D is an amplitude image, B_D an intensity image, and C_D a confidence map providing the reliability of the acquired data. Many commercial products directly supply the confidence map C_D but several fusion schemes compute it from the output data by ad hoc algorithms based on the sensor error model. All these quantities are defined on lattice Λ_D associated with the D-2D reference system.

Various algorithms have been proposed for depth camera and stereo data fusion and an exhaustive recent review can be found in [33]. Similar to stereo vision algorithms, it is possible to divide them into two main families, local methods independently working at each spatial location, and global methods that also include a global optimization step considering the whole depth image.

Notice that like in the previous setup, most methods require calibration and data reprojection to a common reference frame that allows one to spatially relate the 3D geometry information provided by D to the 3D geometry and color information provided by S. The reference frame can be one of the two color cameras or the depth camera, depending on the approach.

Most of the proposed methods are tailored to ToF depth cameras, but some of them can also be adapted to structured light depth cameras.

5.3.1 Local Fusion Methods

Local fusion approaches independently compute the depth value at each location, taking into account the depth measurements and the color data in a spatial neighborhood of the considered point. As discussed in the introduction of this chapter, there are two main possible strategies. The first method, referred to as *late fusion*, estimates the depth from the two color cameras using a suitable stereo vision technique then fuses the two depth maps, eventually exploiting some reliability or confidence information. The second method, called *early fusion*, directly computes the output depth map from depth camera data and color information, typically by using a modified stereo vision algorithm also accounting for the depth camera information.

In the case of late fusion approaches, at each location p^i two different depth measurements are available, i.e., $Z_D(p^i)$ given by the depth camera D, and $Z_S(p^i)$ computed from the stereo system S. The critical issue is how to select the proper depth value at each location p^i from the two hypotheses. To solve this problem most algorithms use some side information representing the reliability of the acquired data, typically:

- a confidence value representing the reliability of the acquired data at the considered location;
- a probability function representing the likelihood of the various depth hypotheses;
- an interval of possible measurements, typically centered on the measured value, with a length depending on the estimated accuracy.

Early fusion approaches typically exploit modified stereo vision algorithms in order to exploit depth camera data. Typical examples of such provisions are:

- a modified cost function also accounting for depth camera data;
- a restricted range for the correspondences search around depth camera estimate;
- depth camera output used as a starting point for the minimization algorithm.

Notice that the two approaches are not mutually exclusive, as there exist techniques that use modified stereo vision algorithms which also use depth camera data.

5.3.1.1 Confidence Based Techniques

The first family of methods exploit confidence values representing the reliability of the acquired data at the considered location. The confidence of depth camera data C_D is typically computed according to error models of the considered sensor. For example, in the case of ToF depth cameras, the confidence can be a function of the standard deviation of the error noise computed from amplitude and intensity data according to (3.47). The confidence of stereo vision data C_S, instead, is typically computed from analysis of the cost function of the stereo algorithm. In particular, a sharp, well-defined peak of the stereo matching cost function is usually a hint of high reliability of the estimated depth. There are several metrics based on this rationale and an exhaustive review can be found in [22]. Other confidence metrics for stereo algorithms use local texture variance or other texture indicators as confidence measures as in [25]. Once one has computed the confidence data, the fusion of the two depth hypotheses can be achieved in various ways, e.g.:

- select the depth hypothesis with the highest confidence

$$Z(p^i) = \begin{cases} Z_D(p^i) & \text{if } C_D(p^i) > C_S(p^i) \\ Z_S(p^i) & \text{otherwise} \end{cases} \tag{5.18}$$

- use a weighted average of $Z_D(p^i)$ and $Z_S(p^i)$ with weights depending on $C_D(p^i)$ and $C_S(p^i)$, e.g., by using the simple weighted average

$$z(p^i) = \frac{Z_D(p^i)C_D(p^i) + Z_S(p^i)C_S(p^i)}{C_D(p^i) + C_S(p^i)} \tag{5.19}$$

  or some more complex combination scheme
- use a threshold on the confidence values to discard one of the two measurements if it is not reliable. If both measurements are reliable then average the two; if neither is reliable then mark the sample as unavailable.

In [25], one of the earliest works where confidence measures are applied to data fusion, a binary confidence map is used to select reliable samples for stereo vision systems. Then, two intervals of possible measurements are associated with D and S and the midpoint of the common region is selected as the final depth measurement. It is also possible to consider small surface regions instead of single samples, for example, [2] fits small planar surfaces (patchlets) over each of the two measurements. The patchlets are then combined in an iterative approach, which takes into account the reliability of data given by the noise estimate.

5.3.1.2 Probabilistic Approaches

As already seen for the single color camera case in Sect. 5.2.2, the fusion of depth information coming from a depth camera D and a stereo vision system S can also be formalized by the probabilistic MAP approach of (5.11), which can be rewritten by Bayes' rule as in (5.12).

In this case the measured data M are the set of data F_D acquired by D and the color images $I_S = (I_L, I_R)$ acquired by the stereo system S. $\mathscr{Z}$ is the random field of the framed scene's depth distribution, defined on Λ_Z, which as in the case of super-resolution can be Λ_D, Λ_S, or another lattice (e.g., a lattice associated with a virtual camera placed between L and D). The choice of lattice Λ_Z is a major design decision when fusing data from S and D. Approaches like [48–50] adopt $\Lambda_Z = \Lambda_S$, while other approaches such as [9] adopt $\Lambda_Z = \Lambda_D$. The choice of Λ_S leads to high resolution depth map estimates, i.e., to depth maps with the resolution of stereo images I_S, while the choice of Λ_D leads to low resolution depth map estimates, although it is also possible to use an interpolated version of Λ_D as seen in [10].

Another fundamental characteristic that defines the random field $\mathscr{Z}$ is the range of values that each random variable $Z^i = \mathscr{Z}(p^i)$ can assume. Each random variable Z^i assumes values in the discrete alphabet $z^{i,j}$, $j = 1, \ldots, J_i$ within a given range $R^i \subset [z_{min}, z_{max}]$ where z_{min} is the nearest measurable distance (e.g., 500 [mm] for a Kinect™ v1) and z_{max} is the farthest measurable distance (e.g., 5000 [mm] for a ToF depth camera with modulation frequency $f_{mod} = 30$ [MHz]). Note that the distances measured by a stereo vision system are also bounded by the choice of a minimum and a maximum disparity value. The choice of z_{min} and z_{max} is the second major design decision which should account for the characteristics of D, S, and the framed scene. Prior knowledge of minimum and maximum depth of the scene to be acquired is a common assumption. It is also worth noting that alphabet values $z^{i,j}$ and alphabet cardinality J_i depend on p^i.

The likelihood term $P(M|Z) = P(I_S, F_D|Z)$ in (5.12) accounts for the depth information acquired by both D and S. Under the common assumption that the measurement likelihood of the two 3D acquisition systems (in this case D and S) are independent, as justified and assumed in [9, 21, 48–50], the likelihood term can be rewritten as

$$P(I_S, F_D|Z) = P(I_S|Z)P(F_D|Z) \tag{5.20}$$

where $P(I_S|Z)$ is the likelihood of the S measurements and $P(F_D|Z)$ is the likelihood of the D measurements. Therefore, in this case the MAP problem of (5.12) can be rewritten as

$$\hat{Z} = \operatorname*{argmax}_{Z \in \mathscr{Z}} P(Z|I_S, F_D) = \operatorname*{argmax}_{Z \in \mathscr{Z}} P(I_S|Z)P(F_D|Z)P(Z). \tag{5.21}$$

Each one of the two likelihood terms $P(I_S|Z)$ and $P(F_D|Z)$ can be computed independently for each pixel $p^i \in \Lambda_Z$.

Let us consider first $P(I_S|Z)$ and observe that by inverting the perspective projection equations, the left camera pixel p^i with coordinates $\mathbf{p}^i = [u^i, v^i]^T$ and the possible depth values $z^{i,j}$ identify a set of 3D points $P^{i,j}$ with coordinates $\mathbf{P}^{i,j} = [x^{i,j}, y^{i,j}, z^{i,j}]^T$, $j = 1, \ldots, J_i$. Such points are projected to the conjugate pairs $(p_L^{i,j}, p_R^{i,j})$. Therefore, the likelihood $P(I_S|Z)$ can be regarded as the set of $P(I_S|P^{i,j}) \triangleq P(I_S|Z^i = z^{i,j}), j = 1, \ldots, J_i$.

The stereo measurement's likelihood for $P^{i,j}$ can be computed by comparing the windows $W_L^{i,j}$ and $W_R^{i,j}$, centered at $p_L^{i,j}$ and $p_R^{i,j}$ in the left and right image, respectively. In principle, the two windows are very similar if $z^{i,j}$ is close to the correct depth value, while they are different if $z^{i,j}$ is different from the correct depth value. The likelihood therefore assumes high values if the windows $W_L^{i,j}$ and $W_R^{i,j}$

have a high similarity score and low values if the windows have a low similarity score. Most stereo vision systems compute a cost function $c_{i,j} = \mathscr{C}(p_L^{i,j}, p_R^{i,j})$ representing the similarity between the two windows, therefore the likelihood of a set of points $P^{i,j}, j = 1, \ldots, J_i$ can be computed as a function of $c_{i,j}$, for instance by an exponential model

$$P(I_S|P^{i,j}) \propto e^{-\frac{c_{i,j}}{\sigma_S^2}} \tag{5.22}$$

where σ_S is a normalization parameter related to the color data variance. The cost function $\mathscr{C}$ can be any of the different functions used in stereo vision techniques [43], such as a simple Sum of Squared Differences (SSD)

$$c_{i,j} = \sum_{p_L^n \in W_L^{i,j}, p_R^n \in W_R^{i,j}} [I_L(p_L^n) - I_R(p_R^n)]^2. \tag{5.23}$$

The windows $W_L^{i,j}$ and $W_R^{i,j}$ can be for instance rectangular windows centered at $p_L^{i,j}$ and $p_R^{i,j}$ respectively. Stereo matching cost computation is pictorially shown in Fig. 5.13 for 3 depth hypotheses $z^{i,1}$, $z^{i,2}$, and $z^{i,3}$. Some examples of the resulting likelihood functions are shown in the third row of Fig. 5.14. Notice how stereo vision data typically lead to accurate measurements, characterized by probability functions with very sharp peaks in highly textured regions and in proximity to edges. Conversely, the likelihood function is quite flat in textureless regions.

Depth camera likelihood $P(F_D|Z)$ for a given point $P^{i,j}$ can be regarded as the set of $P(F_D|P^{i,j}) \triangleq P(F_D|Z^i = z^{i,j}), j = 1, \ldots, J_i$ and can be computed by analyzing the measurement errors of depth cameras. Let us recall that depth camera measurements are generally characterized by random errors distributed along the cameras' optical rays, as shown in Chaps. 2 and 3. For example, in the case of a ToF depth camera, the depth random error can be approximated, at least for points far from depth discontinuities and ignoring multipath and mixed pixel related errors, by a Gaussian distribution along the optical rays with standard deviation σ_P, computed for example from (3.47). Figure 5.15 pictorially shows a Gaussian distribution describing the ToF depth measurement error along the optical ray. For structured light cameras, the error distribution is less regular, but the error can always be assumed to be distributed along the IR camera's optical rays.

Since in both ToF and structured light depth cameras, the error is distributed along the optical rays, the D likelihood terms $P(F_D|Z^i = z^{i,j})$ can be independently considered for each point of lattice $\Lambda_Z = \Lambda_D$. The choice of Λ_D for the output depth field, although not very common, allows one to directly exploit this property.

Let us suppose that the depth estimation error is distributed as a Gaussian function as just described. A depth camera measurement's likelihood for a set of 3D points $P^{i,j}, j = 1, \ldots, J_i$ relative to $p^i \in \Lambda_Z$ can be computed as

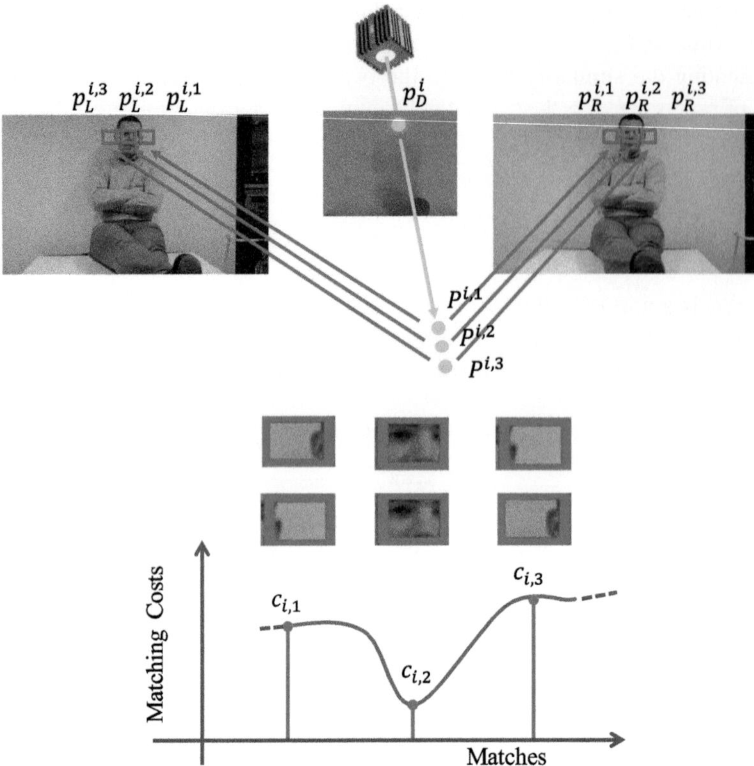

Fig. 5.13 Pictorial illustration of stereo costs computation: 3D points associated with different depth measurements (*green dots*) are projected to different pairs of possible conjugate points according to the distance hypothesis $z^{i,j}$. For each pair of possible conjugate points a matching cost, shown in the plot, is computed by locally comparing the stereo images in windows (*red rectangles*) centered at the relative conjugate points

$$P(F_D|P^{i,j}) \propto e^{-\frac{z^{i,j} - Z_D(p_D^i)}{\sigma_D^2}} \tag{5.24}$$

where $z^{i,j}$ is the candidate depth coordinate of $P^{i,j}$, $Z_D(p_D^i)$ is the measured depth coordinate, and $p_D^i \in \Lambda_D$ is the projection of $P^{i,j}$. Parameter σ_D is a suitable function of depth data variance and noise, e.g., in [9] it depends on both the variance of the 3D point positions σ_P and the local variance of the depth measurements inside a window centered at p_D^i. More refined approaches (e.g. [10]) use Gaussian Mixture Models (GMM) with multiple Gaussians in proximity of edges corresponding to the different surfaces on both sides of the edge. Some examples are shown in the second row of Fig. 5.14; notice how ToF likelihoods have a Gaussian distribution inside the planar regions of the scene, while the behavior in proximity of edges and corners is less regular due to multipath and mixed pixel effects.

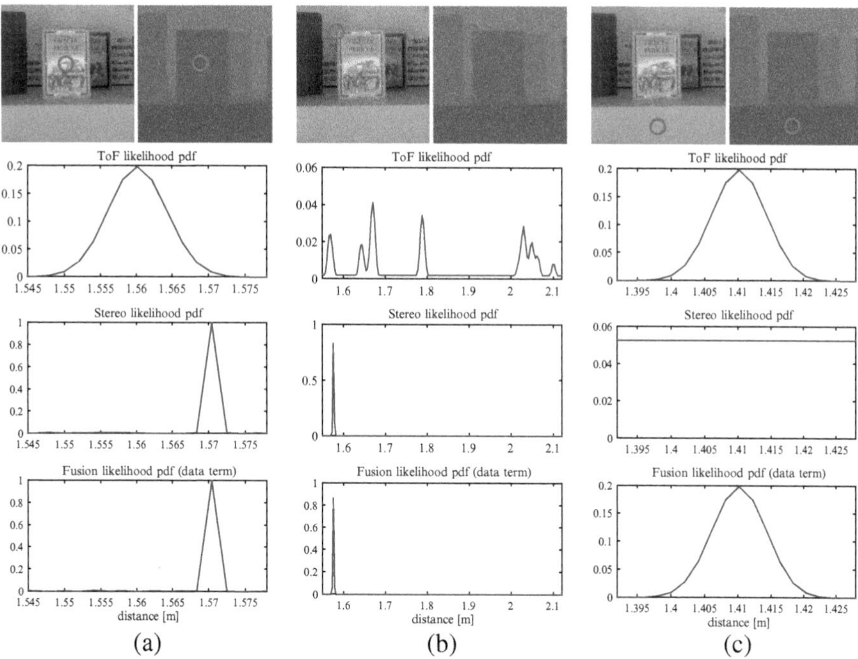

Fig. 5.14 Likelihood functions: Image and depth data with the considered highlighted scene point (*first row*), ToF likelihood (*second row*), stereo likelihood (*third row*) and joint likelihood (*fourth row*) relative to: (**a**) a point far from scene depth discontinuities; (**b**) a point near scene depth discontinuities; (**c**) a point in a textureless area

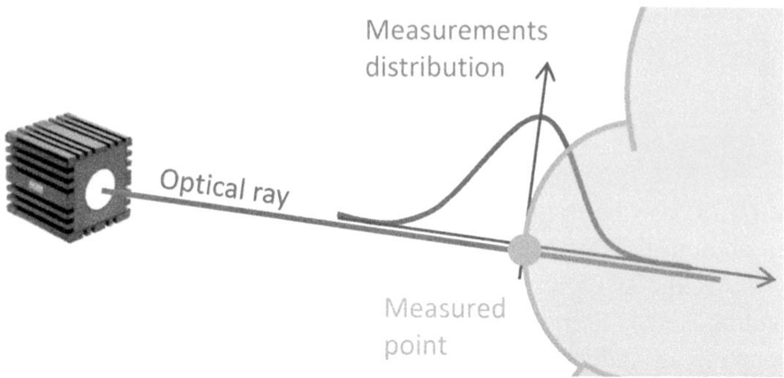

Fig. 5.15 Pictorial illustration of distance measurement error for a ToF depth camera modeled as a Gaussian distribution along the depth camera optical ray

The last term of (5.21) still to be modeled is the prior $P(Z)$. Local approaches, e.g. [9] and [47], model $\mathcal{Z}$ as a juxtaposition of independent random variables $Z^i \triangleq \mathcal{Z}(p^i)$, $p^i \in \Lambda_Z$, where Λ_Z is the considered lattice, that can be Λ_S, Λ_D or another arbitrary lattice. These variables are characterized by prior probability

$P(z^{i,j}) \triangleq P(Z^i = z^{i,j})$, which for instance in [9] is a discrete uniform distribution in $[z_{min}, z_{max}]$. The uniform probability distribution model does not impose any specific structure to the scene depth distribution (e.g., piecewise smoothness). In this case the MAP problem of (5.21) can be simplified and the smoothness term can be removed from the minimization of the energy function $U(Z)$, i.e.,

$$U(Z) = U_{data}(Z) = U_S(Z) + U_D(Z) = \sum_{p^i \in \Lambda_Z} V_S(p^i) + \sum_{p^i \in \Lambda_Z} V_D(p^i). \tag{5.25}$$

Data term U_{data} accounts both for the contribution of the stereo system S, through the energy function $U_S(Z)$ and the corresponding potentials $V_S(p^i)$, and the contribution of the depth camera through $U_D(Z)$ and the corresponding potentials $V_D(p^i)$.

More precisely, $V_S(p^i)$ depends on the cost function $\mathscr{C}$ of the stereo system. For instance, if the model of (5.22) is used for the cost function, $V_S(p^i)$ can be computed as

$$V_S(p^i) = \frac{1}{\sigma_S^2} c_i. \tag{5.26}$$

The potentials $V_D(p^i)$ instead depend on the depth camera measurements, e.g., using the model of (5.24), $V_D(p^i)$ is

$$V_D(p^i) = \frac{[z^i - Z_D(p_D^i)]^2}{\sigma_D^2}. \tag{5.27}$$

Therefore, from (5.26) and (5.27) expression (5.25) can be rewritten as

$$U(Z) = \sum_{p^i \in \Lambda_Z} \left[\frac{c_i}{\sigma_S^2} + \frac{[z^i - Z_D(p_D^i)]^2}{\sigma_D^2} \right]. \tag{5.28}$$

Equation (5.28) can be optimized independently for each random variable of the estimated random field by a Winner-Takes-All (WTA) approach. For each random variable in Λ_Z, the WTA approach, picks the depth value z^i which maximizes the corresponding energy term in (5.28). Local approaches assume that the likelihoods at each location are independent from each other and analyze each random variable of the field independently. This simplifies the model and reduces the computation requirements but does not consider neighborhood information. Some examples of $U(Z)$ are shown in the last row of Fig. 5.14; notice how this method implicitly leads to the selection of the measurement of the most reliable depth source at the considered location. Figure 5.16 reports some results obtained from the local method of [9]. Figure 5.19 instead compares local and global data fusion, described in the next section, with the approach of [10].

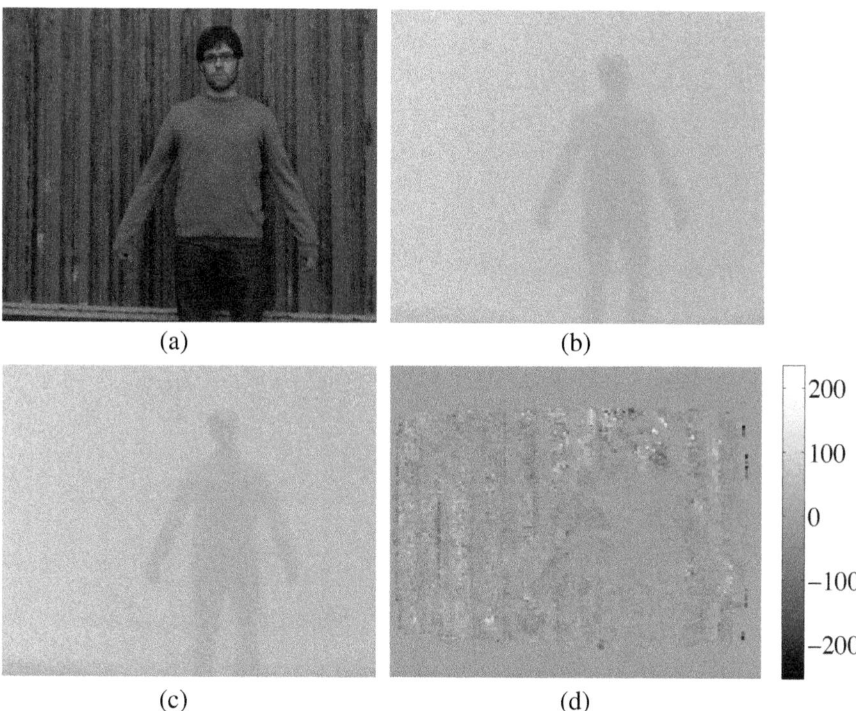

Fig. 5.16 Results of the local fusion approach by [9]: (**a**) image from left color camera L; (**b**) depth map acquired by the ToF depth camera D; (**c**) estimated depth map after data fusion; (**d**) difference between the depth map acquired by the ToF depth camera and the depth map estimated by the fusion approach

5.3.2 Global Optimization Based Approaches

Better results can be obtained with global optimization schemes which exploit neighborhood information and enforce global constraints. The basic idea is similar to that presented in Sect. 5.2.2, i.e., building and optimizing a global energy function accounting for the data from both the depth and color cameras. The data terms in this case include both the measurements from the depth camera and the clues from the stereo vision system, while the smoothness constraints are similar to those presented in Sect. 5.2.2, relying on color information. The most common approach is a probabilistic MAP-MRF formulation [10, 49] as described in the next section; however, other approaches based on graph cuts, total variation, and other clues have been proposed.

5.3.2.1 MAP-MRF Probabilistic Fusion Framework

Similar to the model in Sect. 5.2.2, in this case the prior probability $P(Z)$ can also be modeled as a MRF. As previously explained, this model is used to impose a piecewise smoothness to the estimated depth map. In this case, prior $P(Z)$ can be computed as

$$P(Z) \propto \prod_{p^i \in \Lambda_Z} \prod_{p^n \in N(p^i)} e^{-V_{smooth}(z^i, z^n)} \tag{5.29}$$

where z^i is a realization of $Z^i = \mathscr{Z}(p^i)$ and $N(p^i)$ is a suitable neighborhood of p^i. Also in this case, the MAP problem of (5.21) can be solved by minimizing an energy function $U(Z)$ of the following form

$$\begin{aligned} U(Z) &= U_{data}(Z) + U_{smooth}(Z) \\ &= U_S(Z) + U_D(Z) + U_{smooth}(Z) \\ &= \sum_{p^i \in \Lambda_Z} V_S(p^i) + \sum_{p^i \in \Lambda_Z} V_D(p^i) + \sum_{p^i \in \Lambda_Z} \sum_{p^n \in N(p^i)} V_{smooth}(p^i, p^n) \end{aligned} \tag{5.30}$$

in which $V_S(Z)$ and $V_D(Z)$ can be computed as discussed in the previous section, e.g., it is possible to use (5.26) and (5.27) respectively. The term $V_{smooth}(p^i, p^n)$ instead imposes the piecewise smoothness of depth data, typically by a Potts' model based on the assumption that the depth is a piecewise smooth function, namely by

$$V_{smooth}(p^i, p^n) = \min\left\{[z^i - z^n]^2, T_h\right\} \tag{5.31}$$

where T_h is a truncation threshold that avoids $V_{smooth}\left(p^i, p^n\right)$ to take on too-large values near discontinuities. Note how expression (5.30) differs from (5.14): in (5.14), the prior term depends only on depth information and color information affects the smoothness term, while in (5.30) both depth and color enter the data term, as the comparison of Figs. 5.12 and 5.17 shows.

Equation (5.30) can be rewritten as

$$U(Z) = \sum_{p^i \in \Lambda_Z} \left[\frac{c_i}{\sigma_S^2} + \frac{[z^i - Z_D(p_D^i)]^2}{\sigma_D^2} + \sum_{p^n \in N(p^i)} \min\left\{[z^i - z^n]^2, T_h^2\right\} \right]. \tag{5.32}$$

With respect to (5.28), this energy function takes into account information from both D and S measurements and the smoothness of the estimated depth distribution. Since the depth camera and stereo system have complementary characteristics and responses to different regions of the scene, a key aspect to investigate is how to account for their reliability in the likelihood functions.

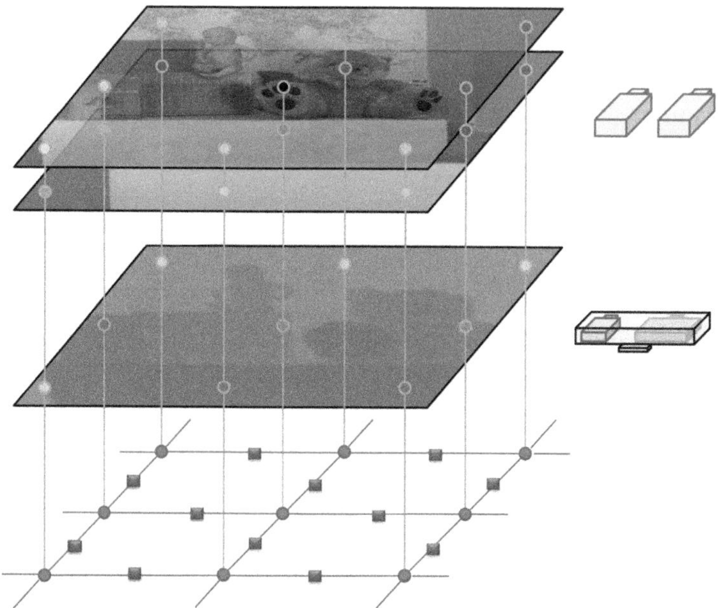

Fig. 5.17 Markov Random Field (MRF) used for the fusion of data from a depth camera and stereo system. The data terms depend on both depth camera and stereo vision measurements. The *blue circles* represent the pixel locations on which the data terms are computed, while *red squares* represent smoothness constraints between neighboring pixels

For depth cameras the simplest models exploit a function of the difference between the measured and considered sample as in (5.32). Different measures can be used, e.g., squared differences, absolute differences, or truncated differences [50]. The resulting term is usually then normalized by a weight depending on the estimated noise level. One can alternatively use a Gaussian function with a standard deviation proportional to the noise [9].

These simple models do not completely capture the behavior of the depth sensors, especially in proximity of edges, where issues like mixed pixels and multipath lead to large errors not captured by the simple Gaussian noise approximation. For this reason more complex likelihood functions can be used; for example, Dal Mutto et al. [10] uses a Mixture of Gaussians model at edge locations with multiple Gaussian functions centered on the measured value and the surfaces on both sides of the edge.

The stereo term implicitly considers data reliability by using the matching cost function, e.g, in unreliable regions the cost function is flat as in Fig. 5.14c, thus limiting its impact on the output depth value. Additional weights based on stereo confidence measures can be used to make the model more robust.

Although thus far we have only made per-frame considerations, another fundamental clue that could be added to (5.32) is the temporal coherence between subsequent frames. The MRF model can be extended by also enforcing smoothness between corresponding locations of subsequent frames. To this end, optical flow information is first computed and used to link each sample with the corresponding one in the previous and subsequent frame, if it exists. Then, a multi-layer MRF can be constructed [49] using the model of (5.32) but with the smoothness terms not only considering the close spatial locations but also the corresponding samples in the previous and (possibly) following frame.

The maximization of (5.32) cannot be computed independently for each random variable of the random field as in local approaches, but it requires the same techniques used for the maximization of (5.14), e.g., loopy belief propagation or graph cuts. Examples of techniques based on MRF in order to fuse data from a ToF camera and stereo vision system are presented in [10, 48–50]. Some results from [50] and [10] are shown in Fig. 5.18 and 5.19, respectively. In particular the increased accuracy of the depth estimation due to the global optimization can be visually appreciated from the comparison between the rows of Fig. 5.19. The global approach reported by [10] leads to a 9 % error reduction. Notice that the maximization of (5.32) is very computationally expensive, and even efficient approaches like loopy belief propagation can require long computation times. One solution is to use hierarchical approaches [49] where the solution is first computed at low resolution, and then used to initialize the optimization at the next resolution level. Another possibility is to restrict the search range by assuming the optimal solution within a given interval near the depth value measured from the depth camera. A modified version of loopy belief propagation with a site-dependent set of possible labels is proposed in [10] and used for the fusion of stereo and ToF depth camera data.

5.3.2.2 Other Global Optimization Based Frameworks

Even if the MAP-MRF probabilistic framework is the most common approach, the global constraints can be enforced by different models. This section offers a brief overview of solutions alternative to the ones seen so far.

A first family of solutions, used for example in [18, 41], leverages graph cuts based models derived from models used in stereo vision [5]. The output depth map is the one that minimizes an energy function containing a data term and a smoothness term as in the previous case. The 3D volume including the scene is modeled by a graph with a node for each possible depth hypothesis at each pixel location of the output lattice. Minimum cut algorithms on the graph are typically used to find the surface corresponding to the minimum of the energy function. In classical stereo vision approaches, the data term depends on local similarity of the image patches according to the matching cost function of the stereo vision algorithm. This approach can be extended by using a linear combination of the stereo vision matching cost with a second metric depending on the difference between the considered depth value and the value measured by the depth camera [18]. A critical issue for this family of approaches is that the volumetric representation has a node for each pixel and each possible depth value on which the search for the minimum

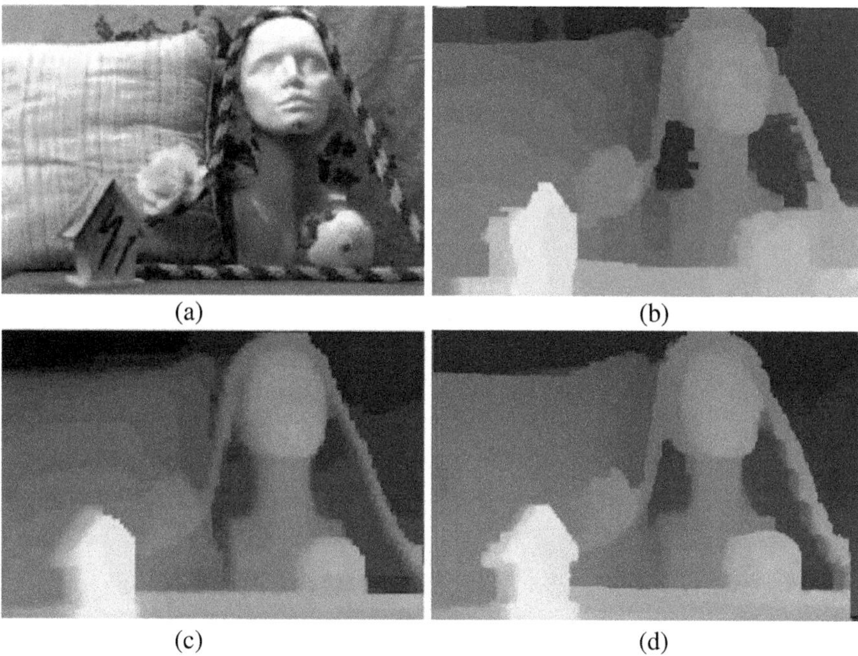

Fig. 5.18 Results of the application of the global fusion approach of [50]: (**a**) acquired left color image; (**b**) depth map acquired by the ToF depth camera and interpolated; (**c**) depth map produced by the stereo vision algorithm; (**d**) depth map obtained by the fusion algorithm (*Courtesy of the authors of [50]*)

cut is performed, leading to large memory usage and computation time. Another advantage of using the depth camera is that the graph can be constructed with only the nodes corresponding to depth values close to the measurements of the depth camera as proposed in [41], thus saving both memory and computation time.

Another framework that can be used to enforce similar constraints is total variation. This technique was first introduced for image denoising by [38] and is based on the minimization of the total variation in the image, simultaneously forcing the solution to not drift too far from the original image. The allowed variation is usually controlled by a parameter of the algorithm. This idea has been adapted to stereo vision and ToF data fusion by [32, 39]. The term controlling the similarity between the original and target depth maps can be modified in order to account for a first component measuring the similarity between the depth map from the depth camera and a second component accounting for the two images from the stereo camera. For example, [32] uses a brightness constancy constraint. The regularization terms instead can be computed on the basis of the norm of the image gradient. The total variation minimization is solved by iterative approaches, typically using the measurements from the depth camera as a starting point.

Fig. 5.19 Comparison of probabilistic local and global data fusion by [10]. *First row* shows five different scenes from the color camera viewpoint. *Second row* shows the estimated depth maps with a local approach. *Third row* shows the estimated depth maps with a global approach. The *arrows* underline some of the larger improvements

5.3.3 Other Approaches

There are a number of approaches for the fusion of data from ToF depth cameras and stereo systems deriving from stereo vision techniques using depth information differently from the previously presented methods. A simple solution is to use standard stereo vision algorithms on the data from the two color cameras with the depth camera data only constraining the search space for the disparity, typically within an interval around the depth camera depth measurements. A variation of this approach [16] adopts a hierarchical stereo vision algorithm and uses the data from a ToF depth camera to initialize the base level of the hierarchy. Another possibility is to simply modify the matching cost function by adding a term dependent on the depth camera measurements as in [47].

Another approach derived from stereo vision techniques [8], after upsampling ToF data to the spatial resolution of the stereo vision images with techniques shown in Sect. 5.2, fuses the two depth fields by enforcing the local consistency of depth data by a technique derived from the approach of [31], originally developed for stereo depth maps' refinement. An example of the depth map produced by this approach is shown in Fig. 5.20.

Fig. 5.20 Fusion of stereo and ToF depth maps by [8]: (*first row*) image acquired by the left camera of the stereo pair; (*second row*) sparse depth data acquired by the ToF depth camera and mapped on the left camera lattice; (*third row*) disparity map computed by a stereo vision algorithm; (*fourth row*) locally consistent disparity map computed from both ToF and stereo data

Another technique, proposed in [12] and [13], relies on region growing approaches that can be used to construct high resolution depth map from the sparse samples of a depth camera reprojected in the high resolution color camera and stereo system.

Another solution [19] exploits confidence measures of ToF data and color image segmentation. Stereo correspondences are computed only in regions where depth data from the ToF depth camera are not reliable. Image segmentation is used to guide the matching search by adaptively fitting the matching window to the color information.

5.4 Conclusions and Further Reading

The fusion of data from sensors of different natures is a classical problem formalized in [20]. This chapter addresses the specific case of data fusion from depth camera and one or two color cameras data. The basic tools available for data fusion are cross bilateral filtering [45] and the Maximum-a-Posteriori Markov-Random-Field (MAP-MRF) framework [4, 26], both widely used in computer vision applications.

In general, the joint use of depth and color cameras improves the quality of each information channel. Data fusion by either local or global approaches improves the accuracy and precision of both depth camera and stereo measurements. Global approaches lead to better results at higher computational costs, therefore local methods are preferred if computation speed is a concern.

A deep understanding of the approaches proposed for the fusion of depth camera and stereo data requires acquaintance with stereo vision methods [40]. A fundamental component of the MRF based methods is the optimization of the global energy function. Such minimization can be carried out by loopy belief propagation [3], graph cuts [5, 6] or other algorithms. A comprehensive survey of optimization algorithms suitable to these kinds of problems can be found in [44].

Although this chapter describes data fusion from depth and color cameras, data fusion for computer vision applications is not limited to imaging sensors. The current integration of depth cameras in portable devices like tablets and smartphones [35] has led to the integration of depth and color data with data from inertial measurement units (IMU), such as accelerometers, gyroscopes, and magnetometers. The information coming from these sensors can be combined with depth sensor data in order to simplify motion tracking and reconstruction problems.

References

1. T.D.A. Prasad, K. Hartmann, W. Wolfgang, S.E. Ghobadi, A. Sluiter, First steps in enhancing 3d vision technique using 2D/3D sensors. O. Chum, V. Franc (eds.), Czech Society for Cybernetics and Informatics, Telc, Czech Republic, pages 82–86 (2006). http://cmp.felk.cvut.cz/cvww2006/
2. C. Beder, B. Barzak, R. Koch, A combined approach for estimating patchlets afrom pmd depth images and stereo intensity images, in *Proceedings of DAGM Conference* (2007)
3. C.M. Bishop, *Pattern Recognition and Machine Learning. Information Science and Statistics.* (Springer, New York, 2007)
4. A. Blake, P. Kohli, C. Rother (eds.), *Markov Random Fields for Vision and Image Processing.* (MIT Press, Cambridge, 2011)
5. Y. Boykov, V. Kolmogorov, An experimental comparison of min-cut/max-flow algorithms for energy minimization in vision. IEEE Trans. Pattern Anal. Mach. Intell. **26**, 359–374 (2001)
6. Y. Boykov, O. Veksler, R. Zabih, Fast approximate energy minimization via graph cuts. IEEE Trans. Pattern Anal. Mach. Intell. **23**, 1222–1239 (2001)
7. A. Buades, B. Coll, J.-M. Morel, A non-local algorithm for image denoising, in *Proceedings of IEEE Conference on Computer Vision and Pattern Recognition (CVPR)*, vol. 2 (2005), pp. 60–65
8. C. Dal Mutto, P. Zanuttigh, S. Mattoccia, G.M. Cortelazzo, Locally consistent tof and stereo data fusion, in *Proceedings of Workshop on Consumer Depth Cameras for Computer Vision* (Springer, Berlin/Heidelberg, 2012), pp. 598–607
9. C. Dal Mutto, P. Zanuttigh, G.M. Cortelazzo, A probabilistic approach to tof and stereo data fusion, in *Proceedings of 3D Data Processing, Visualization and Transmission* (Paris, 2010)
10. C. Dal Mutto, P. Zanuttigh, G.M. Cortelazzo, Probabilistic tof and stereo data fusion based on mixed pixels measurement models. IEEE Trans. Pattern Anal. Mach. Intell. **37**(11), 2260–2272 (2015)

11. J. Diebel, S. Thrun, An application of markov random fields to range sensing, in *Proceedings of Conference on Neural Information Processing Systems* (MIT Press, Cambridge, 2005)
12. G.D. Evangelidis, M. Hansard, R. Horaud, Fusion of range and stereo data for high-resolution scene-modeling. IEEE Trans. Pattern Anal. Mach. Intell. **37**(11), 2178–2192 (2015)
13. V. Gandhi, J. Cech, R. Horaud, High-resolution depth maps based on TOF-stereo fusion, in *Proceedings of IEEE International Conference on Robotics and Automation (ICRA)* (IEEE Robotics and Automation Society, Saint-Paul, 2012), pp. 4742–4749
14. V. Garro, C. Dal Mutto, P. Zanuttigh, G.M. Cortelazzo, A novel interpolation scheme for range data with side information, in *Proceedings of Conference on Visual Media Production* (2009), pp. 52–60
15. V. Garro, C. Dal Mutto, P. Zanuttigh, G.M. Cortelazzo, Edge-preserving interpolation of depth data exploiting color information. Ann. Telecommun. **68**(11–12), 597–613 (2013)
16. S.A. Gudmundsson, H. Aanaes, R. Larsen, Fusion of stereo vision and time of flight imaging for improved 3d estimation. Int. J. Intell. Syst. Technol. Appl. **5**, 425–433 (2008)
17. T. Hach, J. Steurer, A novel RGB-Z camera for high-quality motion picture applications, in *Proceedings of Conference on Visual Media Production* (ACM, New York, 2013), p. 4
18. U. Hahne, M. Alexa, Combining time-of-flight depth and stereo images without accurate extrinsic calibration. Int. J. Intell. Syst. Technol. Appl. **5**(3–4), 325–333 (2008)
19. U. Hahne, M. Alexa, Depth imaging by combining time-of-flight and on-demand stereo, in *Dynamic 3D Imaging*, ed. by A. Kolb, R. Koch. Lecture Notes in Computer Science, vol. 5742 (Springer, Berlin/Heidelberg, 2009), pp. 70–83
20. D.L. Hall, J. Llinas, An introduction to multisensor data fusion. Proc. IEEE **85**(1), 6–23 (1997)
21. C.E. Hernandez, G. Vogiatzis, R. Cipolla, Probabilistic visibility for multi-view stereo, in *Proceedings of IEEE Conference on Computer Vision and Pattern Recognition* (2007)
22. X. Hu, P. Mordohai, A quantitative evaluation of confidence measures for stereo vision. IEEE Trans. Pattern Anal. Mach. Intell. **34**(11), 2121–2133 (2012)
23. B. Huhle, S. Fleck, A. Schilling, Integrating 3d time-of-flight camera data and high resolution images for 3dtv applications, in *Proceedings of Conference on 3DTV* (2007), pp. 1–4
24. B. Huhle, T. Schairer, P. Jenke, W. Strasser, Fusion of range and color images for denoising and resolution enhancement with a non-local filter. Comput. Vis. Image Underst. **114**(12), 1336–1345 (2010)
25. K.D. Kuhnert, M. Stommel, Fusion of stereo-camera and pmd-camera data for real-time suited precise, pages 4780–4785 (2006)
26. S.Z. Li, *Markov Random Field Modeling in Image Analysis*, 3rd edn. (Springer, New York, 2009)
27. M. Lindner, A. Kolb, K. Hartmann, Data-fusion of pmd-based distance-information and high-resolution RGB-images, in *Proceedings of International Symposium on Signals, Circuits and Systems* (2007), pp. 1–4
28. M. Lindner, M. Lambers, A. Kolb, Sub-pixel data fusion and edge-enhanced distance refinement for 2D/3D images. Int. J. Intell. Syst. Technol. Appl., pp. 344–354 (2008)
29. M.Y. Liu, O. Tuzel, Y. Taguchi, Joint geodesic upsampling of depth images, in *Proceedings of IEEE Conference on Computer Vision and Pattern Recognition* (2013), pp. 169–176
30. S. Lu, X. Ren, F. Liu, Depth enhancement via low-rank matrix completion, in *Proceedings of IEEE Conference on Computer Vision and Pattern Recognition* (2014), pp. 3390–3397
31. S. Mattoccia, A locally global approach to stereo correspondence, in *Proceedings of ICCV Workshop, 3D Digital Imaging and Modeling* (Kyoto, 2009), pp. 1763–1770
32. R. Nair, F. Lenzen, S. Meister, H. Schäfer, C. Garbe, D. Kondermann, High accuracy tof and stereo sensor fusion at interactive rates, in *Proceedings of ECCV Workshops and Demonstrations* (Springer, Heidelberg, 2012), pp. 1–11
33. R. Nair, K. Ruhl, F. Lenzen, S. Meister, H. Schäfer, C.S. Garbe, M. Eisemann, M. Magnor, D. Kondermann, A survey on time-of-flight stereo fusion, in *Time-of-Flight and Depth Imaging. Sensors, Algorithms, and Applications*, ed. by M. Grzegorzek, C. Theobalt, R. Koch, A. Kolb. Lecture Notes in Computer Science, vol. 8200 (Springer, Berlin/Heidelberg, 2013), pp. 105–127.

34. J. Park, H. Kim, Y.W. Tai, M.S. Brown, I. Kweon, High quality depth map upsampling for 3d-tof cameras, in *Proceedings of International Conference on Computer Vision* (2011)
35. Project Tango, https://www.google.com/atap/project-tango/
36. Recommendations on uniform color spaces, color difference equations, psychometric color terms. Supplement No. 2 to CIE publication No. 15 (E.-1.3.1) 1971/(TC-1.3.) (1978)
37. R. Reulke, Combination of distance data with high resolution images, in *Proceedings of Image Engineering and Vision Metrology* (2006)
38. L.I. Rudin, S. Osher, E. Fatemi, Nonlinear total variation based noise removal algorithms. Phys. D Nonlinear Phenom. **60**(1), 259–268 (1992)
39. K. Ruhl, F. Klose, C. Lipski, M. Magnor, Integrating approximate depth data into dense image correspondence estimation, in *Proceedings of Conference on Visual Media Production* (ACM, New York, 2012), pp. 26–31
40. D. Scharstein, R. Szeliski, A taxonomy and evaluation of dense two-frame stereo correspondence algorithms. Int. J. Comput. Vis. **47**, 7–42 (2002)
41. Y. Song, C.A. Glasbey, G.W.A.M. van der Heijden, G. Polder, J.A. Dieleman, Combining stereo and time-of-flight images with application to automatic plant phenotyping, in Proceedings of Scandinavian Conference on Image Analysis (Springer, Berlin/Heidelberg, 2011), pp. 467–478
42. J. Sun, N. Zheng, H. Shum, Stereo matching using belief propagation. IEEE Trans. Pattern Anal. Mach. Intell. **25**, 787–800 (2003)
43. R. Szeliski, *Computer Vision: Algorithms and Applications* (Springer, New York, 2010)
44. R. Szeliski, R. Zabih, D. Scharstein, O. Veksler, V. Kolmogorov, A. Agarwala, M. Tappen, C. Rother, A comparative study of energy minimization methods for markov random fields with smoothness-based priors. IEEE Trans. Pattern Anal. Mach. Intell. **30**, 1068–1080 (2008)
45. C. Tomasi, R. Manduchi, Bilateral filtering for gray and color images, in *Proceedings of International Conference on Computer Vision* (1998), pp. 839–846
46. Q. Yang, R. Yang, J. Davis, D. Nistér, Spatial-depth super resolution for range images, in *Proceedings of IEEE Conference on Computer Vision and Pattern Recognition (CVPR)* (2007)
47. Q. Yang, K.H. Tan, B. Culbertson, J. Apostolopoulos, Fusion of active and passive sensors for fast 3D capture, in *Proceedings of IEEE International Workshop on Multimedia Signal Processing* (2010)
48. J. Zhu, L. Wang, R. Yang, J. Davis, Fusion of time-of-flight depth and stereo for high accuracy depth maps, in *Proceedings of IEEE Conference on Computer Vision and Pattern Recognition* (2008)
49. J. Zhu, L. Wang, J. Gao, R. Yang, Spatial-temporal fusion for high accuracy depth maps using dynamic MRFs. IEEE Trans. Pattern Anal. Mach. Intell. **32**, 899–909 (2010)
50. J. Zhu, L. Wang, R. Yang, J.E. Davis, Z. Pan, Reliability fusion of time-of-flight depth and stereo geometry for high quality depth maps. IEEE Trans. Pattern Anal. Mach. Intell. **33**, 1400–1414 (2011)

Part III
Applications of Depth Camera Data

Chapter 6
Scene Segmentation Assisted by Depth Data

Scene segmentation, the detection of the various scene elements inside images and videos, is a well-known and widely studied computer vision and image processing problem. Despite a huge amount of research and many proposed approaches [58], scene segmentation by means of color information alone remains a very challenging task. Even approaches based on graph theory [22, 50] or advanced clustering techniques [12], effective for typical scene imagery, are not able to obtain satisfactory segmentation performance in the most challenging situations. This is due to the lack of information in color data to disambiguate all the scene objects. For example, the segmentation of an object in front of a background with a similar color is still a very difficult task even for the best segmentation algorithms. Objects with complex texture patterns are difficult to segment as well, especially if their texture statistics are similar to that of the background or of other objects.

Furthermore, image segmentation is an ill-posed problem and it is very difficult to define which is the proper solution in many situations. Depth information turns out to be very useful for scene segmentation. Segmentation based on depth data usually provides better results than image segmentation because depth information allows to easily divide near and far objects on the basis of their distance from the depth camera. Depth data also contain relevant information about the shape and size of the framed objects, allowing one to recognize them without ambiguities related to camera distance or perspective distortion. Moreover, since depth information is independent not only from the object and background colors but also from scene illumination, it properly handles objects with complex texture patterns. On the other hand, some scene configurations are critical for depth information, for example the case of two objects touching each other. However it is worth noting that in many cases what is critical for depth information differs from what is critical for color information.

Segmentation by means of depth information has become popular only recently due to the fact that depth data acquisition was rather difficult and expensive

P. Zanuttigh et al., *Time-of-Flight and Structured Light Depth Cameras*,
DOI 10.1007/978-3-319-30973-6_6

before the introduction of consumer depth cameras. Standard image and video segmentation methods can be easily applied to depth maps and generate good results in many situations, although some intrinsic data limitations must be taken into account. First, the relatively low resolution of depth cameras limits the localization accuracy of the edges of segmented objects. Moreover, some scenes are better suited to be segmented by depth information while others are better served by color information. For all of these reasons, the joint use of color and depth data is often the best scene segmentation option. Suitable provisions must also be taken into account to avoid segmentation errors due to noise, missing depths samples, and other artifacts typical of consumer depth cameras.

In this field, there are three main applications for consumer depth cameras (see Fig. 6.1), which are, in order of complexity:

1. **Scene matting:** the task of dividing foreground objects from the scene background. This relatively simple problem has practical relevance in many applications like film-making, special effects, and 3D video.
2. **Scene segmentation:** the partitioning of color images and depth maps into different regions corresponding to the various scene elements.
3. **Semantic segmentation:** the task of segmenting the framed scene and associating each segment to a specific category of objects (e.g., people, trees, buildings etc.). In this task depth data provide valuable information about the object size and shape.

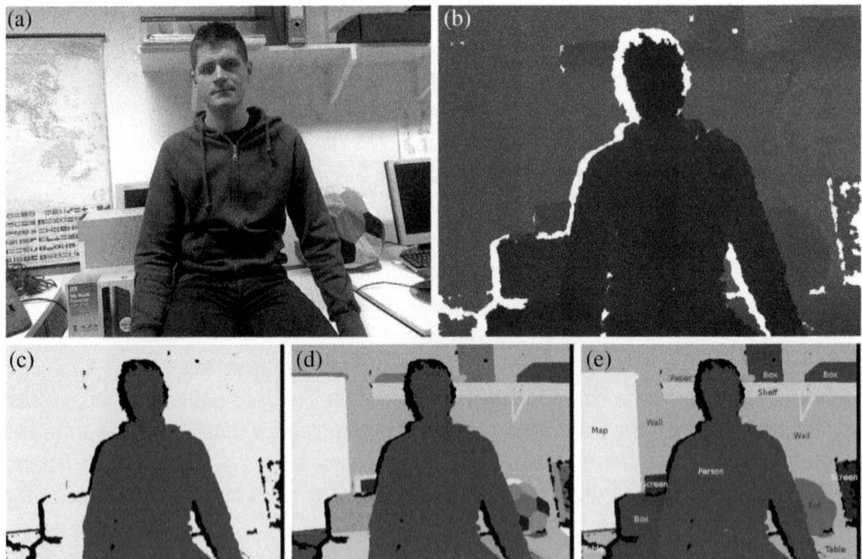

Fig. 6.1 Examples of the considered problems: (**a**) color image, (**b**) depth map, (**c**) scene matting, (**d**) scene segmentation and (**e**) semantic segmentation

Such problems can be either solved from a single depth map possibly associated with an image or a sequence of frames of depth maps associated with a color video. In the latter case, temporal consistency is an important clue for driving the segmentation process.

In this chapter, color and depth data are assumed to have the same resolution and refer to the same viewpoint. This can be achieved by the calibration and interpolation procedures presented in Part II of this book. This chapter presents recent results on the use of depth data for scene matting, segmentation, and semantic segmentation. The chapter is organized in three sections, each devoted to one of these topics.

6.1 Scene Matting with Color and Depth Data

Video matting traditionally represents an important task in film-making and video production. It remains a challenging task despite considerable research and the existence of various commercial products. Accurate foreground extraction is achieved by using a cooperative background, but if the background is not controllable and, for instance, it includes moving objects or colors similar to those in the foreground, video matting becomes difficult. Depth data represent the distance of each single pixel from the camera and thus provide valuable information to recognize the foreground objects, independently from their color and illumination artifacts like shadows.

The video matting problem can be formalized [45] by representing each pixel p^i in the observed color image $I_C(t)$ at time t as the linear combination of a background color $B(p^i)$ and foreground color $F(p^i)$ weighted by the opacity value $\alpha(p^i)$, i.e.

$$I_C(p^i) = \alpha(p^i)F(p^i) + [1 - \alpha(p^i)]B(p^i) \tag{6.1}$$

where $\alpha(p^i) = 1$ for foreground pixels and $\alpha(p^i) = 0$ for background pixels. If α is constrained to assume values of only 0 or 1, the problem is reduced to a binary segmentation task. Many methods also allow fractional α values in order to handle transparent objects and pixels close to edges that include both foreground and background elements in their area.

The set made by the $\alpha(p^i)$ values of the whole image is called *alpha matte* and its estimate is the *matting problem*. For video sequences, one could independently solve the problem for each frame of the sequence or also exploit temporal constraints between subsequent frames (in this case, an estimate of the motion field is typically useful).

Matting is an underconstrained problem: assuming a standard three-dimensional color space, for each pixel there are seven unknowns (the α value and the three components of the foreground and background color vectors) and only three known values (i.e., the color components of the pixel in the observed image). Standard matting approaches make assumptions on background and foreground image statistics, spatial and temporal coherence (the latter only when considering video sequences), and user input in order to further constrain the problem. Novel matting methods based on depth data instead leverage the foreground object closer proximity to the camera compared to background objects, making their separation easier to distinguish.

The analysis of an object depth allows to easily separate background and foreground within the already underlined limitations of depth cameras. The simplest approach consists of thresholding the depth values in order to associate all pixels closer than a suitable threshold T_{fg} to the foreground and the remaining ones to the background, i.e.,

$$p^i \in \begin{cases} \text{foreground} & \text{if } Z(p^i) \leq T_{fg} \\ \text{background} & \text{otherwise} \end{cases} \tag{6.2}$$

where $Z(p^i)$ is the depth value corresponding to pixel p^i. Although (6.2) gives a relatively accurate foreground in many situations, it has limitations. The first is the so called *depth camouflage* problem (Fig. 6.2b), i.e., when a moving object or person belonging to the foreground gets close to the background. In this case, the depth values of the foreground and background become similar and it is difficult to separate them given only depth information. The second is that the simple thresholding of (6.2) assigns static objects close to the camera to the foreground, a behavior not suited to some applications like video surveillance. It also provides a binary assignment decision between background or foreground, without the possibility of a soft transition as in the model of (6.1). In particular, it is difficult to precisely locate the edges between the two regions from depth data alone, because of the limited spatial resolution of most ToF cameras and of the edge artifacts introduced by many ToF and structured light depth cameras. If the depth cameras are assisted by standard cameras, another critical issue is the limited accuracy of the calibration between the depth camera and the high quality color camera used for video acquisition (the target is typically the matting of the video stream from the color camera). Moreover, as already shown in Chap. 5, depth and color cameras usually have slightly different optical centers, therefore, some background regions visible from the color camera may be occluded from the depth camera point of view and a depth-to-color value association may not be feasible. On the other hand, the amplitude of the signal received by ToF depth cameras is also a useful source of information, since the pixel amplitude values depend on the combination of the point distance and the surface reflectivity, thus combining depth and color dependent properties. Table 6.1 summarizes which sources of information are best suited to different situations, and also shows how combining different clues can be beneficial.

A common approach to address the above matting issues is to build a *trimap* from depth information, i.e., dividing the image into a foreground, a background, and an uncertainty region. The uncertainty region usually corresponds to areas near the edges between foreground and background (as shown in Fig. 6.3). Standard matting algorithms typically build a trimap under human supervision, but depth information

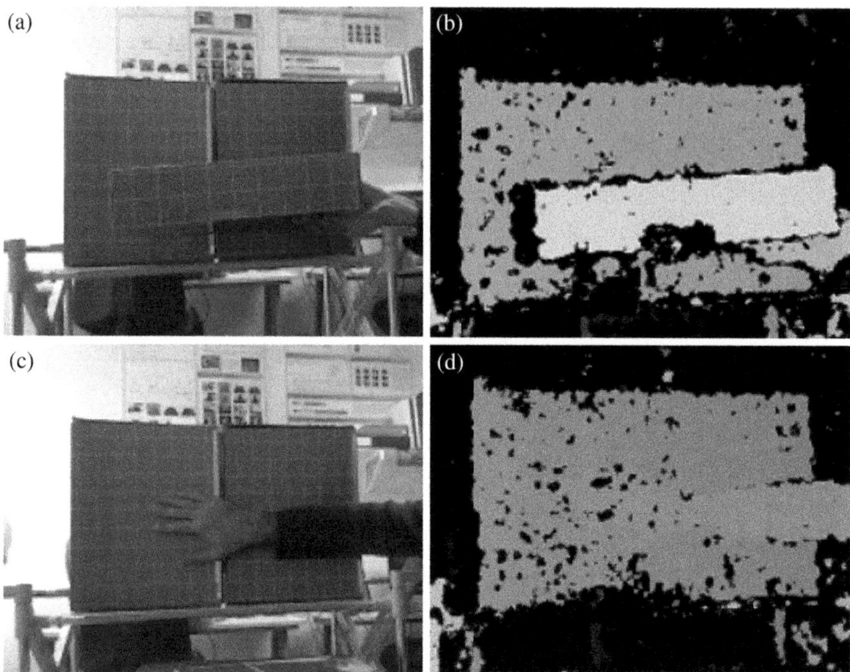

Fig. 6.2 Examples of matting with color and depth information: the foreground and background have similar colors (**a**) but different depths (**b**); the foreground and background have different colors (**c**) but similar depths (**d**) (depth camouflage). Note how the first situation can be properly handled by depth data while the second is better solved by color data

Table 6.1 Various scene characteristics and possible related issues for scene matting

Scene characteristics	Color data	Depth data	Amplitude data (ToF)
Similar colors in background and foreground	Issues	Ok	Ok[a]
Foreground object close to background	Ok	Issues	Ok[a]
Illumination changes	No	Ok	Ok
Shadows	No	Ok	Ok
Edges accuracy	Good	Low	Low
Highly textures regions	Issues	Ok	Issues

[a] The amplitude value is a combination of the distance and color values: different values of distance and color can lead to the same amplitude value

allows one to solve this task automatically, with advantages in many commercial applications. The simplest way to build a trimap is to threshold the depth map, then erode the foreground and background regions to remove pixels near their boundary. The uncertainty region is usually given by eroded samples and by samples without depth values, either due to occlusions or because they were not provided by the depth camera. The trimap computation can also take advantage of other clues, if

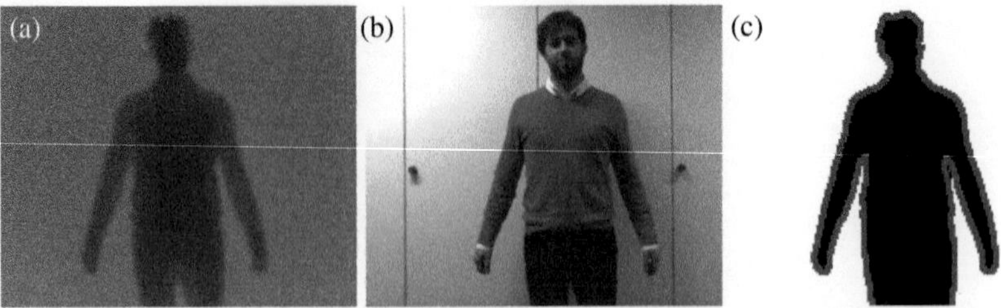

Fig. 6.3 Example of a matting trimap: (**a**) depth map acquired by the ToF camera; (**b**) color image acquired by the video camera; (**c**) trimap computed from depth information with *black*, *white* and *gray* pixels referring to foreground, background and uncertainty region respectively

available: for instance, one can use the ToF camera confidence map to include pixels with low confidence depth values in the uncertainty region.

6.1.1 Single Frame Matting with Color and Depth Data

Frame matting based on the joint use of color and depth information has been the subject of a considerable amount of research. This section presents the approaches focusing on a single frame only, without taking into account the temporal coherence between subsequent frames.

A first approach for computing a trimap [14] is assigning to each depth map pixel p^i the probability of belonging to the foreground. This probability does not only depend on the considered pixel depth but also on the depth of the pixels in the neighborhood of p^i. Namely, a probability $P_{fg}(p^i)$ of being the foreground is assigned to each pixel p^i on the basis of its corresponding depth value $Z(p^i)$. Pixels without an associated depth value usually receive a low P_{fg} score since occluded pixels usually belong to the background (though the lack of a depth measurement may be due to other causes). Then, the likelihood of each pixel being the foreground or background is computed as a weighted average of the probabilities of the pixels around it, according to

$$P_{fg}(p^i|W^i) = \frac{1}{\sum_k w(p^k)} \sum_{p^n \in W^i} w(p^n) P_{fg}(p^n) \tag{6.3}$$

where W^i is a rectangular window around p^i and $w(p^n)$ are weights usually giving more relevance to the contributions of the pixels closer to p^i. A common choice for modeling $w(p^n)$ is a Gaussian function. Equation (6.3) assigns likelihood values close to 0 or 1 for pixels in the background or foreground regions, respectively, and intermediate likelihood values to pixels near edges. Two thresholds T_l and T_h can be used to assign the samples to the foreground, background, or uncertainty region, namely

$$p^i \in \begin{cases} \text{foreground} & \text{if } P_{fg}(p^i|W) > T_h \\ \text{background} & \text{if } P_{fg}(p^i|W) < T_l \\ \text{uncertainty region} & \text{if } T_l \leq P_{fg}(p^i|W) \leq T_h. \end{cases} \tag{6.4}$$

The critical issue for this family of approaches is how to assign the pixels of the uncertainty region to the background or foreground. A possible solution is to apply cross bilateral filtering to the alpha matte, as for the super resolution case [14] seen in Sect. 5.2.1, according to

$$\alpha_f(p^i) = \frac{1}{n_f} \sum_{p^n \in W^i} G_s(p^i, p^n) G_r(I(p^i), I(p^n)) \alpha(p^n) \tag{6.5}$$

where W^i is a rectangular window around p^i, n_f is a normalization factor defined in (5.3), and G_s and G_r are the spatial and range Gaussian weighing functions introduced in Sect. 5.2.1. Note that this approach can be used either to filter the existing alpha matte or to assign an α value to the samples without a depth value because of occlusions or missing data in the depth camera acquisitions. In order to handle missing data, W^i can be defined as the set of window samples with a *valid* α value (i.e., if the trimap is computed by thresholding depth information, valid pixels are the those with an associated depth value). Figure 6.4 shows an example of alpha matte computed by the method of [14] based on this approach.

Another possibility is to assign to each image pixel a 4D vector $I_{C,Z}(p^i) = [R(p^i), G(p^i), B(p^i), Z(p^i)]^T$ with depth $Z(p^i)$ as a fourth channel to be added to the three standard RGB color channels, and then apply standard matting techniques originally developed for color data to $I_{C,Z}(p^i)$. For example, Wang et al. [62] extend the Bayesian matting scheme originally proposed in [11] for color images by introducing the depth component in the probability maximization scheme. Another interesting idea, also proposed in [62], is weighting the confidence of the depth channel on the basis of the estimated α value in order to give pixels with α close to 0 or 1 more relevance and less relevance to the depth of the pixels in the uncertainty region. This provision takes into account the fact that depth data are less reliable on the boundary regions between foreground and background due to low resolution and edge artifacts typical of consumer depth cameras. The approach of [38] also gives a lower depth weight to pixels close to boundaries and at the same time increases the depth weights when the foreground and background color are similar, a situation in which color data are not very reliable.

Poisson matting [57] can be similarly extended to include depth information. In this approach, the color image is first converted into a single channel representation. Although, as in the previous case, depth data can be added as a fourth channel for the color image, the sharp edges typical of depth maps have a very strong impact on the gradients used by Poisson matting and the results are similar to those obtained from depth information alone. In [62] a confidence map is first built from the depth map and then used to derive a second alpha matte to be combined with the first

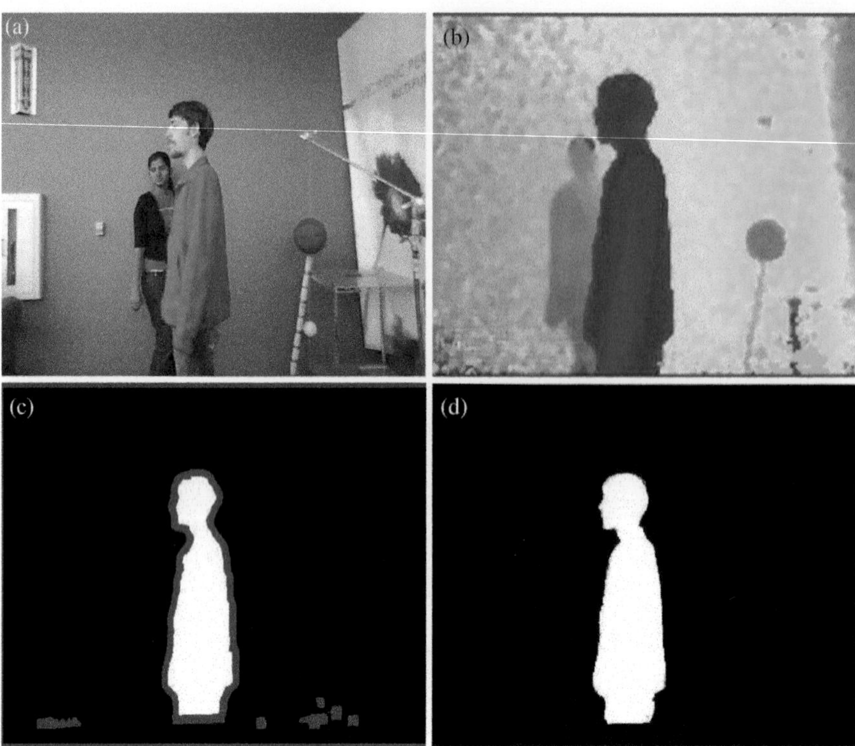

Fig. 6.4 Example of matting with the method of [14] based on bilateral filtering on color and depth data: (**a**) color image; (**b**) depth map; (**c**) trimap computed on the basis of depth information; (**d**) alpha matte computed by joint bilateral filtering on depth and color data. (*Courtesy of the authors of [14]*)

one from Poisson matting. Finally, a multichannel extension of Poisson matting is proposed in [61] in order to jointly consider the three color channels and the depth channel.

A different class of approaches for the joint segmentation of depth and color data extends the graph-cut segmentation framework to this particular task and represents the image as a graph $G = \{V, E\}$ where the graph nodes correspond to pixels and the edges to relationships between neighboring pixels (see Fig. 6.5). The matting problem can be formalized as the identification of the labeling minimizing the energy functional

$$U(\alpha) = \sum_{p^i \in I} V_{data}(\alpha(p^i)) + \sum_{p^i \in I} \sum_{p^n \in N(p^i)} V_s(\alpha(p^i), \alpha(p^n)). \tag{6.6}$$

Such approaches do not allow fractional α values and $\alpha(p^i) \in \{0, 1\}$ are binary labels assigning each pixel to the foreground or background. The data term V_{data} of

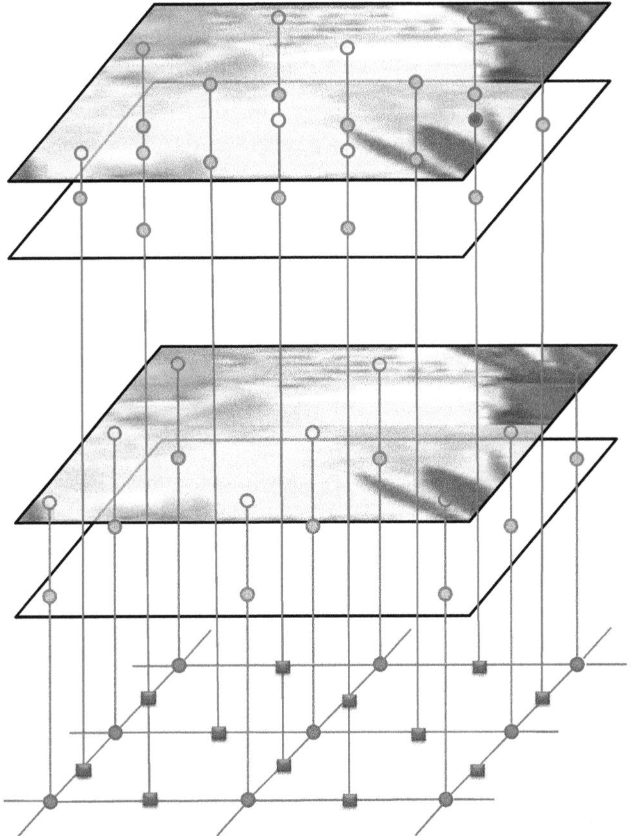

Fig. 6.5 Structure of the Markov Random Field used in joint depth and color matting

(6.6) models the likelihood that a pixel belongs to the foreground or background, and is typically the sum of two terms, one depending on color and one depending on depth, i.e.,

$$V_{data}(\alpha(p^i)) = V_{color}(\alpha(p^i)) + \lambda V_{d,depth}(\alpha(p^i)) \tag{6.7}$$

where the color term V_{color} can simply be the distance between the pixel color and the mean foreground or background color as in [3]. The depth term V_{depth} can be modeled in a similar way. The term V_{depth} must, however, take into account that foreground pixels usually lie at a similar distance from the camera, while background pixels depth may vary sensibly, especially in complex scenes. In [3] this issue is handled by considering the depth term for foreground pixels only.

Better results can be obtained by using more complex models of the foreground and background likelihoods. Figure 6.6 shows some results obtained by the approach of [60] that models the two likelihoods using Gaussian Mixture Models (GMM).

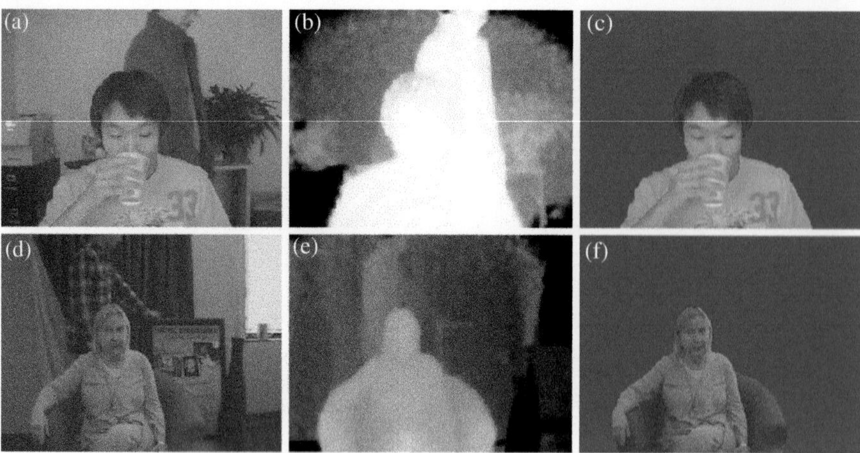

Fig. 6.6 Joint depth and color matting by the graph-cut based method of [60] on two different scenes: (**a**)–(**d**) color image; (**b**)–(**e**) depth map; (**c**)–(**f**) extracted foreground. (*Courtesy of the authors of [60]*)

Another key issue is that color and depth lie in completely different spaces and it is necessary to adjust their mutual relevance. The proper setting of the weight constant λ is a challenging task that can be solved by adaptive weighting strategies. In [60] the weights are updated on the basis of the foreground and background color histograms along the spatial and temporal dimensions.

The smoothness term V_{smooth} of (6.6) can be built in different ways. In standard graph-based segmentation and matting approaches based on color information only, V_{smooth} usually forces a smoothness constraint in the color domain within each segmented region. In the case of color and depth information, the same constraint can be adapted to the depth domain, for instance by an exponential function of the color and depth differences such as

$$V_{smooth}(\alpha(p^i), \alpha(p^n)) = \\ = \left|\alpha(p^i) - \alpha(p^n)\right| e^{-\frac{\delta(I_C(p^i), I_C(p^n))^2}{2\sigma_c^2}} e^{-\frac{[Z(p^i) - Z(p^n)]^2}{2\sigma_z^2}} \tag{6.8}$$

where σ_c and σ_z are the standard deviations of the color and depth data noise, respectively, $I_C(p^i)$ and $I_C(p^n)$ are the color values of the considered samples, and $Z(p^i)$ and $Z(p^n)$ their depth values. The δ function can be any suitable measure of the color difference between the two samples.

The energy functional $U(\alpha)$ of (6.6) can be minimized by efficient graph-cuts optimization algorithms [6] with the methods seen in Sect. 5.2.2.

6.1.2 Video Matting with Color and Depth Data

The matting problem is of great interest in video processing. When dealing with video sequences the temporal consistency between subsequent frames is another fundamental clue to be exploited in matting. There are two main situations to consider in this case, i.e, the simplest case of fixed cameras and the more challenging case of moving cameras.

When the depth camera is fixed, that is, it does not move or rotate, it is possible to assume that the background location is also fixed. Namely, the depth values of the background pixels do not change because of motion and consequently, depth variations are only due to the movement of people and objects in the foreground. Background subtraction techniques developed for standard videos can easily be adapted to depth data and used to recognize foreground and background regions. Since depth is not affected by illumination changes, shadows, or reflections, the depth values of the background are more stable than the corresponding color values and even the simplest background subtraction schemes are rather effective. However, there can still be depth values changes due to sensor noise and the various depth camera acquisition artifacts described in Chaps. 2 and 3. Another relevant difference is that while color pixels are defined at each raster location, depth data may not exist in some locations due to the various issues described in the first part of the book (e.g., objects too close or too far or occluded from the point of view of the projector in structured light color cameras). All of these reasons suggested the development of ad hoc approaches for video matting from color and depth data.

In particular, since a moving foreground object is typically associated with a change in both color and depth information, the combination of the two clues gives better results. A simple solution proposed in [36] is an independent foreground extraction from depth and color data followed by the combination of the results by the logical OR operator. Background subtraction algorithms can be easily modified in order to handle both color and depth data, typically by considering depth as an additional color channel. The approach proposed by Stauffer and Grimson [55] based on a Gaussian Mixture Model (GMM), for example, is one of the most successful solutions for video background subtraction and has been adapted to jointly account for depth and color data in [26] and [35]. The approach of [26] basically adds a new component representing depth data to the three standard color components in the background model. In this method the criteria used in [55] to accept a match between the input and model distribution has been modified in order to require that the condition is satisfied on both depth and color data. The method of [35] considers five different components, namely the three color components, depth information and the amplitude data from the ToF depth camera. Amplitude information is an interesting clue since it depends on both the distance and reflectivity of the framed scene point. A similar but more refined work [24] uses two distinct Gaussian Mixture Models for the foregrounds of depth and color spaces and combines them in a Bayesian Framework. The approaches of [24, 26], and [36] also consider temporal constraints in depth and color video streams.

Another possibility for extending background subtraction schemes for color data to joint color and depth matting is building two different classifiers, one working on the color data and one on depth information, and then combining the results. The method of [8] follows this rationale and adopts two different classifiers (also based on a Gaussian Mixture Model approach inspired by [55]) independently working on the depth and color data. The final result is a weighted average of the two outputs where the adaptive weights depend on two main clues. The first clue is the sample distance from color and depth edges: since depth data have lower resolution and less accurate edges, the idea is to give more relevance to color information in proximity of edges. The second clue is only applied to pixels detected as foreground in the previous frame and incorporates the difference between the depth of the considered foreground pixel and the corresponding pixel in the background model. The larger such a difference is, the higher the weight associated with the depth data. If a moving object gets closer to the background (i.e., the depth difference becomes small), its detection with depth data becomes less reliable due to the depth camouflage problem. Figure 6.7 exemplifies the results of [8] by comparing matting based on color only, depth only, and the two combined. In [9] and [19] this approach has been combined with a Bayesian network that predicts the background and foreground regions across subsequent times.

Some applications also require subdividing the foreground into various moving objects or people, for example, [56] analyzes the gradient of depth data for this task. Another correlated problem is people detection, where various classification schemes are used to detect people in regions associated with the foreground.

When the depth camera moves, standard background subtraction approaches cannot be applied, although all of the single frame approaches from the previous section can still be utilized, possibly reinforced by temporal constraints. Furthermore, as described in Chap. 7, depth data allow one to estimate and compensate camera motion in order to switch back to the static camera case.

Fig. 6.7 Example of the results of [8]: (**a**) color view; (**b**) depth map; (**c**) segmentation from color data only; (**d**) segmentation from depth data only; (**e**) segmentation jointly exploiting color and depth data. (*Courtesy of the authors of [8]*)

6.2 Scene Segmentation from Color and Depth Data

Scene segmentation is a more general problem than matting, targeting the extraction of the different scene regions from the framed scene. The literature presents several approaches for this task, either working on a single frame or a video stream of color and depth data. As already stated, depth data allow to achieve better results than color information alone, but the improvement also depends on the depth information accuracy. This is exemplified in Fig. 6.8, which shows segmentation results of [17] obtained from depth data acquired by a ToF depth camera, a structured light depth camera (i.e., a Kinect™ v1), and a passive stereo vision system.

Recall that a critical issue in segmenting ToF depth data registered with color images is their different spatial resolutions. For clarity sake, all the methods presented in this section assume the availability of a depth map and color image of the same resolution. It is possible to obtain such a pair by interpolating the depth data to the (high) resolution of the color image eventually using also the additional information provided by color data with any of the methods described in Chap. 5. Alternatively it is possible to just subsample the color image to the (low) depth map resolution.

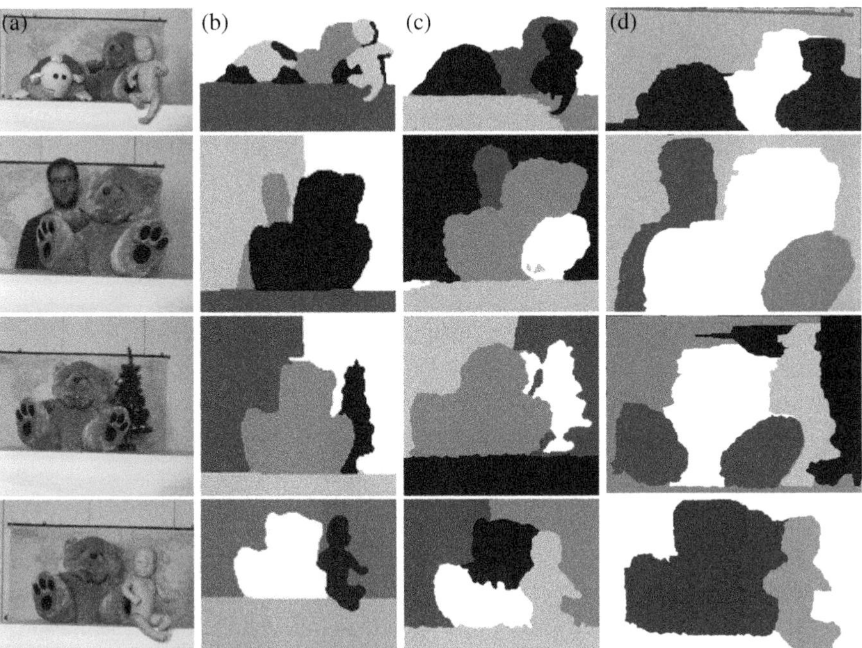

Fig. 6.8 Examples of segmentation by the approach of [17] based on color and depth data coming from different acquisition systems: (**a**) color image; (**b**) segmentation using data from a ToF depth camera; (**c**) segmentation using data from a Kinect™ v1 sensor; (**d**) Segmentation using data from a stereo vision system

There are also segmentation techniques which directly handle the different resolution issue without scaling the depth map to the color image spatial resolution beforehand. For instance, the method of [20] first performs a multi-resolution oversegmentation of the color image and then tries to fit the sparse depth samples inside each segment to a surface shape model (e.g., planar or quadratic). The segments minimizing the fitting error are then selected and used as an initial segmentation. A region merging algorithm and a growing stage are then used for the final segmentation which also exploits the sparse depth samples.

6.2.1 Single Frame Segmentation from Color and Depth Data

This section considers the segmentation of a single frame made by an image and a depth map corresponding to the same time instant. This problem can be solved by color data, depth data, or both types of data jointly considered. Segmentation from color images is a well-known research topic with a vast literature but is beyond the scope of this book. The interested reader is referred to [2, 58] for an extensive treatment of state of the art methods. Concerning depth data, the simplest approach is to treat the depth map as a grayscale image and apply suitable standard image segmentation techniques, e.g., [22] or [12].

Depth data are simpler to segment than color images since edges are sharp and well defined and there are no issues due to complex color and texture patterns or illumination changes. However, there are situations that cannot be easily solved only with depth data, like in the case of close or touching objects or the already introduced depth camouflage problem. Figure 6.9 shows a comparison between standard color segmentation and depth segmentation performed by representing depth maps as grayscale images and applying standard image segmentation techniques to them (the state of the art method of [12] has been used). Figure 6.9 confirms that depth segmentation is not affected by complex texture patterns and in general gives reasonable results, but some critical issues remain. For example, in Fig. 6.9c depth segmentation cannot divide the two people who are easily distinguishable by the color segmentation in Fig. 6.9b. Moreover, image segmentation techniques do not consider the three-dimensional structure behind depth data with various consequent artifacts. For instance, long uniform structures spread across a wide range of depth values are typically divided into several regions instead of being recognized as a single object (e.g., the slanted wall in Fig. 6.9d).

Better results can be achieved by using both depth and color information, as shown in Fig. 6.10. An associated color camera or alternatively the ToF amplitude image can be used to assist the depth data with a color image for segmentation.[1] There are different ways of jointly exploiting color and depth information for scene

[1]In the case of structured light depth cameras the amplitude image is not very informative about the scene's color, since it is dominated by the projected pattern.

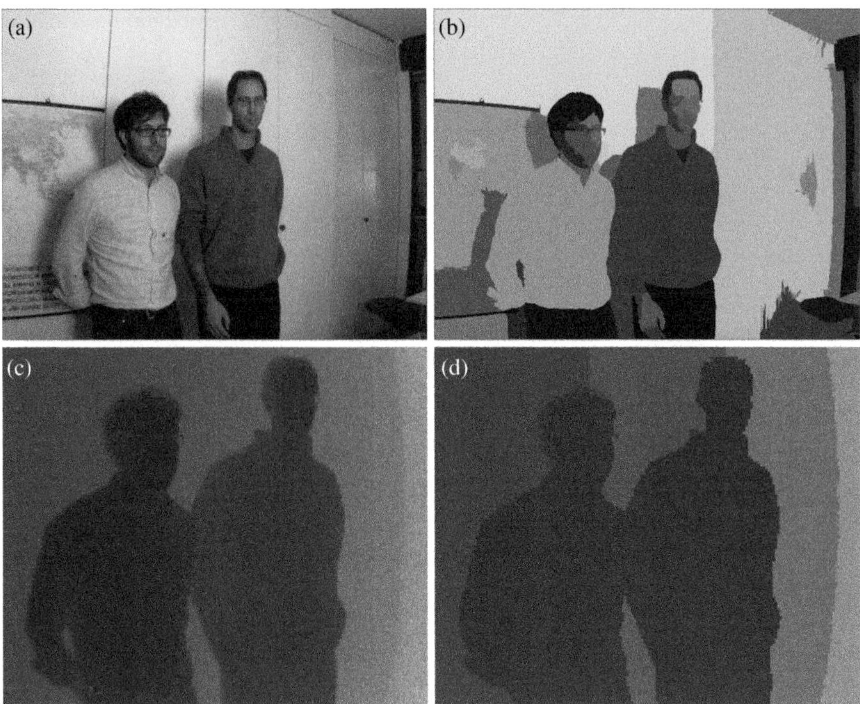

Fig. 6.9 Comparison between color and depth data segmentation: (**a**) color image from a standard camera; (**b**) color image segmentation by [12]; (**c**) depth map from a ToF depth camera; (**d**) depth map segmentation by [12]

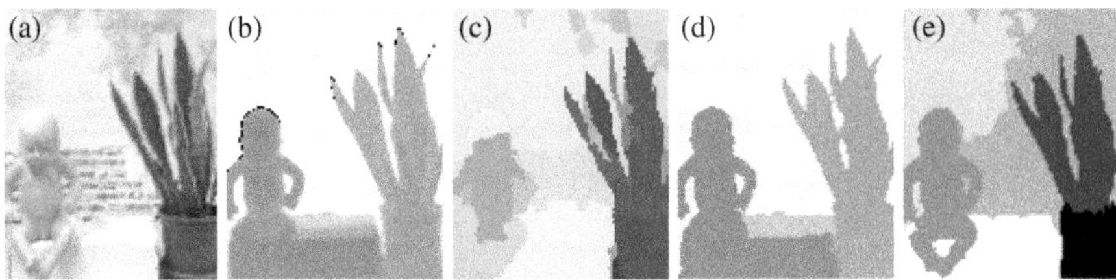

Fig. 6.10 Segmentation based on joint color and depth data information vs. segmentation based on color or depth information only from [17]: (**a**) color image from a standard camera; (**b**) depth map from a ToF depth camera; (**c**) segmentation based on color information only; (**d**) segmentation based on depth data only; (**e**) segmentation based on joint color and depth data

segmentation. A simple option is to first perform segmentation on color and depth data separately, and then combine the results according to some suitable criteria. For instance, in [7] the two segmentations are fused together in an iterative cooperative region merging process. Similarly, the approach of [36] first segments the scene based on each specific clue, as well as on the basis of the ToF amplitude and intensity images, then combines the results of the separate segmentations together.

More refined methods which use both clues into a single segmentation operation can be divided into three main families:

- **Clustering-based approaches:** each scene point is represented by a multidimensional vector containing color or geometric properties (Sect. 6.2.2).
- **Graph-based approaches:** the scene is represented by a graph with weights associated with the edges and nodes modeling the color and depth properties of the scene samples (Sect. 6.2.3).
- **Geometric approaches:** the scene is segmented by detecting specific geometric structures like planes or parametric surfaces (Sect. 6.2.4).

6.2.2 *Single Frame Segmentation: Clustering of Multidimensional Vectors*

A first possible solution to jointly segmenting a scene on the basis of color and depth data is to represent the scene as a set of multidimensional vectors containing color and geometry information for each point, then cluster them by a suitable algorithm [5, 16, 17].

Many color image segmentation methods (e.g., [12]) represent each image pixel as a 5D vector, with three components corresponding to the color coordinates in the selected color space and the remaining two components corresponding to the 2D pixel coordinates in the camera image plane, and cluster them by state of the art techniques. This approach can easily be extended to handle both depth and color data. The basic idea, used in [5] and [17] for instance, is to replace the 2D coordinates of the image pixels with the corresponding 3D coordinates obtained from the depth data and calibration information, as seen in Chap. 4. In this case, each pixel is associated with a 6D vector instead of a 5D vector like in image segmentation. It is worth noting that the three vector components representing scene geometry need to be expressed in a consistent way, and the 3D point coordinates $\mathbf{P}^i = [x^i, y^i, z^i]^T$ are a viable solution. It is also necessary to represent color in a uniform color space to ensure the consistency of the three color components, and the CIELab or CIELuv color spaces [47] are the most common choices. In any case, as geometry and color information are defined on different spaces, they must be normalized for a consistent representation. Assuming the use of the CIELuv color space and the 3D spatial coordinates (x, y, z), each pixel p^i can be associated with a 6D vector

$$V^i = \left[\frac{L(p^i)}{n_c}, \frac{u(p^i)}{n_c}, \frac{v(p^i)}{n_c}, \frac{x^i}{n_g}, \frac{y^i}{n_g}, \frac{z^i}{n_g} \right]^T \tag{6.9}$$

where $L(p^i)$, $u(p^i)$, and $v(p^i)$ are the L, u, and v components of p^i in the CIELuv color space. The normalization factors n_c and n_g are critical to properly comparing color and depth data and balancing their mutual relevance. It was experimentally found that segmentation performance strongly depends on the weights associated with the two types of clues.

A possible solution consists of normalizing both color and depth data with respect to their standard deviations σ_c and σ_g [17], but this does not allow one to modulate the contribution of color or depth information in the segmentation process. For this reason, the task of representing the relevance of depth information is often assigned to a further weighting term λ applied to the geometric components. In this case, the 6D vectors V^i of (6.9) can be rewritten as

$$V^i = \left[\frac{L(p^i)}{\sigma_c}, \frac{u(p^i)}{\sigma_c}, \frac{v(p^i)}{\sigma_c}, \lambda \frac{x^i}{\sigma_g}, \lambda \frac{y^i}{\sigma_g}, \lambda \frac{z^i}{\sigma_g} \right]^T. \tag{6.10}$$

The proper setting of λ is critical and its optimal value depends on scene characteristics, as shown by examples in Figs. 6.11, 6.12 and 6.13. An unsupervised segmentation metric for the joint segmentation of depth and color is proposed in [17] and used to evaluate the quality of the resulting segmentation (see the examples of Figs. 6.11 and 6.12). In this work the λ parameter is iteratively adjusted until an optimal segmentation is obtained with respect to the chosen metric. Even though this approach allows one to automatically tune the relevance of the depth and color clues, it is computationally expensive since it requires one to perform multiple segmentations.

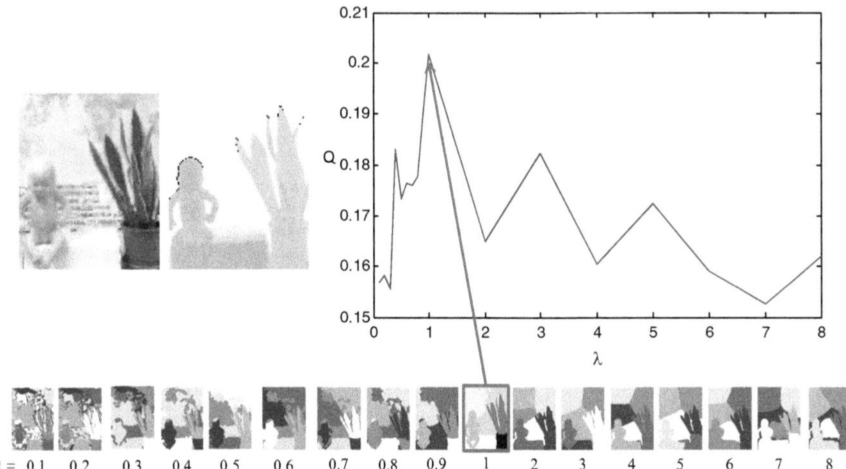

Fig. 6.11 Segmentation of a sample scene acquired by a ToF depth camera with the clustering-based approach of [17] for different values of the λ parameter. The segmentation quality is measured with the Q metric introduced in [17]

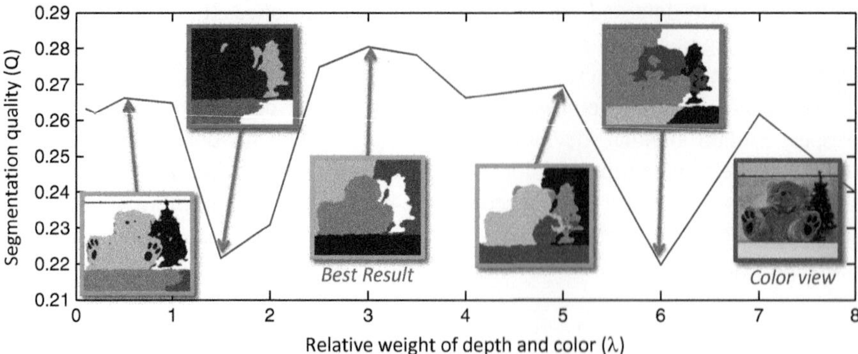

Fig. 6.12 Segmentation of a sample scene acquired by a Kinect™ v1 with the clustering-based approach of [17] for different values of the λ parameter. The best segmentation according to the Q metric [17] is highlighted in *green*, good segmentations (in *blue*) correspond to high Q values and the bad segmentations (in *red*) to low Q values

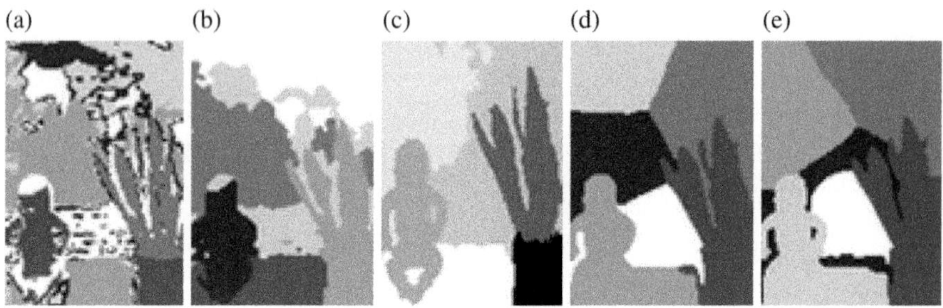

Fig. 6.13 Segmentation of the scene of Fig. 6.10 by the approach of [17] for different values of the λ parameter in (6.10): (**a**) $\lambda = 0.1$, (**b**) $\lambda = 0.5$, (**c**) $\lambda = 1$, (**d**) $\lambda = 4$, (**e**) $\lambda = 8$

Any suitable clustering technique can be used on the vector set of (6.10) to segment the scene. Although simpler clustering methods like K-means can be used to reduce computation time, more refined techniques lead to better results. The mean-shift algorithm [12], for instance, has been widely used for image segmentation and also gives good results with the vectors of (6.10). Another high performance method is *normalized cuts spectral clustering* [50], although its computational requirements are high. Note how the computational demand can be reduced by approximations like the Nyström method [23]. Figure 6.14 shows the segmentation results on a sample scene with various clustering-based approaches using color only, geometry only, and the fusion of the two clues. The figure exemplifies how more advanced clustering techniques, like mean-shift and normalized cuts spectral clustering, lead to more accurate segmentations.

One can consider a number of variations of this framework. For instance it is possible to use the 2D pixel coordinates $[u^i, v^i]^T$ together with the corresponding depth value z^i, but this requires a further normalization between vector $[u^i, v^i]^T$

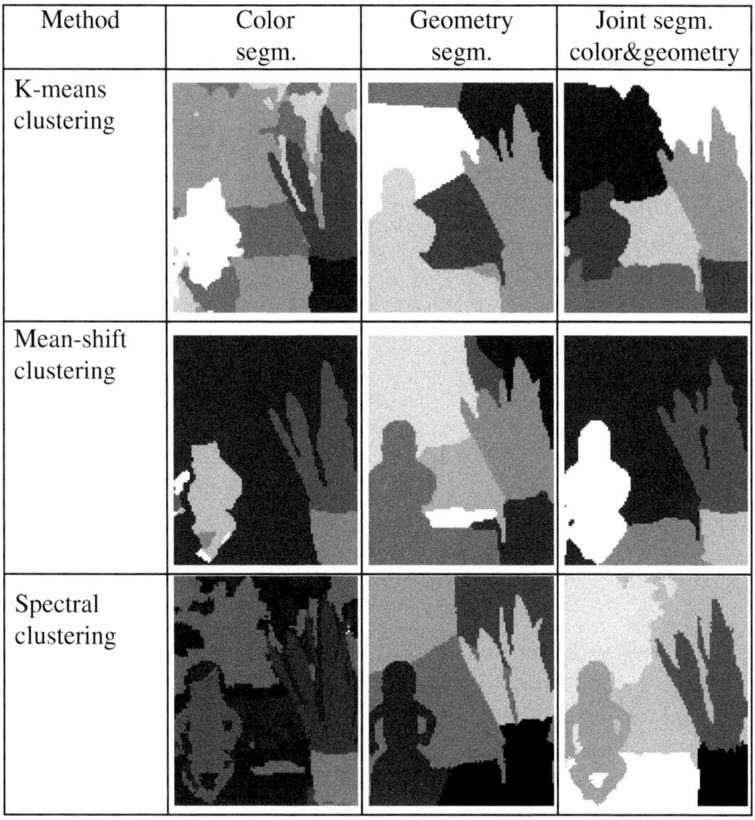

Method	Color segm.	Geometry segm.	Joint segm. color&geometry
K-means clustering			
Mean-shift clustering			
Spectral clustering			

Fig. 6.14 Segmentation of a scene of Fig. 6.11 with different clustering algorithms on color, geometry, and the fusion of color and geometry from [17]

and z^i since they lie in different spaces. Another possibility is to use derivatives or gradients of depth data [10, 59] as segmentation clues. Such methods are advantageous since they can discriminate between close surfaces with different orientations (notice that the gradient of depth data is equivalent to the normals to the surface). On the other hand, the computation of the normals requires one to use a window centered on the considered sample, with a smoothing effect close to edges that can create small segments aligned with edge regions. Moreover, the gradients used to compute the normals are sensitive to noise, and the high noise level of ToF depth cameras can become an issue. Finally, normal-based segmentation cannot distinguish between distant surfaces with the same orientation, although this problem can be easily solved by a 9-dimensional vector representation including spatial positions along with normal information.

When the acquisition setup contains multiple color cameras, e.g., stereo setups or setups considered for the fusion algorithms of Chap. 5, each pixel can be associated with multiple color values [17]. For example, in the case of a stereo setup it is possible to define a 9-dimensional representation with two CIELuv vectors and the 3D position at each location. It is finally worth noting that the previously introduced clustering-based segmentation approaches can be applied to any general multi-channel representation, namely, it is possible to consider not only depth or 3D information along with color, but also any other type of additional information provided by depth cameras. For instance, in the case of ToF depth cameras, the multidimensional representation of (6.10) can be extended with additional components related to the amplitude A_T or intensity B_T, which become additional components of the multidimensional vectors used in the clustering process. Recall that the values of the intensity A_T and amplitude B_T are inversely proportional to the distance of the acquired points from the camera, that is, their values are greater for closer points. This is an interesting property for segmentation purposes since it allows one to disambiguate objects of the same color placed at different distances, which is a critical issue for color-based segmentation techniques (e.g., the case of an object in front of a background of similar color).

6.2.3 *Single Frame Segmentation: Graph-Based Approaches*

Graph cut segmentation can also be adapted to joint color and depth segmentation [15]. These approaches work similarly to the graph-based methods considered for scene matting in Sect. 6.1. The scene is represented with a graph $G = \{V, E\}$ with nodes associated with the pixels and the edges representing the relationships between neighbor pixels (see Fig. 6.15). The segmentation problem can be expressed as the identification of the labeling that minimizes an energy functional of the form

$$U(S) = \sum_{p^i \in I} V_{data}(S(p^i)) + \sum_{p^i \in I} \sum_{p^n \in N(p^i)} V_{smooth}(S(p^i), S(p^n)) \tag{6.11}$$

where the function $S(p) \in \{0, N_s\}$ assigns each pixel to the corresponding segment (with N_s representing the total number of segments). Note how the labels are no longer binary as in the matting case. The data term V_{data} of (6.11) models the likelihood that a pixel p^i belongs to segment $S(p^i)$ and is typically the sum of two terms, one depending on color and one depending on depth, i.e.

$$V_{data}(S(p^i)) = V_{color}(S(p^i)) + \lambda V_{depth}(S(p^i)). \tag{6.12}$$

The color term V_{color} can simply be the distance between the pixel color and the mean segment color or it can take into account more complex models as in [15] where Gaussian Mixture Models (GMM) are used. The depth term V_{depth} can be modeled in a similar way. A remaining critical issue is the mutual relevance of color and depth, namely, setting the λ parameter. Some approaches, e.g. [15], combine the two measures by taking the maximum instead of the sum of the two components.

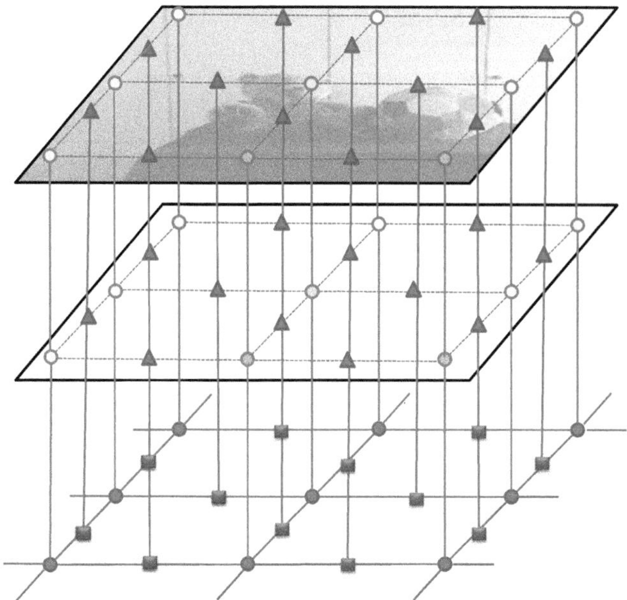

Fig. 6.15 Structure of the graph used in joint depth and color segmentation. The *circles* correspond to data terms computed over the color and depth values of the pixel, while *triangles* correspond to smoothness terms computed from the difference of the two close pixels along the *dotted lines*

The smoothness term V_{smooth} of (6.11) accounts for the color and depth difference between neighboring pixels and typically enforces a smoothness constraint, e.g., by employing an exponential function of the color and depth differences such as

$$V_{smooth}(S(p^i), S(p^n)) = \\ = \begin{cases} 0 & \text{if } S(p^i) - S(p^n) = 0 \\ e^{-\frac{\delta(I_C(p^i), I_C(p^n))^2}{2\sigma_c^2}} e^{-\frac{[Z(p^i) - Z(p^n)]^2}{2\sigma_z^2}} & \text{if } S(p^i) - S(p^n) \neq 0 \end{cases} \quad (6.13)$$

where σ_c and σ_z can be used to respectively tune the color and depth data relevance, $I_C(p^i)$ and $I_C(p^n)$ are the color values of the considered samples and $Z(p^i)$ and $Z(p^n)$ their depth values. The δ function can be any suitable measure of the color difference between the two samples. Other approaches, instead, directly use the difference without the exponential weighting or take the maximum of the two weights in place of the product [15]. The energy functional $U(S)$ of (6.11) can be minimized by efficient graph-cut optimization algorithms [6] as seen in Sect. 5.2.2.

6.2.4 Single Frame Segmentation Based on Geometric Clues

Another interesting option is segmenting the scene based on an explicit representation of the 3D information associated with depth data. There are a number of strategies based on this rationale.

The simplest idea is to compute the 3D point cloud from the depth map by using calibration information and then apply one of the many segmentation techniques developed for this kind of data. For a recent review of these techniques see [40]. Even though such approaches in principle are feasible with consumer depth camera data, better results can be obtained by ad hoc procedures.

One commonly exploited geometric clue is that man-made objects usually feature sets of planar surfaces, so one can recognize the different planes of the acquired point cloud and use this information for segmentation. The planes can be computed by clustering the 3D points on the basis of their surface normal information [31], similar to the clustering-based approaches of Sect. 6.2.2. As already pointed out, such an approach cannot discriminate different planes having the same orientation but different 3D positions; all the 3D points sharing the same orientation are considered part of a unique plane.

The method of [31] solves the problem with a two-phase procedure. In the first phase the 3D vectors are clustered according to their surface normal orientations by a region growing approach, although any other clustering technique (e.g., K-means, mean-shift [10] or normalised cuts) could be applied. The second phase separates different planes with the same orientation by grouping points in the same cluster with similar distances from the plane representing the cluster.

In a more refined approach [21], an initial over-segmentation by the method of [22] is followed by a linear plane fitting on the segmented super-pixels and finally a Markov Chain Monte Carlo is used to obtain the final solution. This method was extended in [54] to segment multiple images and depth maps framing the same still scene.

Although planar surfaces are very common, especially in indoor scenes, this family of approaches fails when scene objects are not described by perfectly planar surfaces. In order to overcome this limitation one can fit a parametric NURBS model [42] over the surface instead of the simpler planar one, and therefore handle scenes with more complex geometry. This work adopts a recursive tree-structured strategy, where at each step a segment is divided in two parts by a clustering-based scheme similar to [17]. The key idea is to fit a NURBS surface on each computed segment and compare the accuracy of the fitting before and after each segmentation (splitting) step. If the fitting accuracy increases (different fitting metrics are considered in [43]) the segment subdivision into two parts is accepted, and the two sub-segments are recursively considered for further splitting. Otherwise the recursion on the considered segment stops and the latter is considered part of the final solution. Figure 6.16 exemplifies the results of this procedure.

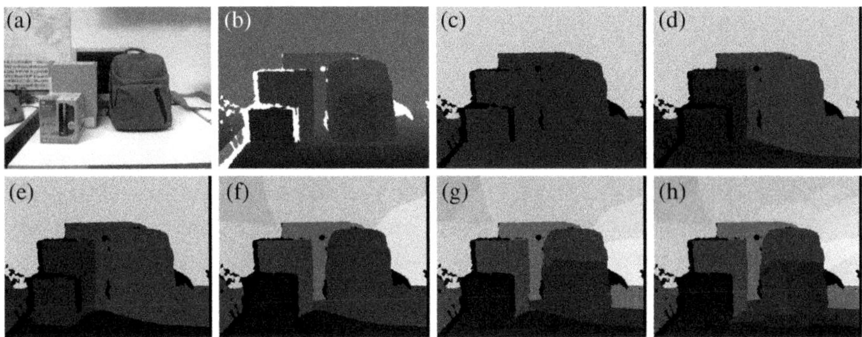

Fig. 6.16 Results of [42]: color image (**a**) and associated depth map (**b**); segmentation results after 1 (**c**), 2 (**d**), 3 (**e**), 10 (**f**) and 20 (**g**) iterations, and the final result (**h**)

6.2.5 Video Segmentation from Color and Depth Data

The segmentation of video sequences allows one to introduce temporal constraints in the segmentation framework. On one hand, temporal consistency constraints simplify the search for the final solution since segmentation results of previous frames can be used to initialize segmentation of the current frame. On the other hand, ensuring the consistency between segmentations of subsequent frames is not trivial. The optical flow, i.e., the apparent motion between the camera and objects in subsequent frames, can play a fundamental role for the solution of this problem since it allows one to propagate the segmentation across the various frames in order to enforce the temporal constraints.

In segmentation algorithms based on iterative optimization schemes, starting from an initial state and converging to the optimal solution (as is the case of several clustering algorithms), optical flow can be used to *warp* the previously segmented frame to the current frame in order to initialize the optimization procedure. This idea is adopted in [1], where super-paramagnetic clustering is used for segmentation. The Metropolis algorithm used by [1] is initialized at each step with the solution of the previous frame warped according to optical flow information.

In the approach of [29], also aiming to maintain temporal consistency between subsequent frames, each frame is segmented by a modified version of the graph cut approach of [22] that also accounts for depth information. Then, the optical flow is estimated and the consistency between subsequent frames is enforced by representing the segmentation as a bipartite graph matching problem over a graph built from the current and eight previous frames.

For dynamic scenes acquired by a moving camera, the segmentation problem can be solved along with the 3D reconstruction problem of Chap. 7. For instance, [27] tries to locate the moving objects while reconstructing the scene structure.

6.3 Semantic Segmentation from Color and Depth Data

The last problem considered in this chapter is semantic segmentation, whose goal is not only segmenting the scene into its various components, but also assigning each segmented region a label identifying the object category the segment belongs to. Semantic segmentation from color data is a well established and challenging problem with a rich literature. Depth information provides many useful clues like object size, their relative positions in 3D space, or surface normals that can improve the segmentation results on color data.

Figure 6.17 shows an example of semantic segmentation input and output from the NYUv2 dataset [52], commonly used for evaluating the performance of semantic segmentation algorithms on joint color and depth data. The algorithm input is a color view with the corresponding depth map (or a depth map stream and a color video in the dynamic case) and the output is a segmentation map associating each pixel to a segmented and labeled region, with labels indicating what class each region belongs to (the classes are typically associated with the segmented regions). For instance, in the NYUv2 dataset there are around 900 different labels corresponding to various object and surfaces types (e.g., window, table or wall). The labels are typically grouped in a smaller number of more general categories to make the problem more tractable.

Although semantic segmentation from color and depth data can be solved with different approaches, they typically share most blocks of the pipeline of Fig. 6.18. As shown by the figure, there are two fundamental steps. First, relevant features are extracted from the color and depth data. Then, they are fed to a classifier based on a suitable machine learning technique. Notice how this pipeline resembles that of gesture recognition which will be presented in Chap. 9. Table 6.2 reports various proposed solutions and the adopted techniques.

The first step of the pipeline of Fig. 6.18 is pre-processing the input data, like constructing a refined 3D representation of the scene from a single frame or multiple frames (e.g., [28]), computing the surface normals, or estimating of the main planes of the scene (e.g., [52]). In particular, surface orientations and the identification of planar surfaces supply useful clues for segmentation and classification algorithms.

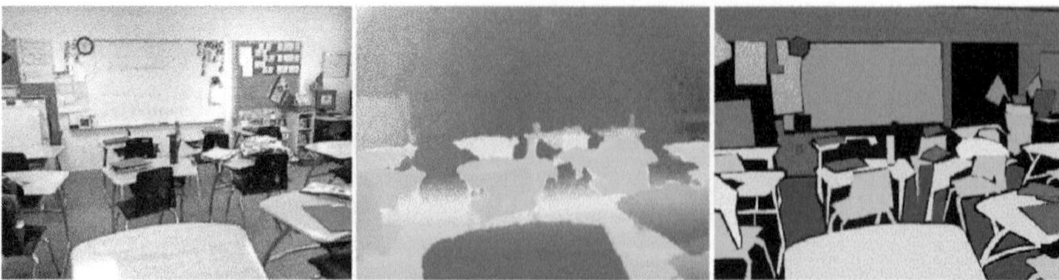

Fig. 6.17 Sample scene from the NYUv2 dataset [52]. The image shows the color view, the depth map and the semantic segmentation. In the latter, each different color corresponds to a different label (e.g., *green* to table, *light blue* to chair or *red* to wall)

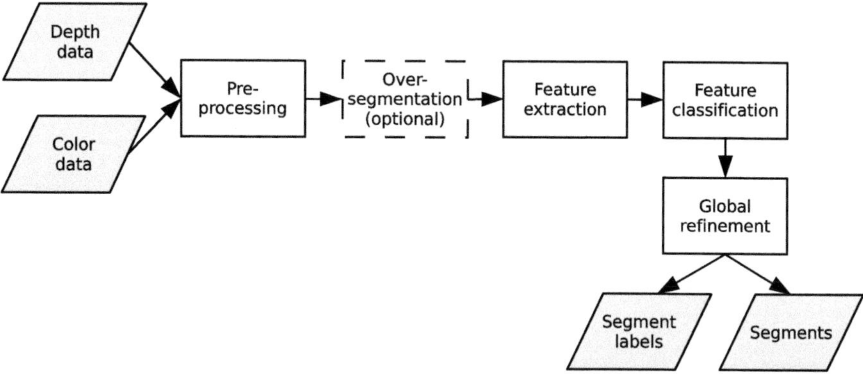

Fig. 6.18 General pipeline of semantic segmentation schemes. The figure shows blocks common to most approaches, but there exist also techniques based on different pipelines

A first key difference between the various approaches is the type of input given to the feature extractor and classification algorithms. While some methods directly extract the selected features on the image or depth map pixels, other approaches perform a prior image *oversegmentation* in several small regions (named *superpixels*) using any suitable segmentation technique from Sect. 6.2. Oversegmentation typically produces a much larger number of segments (superpixels) than the final output segmentation, which will be merged according to proper adjacency criteria after segment labeling. For instance, [52] starts with an oversegmentation computed from surface normals and color data then merges the segments by a hierarchical segmentation scheme.

The next step in the pipeline of Fig. 6.18 is feature extraction from the pre-processed pixels or superpixels generated by the oversegmentation. Several types of descriptors can be extracted both from color and depth data by one of the following characteristics.

- Differences between color and depth values in the selected sample neighborhood [28, 39, 48], e.g., simple color and depth differences between a few pixels within a square patch surrounding the considered sample [28]. Color differences are computed in the Lab space in order to give a perceptual significance to the Euclidean distance and the size of the patch is normalized with respect to the distance from the sensor in order to make the approach invariant to the object's distance from the camera.
- Information provided by the *surface normals* [39, 48, 52]. Note how the normal directions are an alternative representation of depth differences or gradients.
- 2D or 3D bounding boxes enclosing the segmented superpixels, the average superpixel depths or their color or depth histograms [52]. Another approach [39] instead models the directed angle between superpixel centers in the 2D image plane in order to get the relative object positions.

Table 6.2 Semantic segmentation methods using color and depth data

Method	Superpixel segm.	Graph. model	CRF	SIFT	LBP	Spin images	HOG (HOD)	3D bbox	Color diff	Depth diff	Classification			
											SVM	RDF	CNN	Other
Silberman and Fergus [51]	X		X	X		X			X	X				
Silberman et al. [52]		X		X			X	X	X	X				X
Ren et al. [48]		X		X	X				X		X			
Hermans et al. [28]		X	X						X	X		X		
Couprie et al. [13]	X												X	
Hoft et al. [30]							X						X	
Muller et al. [39]		X	X						X	X	X	X		
Banica et al. [4]	X	X		X	X	X							X	X
Gupta et al. [25]							X				X	X		

The table shows the adopted model, features, and machine learning techniques

- Scale-Invariant Feature Transform (SIFT) descriptors [37] extracted from color or depth data [4, 51, 52]. In particular, [51] computes SIFT descriptors on both color and depth data and also proposes an extended SIFT descriptor made by the juxtaposition of separate descriptors extracted from color and depth data.
- Local Binary Patterns (LBP) [41] descriptors on the input data [4, 48].
- Histogram of Oriented Gradients (HOG) [18] descriptors from the input data. Depth data can also be analyzed by the Histogram of Oriented Depths (HOD) [53], extending the HOG feature extraction approach to depth data. HOD have been used for semantic segmentation in [30]. Another variant of this approach concerns the use of histograms of surface normals as in [25, 52]. As reported in [25], the depth gradient magnitude depends on the distance from the camera, thus requiring proper normalization.
- Ad hoc descriptors for geometric information, such as the spin-images [32], originally introduced for 3D surface registration but widely used for several other applications due to their invariance to the viewpoint and to the object position and orientation in 3D space. By changing the region of support, spin-images can be used both as a local or global descriptor [4, 48, 51].

It is also possible to avoid feature extraction and directly use the image and depth data as input for the classification algorithm. This is typically the case for the approaches based on deep learning techniques such as [13].

There are various possibilities to classifying scene samples and producing the final segmentation. One possibility is to first independently classify each single pixel and then apply suitable global optimization schemes enforcing spatial consistency among the segments in order to build a set of labeled regions. For instance, the approaches of [25, 28, 39] use Randomized Decision Forests (RDF) as pixel classifiers. In particular, [28] classifies each single pixel independently and uses simple color and depth differences as features in the considered sample neighborhood, while [39] works on a superpixel basis and extracts features describing segment properties. Support Vector Machines (SVM) can also be used for this task [25, 39, 48]. Note how, since SVM is a binary classifier, multi-class classification has to be performed with the one-vs-one (i.e., each class is tested against each other class) or the one-vs-all (i.e., each class is tested against all the others) approaches.

The independent classification of each single pixel or superpixel typically produces a very noisy label assignment which is typically fixed by a global optimization step. Among the various methods, Conditional Random Fields (CRF) are commonly used. CRFs model both the hypothesis on each single pixel or superpixel of the initial segmentation and the bounds between close pixels. CRFs can either be applied over the output of a previous label assignment $l_1(p^i)$ or to directly compute the final label assignment $l_2(p^i)$. For instance, [28] uses CRF to refine the initial classification performed by RDFs, while in [39, 51] CRFs are directly used as classifier. The energy function is based on the previously presented standard MRF formulation:

$$
\begin{aligned}
U(l_2) &= \sum_{p^i \in I} V_d(p^i, l_2(p^i)) + \\
&+ \sum_{p^i \in I} \sum_{p^n \in N(p^i)} V_s(p^i, p^j, l_2(p^i), l_2(p^j)) \\
U(l_1, l_2) &= \sum_{p^i \in I} V_d(p^i, l_1(p^i), l_2(p^i)) + \\
&+ \sum_{p^i \in I} \sum_{p^n \in N(p^i)} V_s(p^i, p^j, l_1(p^i), l_1(p^j), l_2(p^i), l_2(p^j)).
\end{aligned}
\tag{6.14}
$$

The first equation of (6.14) is used in case there are no initial assignments, while the second also accounts for an initial labeling $l_1(p^i)$ given as input to the CRF model. The corresponding graphical model (Fig. 6.19) has a node for each single pixel (superpixel) to be labeled, enumerated by the set I, and an edge between each couple of adjacent pixels (superpixels) modeling their relationship. The unary energy function V_d is either learned from the features associated with each pixel (superpixel) or directly from the output of a previous segmentation classifying each pixel separately. The pairwise energy function V_s represents the relationship

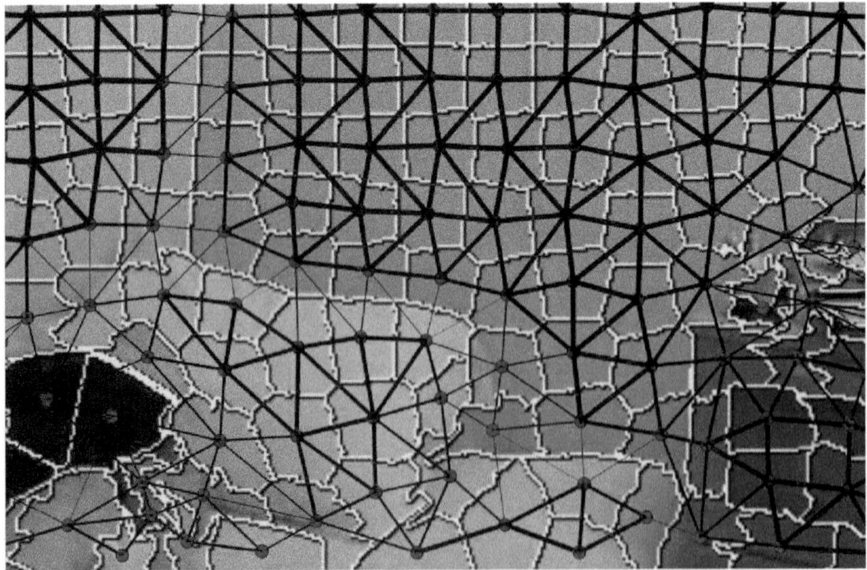

Fig. 6.19 Example of a CRF applied on a superpixel segmentation. The image shows a detail of a bedroom scene with the superpixel segmentation overlaid, the lines connect adjacent superpixels and have a color proportional to the strength of the link between the superpixels (the employed criterion in this case is the similarity between surface normals). (*Courtesy of the authors of [39]*)

between the label assignment of close pixels and is typically a function of the color, depth or normals difference between two pixels (superpixels). In [39], where the CRF is applied over the superpixel segmentation, the potentials V_s are computed from color, depth and normal differences. The CRF can be trained in various ways, e.g., by a structural support vector machine (SSVM) [39].

Another possibility to solve the classification step and enforce the global consistency at the same time is to use deep learning techniques, in particular Convolutional Neural Networks (CNN) as in [13, 30]. Finally, in order to summarize the various approaches presented in this section, Table 6.2 presents an overview of the solutions in the literature.

6.4 Conclusions and Further Reading

Image segmentation is a classical image processing and computer vision problem treated by a vast literature. For a review of the early image segmentation methods see for instance [44], while for an overview of current advancements see [58]. Among the wide variety of proposed techniques, clustering based approaches (e.g., [12]) and methods based on graph representations (e.g., [50] and [22]) have been widely used with excellent resulting performance. Clustering and graph-based approaches can be easily extended to exploit depth data or additional information like amplitude and intensity coming from ToF depth cameras.

The data acquired by depth cameras can be represented as colored point clouds, therefore current methods developed in the 3D scanning and photogrammetry fields for colored point cloud segmentation, e.g., [46] and [49], can also be adapted to this task.

The approaches presented in this chapter can be applied to depth data coming from different sources, e.g., stereo vision systems or structured light and laser scanners. Stereo vision systems are a widely used alternative to depth cameras, especially for outdoor environments and long range acquisitions. Some recent works, e.g., [16, 33] and [34] use depth data from stereo in order to improve segmentation results or try to jointly solve the stereo depth estimation and the segmentation problems. It is possible to replace the depth data estimated by stereo vision with depth data obtained by ToF or structured light depth cameras and use most of the ideas and techniques presented in those works for joint color and depth segmentation. Finally, prior knowledge of the objects that are going to be segmented can be used in order to develop ad hoc procedures for specific tasks, e.g., in the case of human body or hand segmentation.

References

1. A. Abramov, K. Pauwels, J. Papon, F. Worgotter, B. Dellen, Depth-supported real-time video segmentation with the kinect, in *Proceedings of IEEE Workshop on Applications of Computer Vision* (2012), pp. 457–464
2. P. Arbelaez, M. Maire, C. Fowlkes, J. Malik, Contour detection and hierarchical image segmentation. IEEE Trans. Pattern Anal. Mach. Intell. **33**(5), 898–916 (2011)
3. O. Arif, W. Daley, P.A. Vela, J. Teizer, J. Stewart, Visual tracking and segmentation using time-of-flight sensor, in *Proceedings of IEEE International Conference on Image Processing* (2010), pp. 2241–2244
4. D. Banica, C. Sminchisescu, Second-order constrained parametric proposals and sequential search-based structured prediction for semantic segmentation in RGB-D images, in *Proceedings of IEEE Conference on Computer Vision and Pattern Recognition* (2015), pp. 3517–3526
5. A. Bleiweiss, M. Werman, Fusing time-of-flight depth and color for real-time segmentation and tracking, in *Proceedings of DAGM Workshop, Dynamic 3D Imaging* (2009), pp. 58–69
6. Y. Boykov, O. Veksler, R. Zabih, Fast approximate energy minimization via graph cuts. IEEE Trans. Pattern Anal. Mach. Intell. **23**, 1222–1239 (2001)
7. F. Calderero, F. Marques, Hierarchical fusion of color and depth information at partition level by cooperative region merging, in *Proceedings of IEEE International Conference on Acoustics, Speech and Signal Processing* (2009), pp. 973–976
8. M. Camplani, M. Salgado, Background foreground segmentation with RGB-D kinect data: An efficient combination of classifiers. J. Vis. Commun. Image Represent. **25**(1), 122–136 (2014). Visual Understanding and Applications with RGB-D Cameras
9. M. Camplani, C.R. Del Blanco, L. Salgado, F. Jaureguizar, N. García, Advanced background modeling with RGB-D sensors through classifiers combination and inter-frame foreground prediction. Mach. Vis. Appl. **25**(5), 1197–1210 (2014)
10. Y. Cheng, Mean shift, mode seeking, and clustering, IEEE Trans. Pattern Anal. Mach. Intell. **17**(8), 790–799 (1995)
11. Y.Y. Chuang, B. Curless, D.H. Salesin, R. Szeliski, A bayesian approach to digital matting, in *Proceedings of IEEE Conference on Computer Vision and Pattern Recognition* (2001), p. 264
12. D. Comaniciu, P. Meer, Mean shift: a robust approach toward feature space analysis. IEEE Trans. Pattern Anal. Mach. Intell. **24**, 603–619 (2002)
13. C. Couprie, C. Farabet, L. Najman, Y. LeCun, Convolutional Nets and Watershed Cuts for Real-Time Semantic Labeling of RGBD Videos. J. Mach. Learn. Res. **15**(Oct), 3489–3511 (2014)
14. R. Crabb, C. Tracey, A. Puranik, J. Davis, Real-time foreground segmentation via range and color imaging, in *Proceedings of IEEE Conference on Computer Vision and Pattern Recognition Workshops* (2008), pp. 1–5
15. M.J. Dahan, N. Chen, A. Shamir, D. Cohen-Or, Combining color and depth for enhanced image segmentation and retargeting. Vis. Comput. **28**(12), 1181–1193 (2012)
16. C. Dal Mutto, P. Zanuttigh, G.M. Cortelazzo, Scene segmentation assisted by stereo vision, in *Proceedings of International Conference on 3D Imaging, Modeling, Processing, Visualization and Transmission 2011* (Hangzhou, 2011)
17. C. Dal Mutto, P. Zanuttigh, G.M. Cortelazzo, Fusion of geometry and color information for scene segmentation. Proceedings of IEEE J. Sel. Top. Signal Process. **6**(5), 505–521 (2012)
18. N. Dalal, B. Triggs, Histograms of oriented gradients for human detection, in *Proceedings of IEEE Conference on Computer Vision and Pattern Recognition* (2005), pp. 886–893
19. C.R. Del-Blanco, T. Mantecón, M. Camplani, F. Jaureguizar, L. Salgado, N. García, Foreground segmentation in depth imagery using depth and spatial dynamic models for video surveillance applications. Sensors **14**(2), 1961–1987 (2014)
20. B. Dellen, G. Alenyá, S. Foix, C. Torras, Segmenting color images into surface patches by exploiting sparse depth data, in *Proceedings of Winter Vision Meeting: Workshop on Applications of Computer Vision* (2011), pp. 591–598

21. C. Erdogan, M. Paluri, F. Dellaert, Planar segmentation of RGBD images using fast linear fitting and markov chain monte carlo, in *Proceedings of Conference on Computer and Robot Vision* (Toronto, 2012), pp. 32–39
22. P.F. Felzenszwalb, D.P. Huttenlocher, Efficient graph-based image segmentation. Int. J. Comput. Vis. **59** (2004)
23. C. Fowlkes, S. Belongie, Fan Chung, J. Malik, Spectral grouping using the nystrom method. IEEE Trans. Pattern Anal. Mach. Intell. **26**(2), 214–225 (2004)
24. J. Gallego, M. Pardàs, Region based foreground segmentation combining color and depth sensors via logarithmic opinion pool decision. J. Vis. Commun. Image Represent. **25**(1), 184–194 (2014)
25. S. Gupta, P. Arbeláez, R. Girshick, J. Malik. Indoor scene understanding with RGB-D images: Bottom-up segmentation, object detection and semantic segmentation. Int. J. Comput. Vis. **112**(2), 133–149 (2015)
26. M. Harville, G. Gordon, J. Woodfill, Foreground segmentation using adaptive mixture models in color and depth, in *Proceedings of IEEE Workshop on Detection and Recognition of Events in Video* (2001)
27. E. Herbst, P. Henry, D. Fox, Toward online 3-D object segmentation and mapping, in *Proceedings of IEEE International Conference on Robotics and Automation* (2014)
28. A. Hermans, G. Floros, B. Leibe, Dense 3d semantic mapping of indoor scenes from RGB-D images, in *Proceedings of 2014 IEEE International Conference on Robotics and Automation* (IEEE, Hong Kong, 2014), pp. 2631–2638
29. S. Hickson, S. Birchfield, I. Essa, H. Christensen, Efficient hierarchical graph-based segmentation of RGBD videos, in *Proceedings of IEEE Conference on Computer Vision and Pattern Recognition* (Columbus, 2014)
30. N. Höft, H. Schulz, S. Behnke, Fast semantic segmentation of RGB-D scenes with gpu-accelerated deep neural networks, in *Proceedings of Conference on Advances in Artificial Intelligence* (Springer, Cham, 2014), pp. 80–85
31. D. Holz, S. Holzer, R. Bogdan Rusu, S. Behnke, Real-time plane segmentation using RGB-D cameras, in *Proceedings of RoboCup International Symposium*, Istanbul (2011)
32. A.E. Johnson, M. Hebert, Using spin images for efficient object recognition in cluttered 3d scenes. IEEE Trans. Pattern Anal. Machine Intell. **21**(5), 433–449 (1999)
33. V. Kolmogorov, A. Criminisi, A. Blake, G. Cross, C. Rother, Bi-layer segmentation of binocular stereo video, in *Proceedings of IEEE Conference on Computer Vision and Pattern Recognition* (2005), p. 1186
34. L. Ladicky, P. Sturgess, C. Russell, S. Sengupta, Y. Bastanlar, W. Clocksin, P. Torr, Joint optimisation for object class segmentation and dense stereo reconstruction. International Journal of Computer Vision, Springer US **100**(2), 122–133 (2012)
35. B. Langmann, S.E. Ghobadi, K. Hartmann, O. Loffeld, Multi-modal background subtraction using gaussian mixture models, in *Proceedings of ISPRS Technical Commission III Symposium on Photogrammetry Computer Vision and Image Analysis* (2010), pp. 61–66
36. J. Leens, S. Pierard, O. Barnich, M. Van Droogenbroeck, J.M. Wagner, Combining color, depth, and motion for video segmentation, in *Proceedings of Conference on Computer Vision Systems* (2009)
37. D.G. Lowe, Distinctive image features from scale-invariant keypoints. Int. J. Comput. Vis. **60**(2), 91–110 (2004)
38. T. Lu, S. Li, Image matting with color and depth information, in *Proceedings of 2012 International Conference on Pattern Recognition* (2012), pp. 3787–3790
39. A.C. Muller, S. Behnke, Learning depth-sensitive conditional random fields for semantic segmentation of RGB-D images, in *Proceedings of 2014 IEEE International Conference on Robotics and Automation* (2014), pp. 6232–6237
40. A. Nguyen, B. Le, 3d point cloud segmentation: a survey, in *Proceedings of 2013 IEEE Conference on Robotics, Automation and Mechatronics* (2013), pp. 225–230
41. T. Ojala, M. Pietikainen, T. Maenpaa, Multiresolution gray-scale and rotation invariant texture classification with local binary patterns. IEEE Trans. Pattern Anal. Mach. Intell. **24**(7), 971–987 (2002)

42. G. Pagnutti, P. Zanuttigh, Scene segmentation from depth and color data driven by surface fitting, in *Proceedings of International Conference on Image Processing* (2014)
43. G. Pagnutti, P. Zanuttigh, Scene segmentation based on nurbs surface fitting metrics, in *Proceedings of Smart Tools and Apps in computer Graphics* (2015)
44. N.R. Pal, S.K. Pal, A review on image segmentation techniques. Pattern Recogn. **26**(9), 1277–1294 (1993)
45. T. Porter, T. Duff, Compositing digital images, in *Proceedings of ACM SIGGRAPH* (New York, 1984), pp. 253–259
46. T. Rabbani, F. Van Den Heuvel, G. Vosselmann, Segmentation of point clouds using smoothness constraint, in *Proceedings of International Archives of Photogrammetry, Remote Sensing and Spatial Information Sciences*, vol. 36 (Dresden, 2006)
47. Recommendations on uniform color spaces, color difference equations, psychometric color terms. Supplement No. 2 to CIE publication No. 15 (E.-1.3.1) 1971/(TC-1.3.) (1978)
48. X. Ren, L. Bo, D. Fox, RGB-(D) scene labeling: Features and algorithms, in *Proceedings of IEEE Conference on Computer Vision and Pattern Recognition* (2012), pp. 2759–2766
49. R. Schnabel, R. Wahl, R. Klein, Efficient ransac for point-cloud shape detection. Comput. Graph. Forum **26**(2), 214–226 (2007)
50. J. Shi, J. Malik, Normalized cuts and image segmentation. IEEE Trans. Pattern Anal. Mach. Intell. **22**, 888–905 (2000)
51. N. Silberman, R. Fergus, Indoor scene segmentation using a structured light sensor, in *Proceedings of IEEE International Conference on Computer Vision Workshops* (Barcelona, 2011), pp. 601–608
52. N. Silberman, D. Hoiem, O. Kohli, R. Fergus, Indoor segmentation and support inference from RGBD images, in *Proceedings of European Conference on Computer Vision* (2012)
53. L. Spinello, K.O. Arras, People detection in RGB-D data, in *Proceedings of IEEE/RSJ International Conference on Intelligent Robots and Systems* (San Francisco, 2011), pp. 3838–3843
54. N. Srinivasan, F. Dellaert, A rao-blackwellized mcmc algorithm for recovering piecewise planar 3d model from multiple view RGBD images, in *Proceedings of International Conference on Image Processing* (2014)
55. C. Stauffer, W.E.L. Grimson, Learning patterns of activity using real-time tracking. IEEE Trans. Pattern Anal. Mach. Intell. **22**(8), 747–757 (2000)
56. A. Störmer, M. Hofmann, G. Rigoll, Depth gradient based segmentation of overlapping foreground objects in range images, in *Proceedings of 2010 Conference on Information Fusion* (Edinburgh, 2010), pp. 1–4
57. J. Sun, J. Jia, C. Tang, H. Shum, Poisson matting. ACM Trans. Graph. **23**, 315–321 (2004)
58. R. Szeliski, *Computer Vision: Algorithms and Applications* (Springer, New York, 2010)
59. M. Wallenberg, M. Felsberg, P. Forssen, B. Dellen, Channel coding for joint colour and depth segmentation. in *Proceedings of Annual Symposium of the German Association for Pattern Recognition*, vol. 6835 (Springer, Heidelberg, 2011), pp. 306–315
60. L. Wang, C. Zhang, R. Yang, C. Zhang, Tofcut: Towards robust real-time foreground extraction using time-of-flight camera, in *Proceedings of 3D Data Processing, Visualization and Transmission* (Paris, 2010)
61. L. Wang, M. Gong, C. Zhang, R. Yang, C. Zhang, Y.-H. Yang, Automatic real-time video matting using time-of-flight camera and multichannel poisson equations. Int. J. Comput. Vis. **97**, 1–18 (2011)
62. O. Wang, J. Finger, Q. Yang, J. Davis, R. Yang, Automatic natural video matting with depth, in *Proceedings of Pacific Conference on Computer Graphics and Applications* (2007), pp. 469–472

Chapter 7
3D Scene Reconstruction from Depth Camera Data

Among the applications of consumer depth cameras, 3D reconstruction is one of the most important, and it presents several key challenges. This task, also called *free-form surface reconstruction* or *3D modeling*, is well established in several fields, e.g., the automotive industry, movies, special surgery and industrial mechanics, and is traditionally solved with lasers and structured light scanners [6, 33, 44]. These devices are very expensive though, and the modeling tools used to build 3D models from their data require intensive manual work from skilled users.

In these applications the acquired depth maps, also called *range data* or *3D views*, are turned into point clouds, generic 3D point sets not characterized by a lattice domain like depth maps. The 3D surface model is first obtained as a point cloud, but is typically delivered as a mesh of polygons, typically triangles, or as a parametric surface in some cases. The standard procedure followed to reconstruct an object or scene from multiple 3D views is called the *3D modeling pipeline* [5], and it encompasses the following steps:

1. pairwise registration (i.e., alignment) of all the 3D views;
2. global alignment refinement of all the 3D views;
3. fusion of the 3D data describing the surface as a cloud of points into a single 3D surface;
4. post-processing, such as hole filling, surface simplification, or inclusion of color information.

The first step is usually split in two phases: a coarse alignment and a subsequent fine registration. Traditionally, the rough registration phase was performed by manually selecting corresponding points in the two 3D views, then finding the roto-translation $(\mathbf{R}, \mathbf{t})$ aligning them with respect to the coordinate system of one of the 3D views using the Horn algorithm [19] presented in Sect. 4.4.1. In some recent approaches, this has been replaced by either the use of 3D features [20, 22, 48], or assuming that subsequent views in a "video" stream of depth maps are very close to each other. Note how in the latter case, the coarse alignment step becomes unnecessary.

P. Zanuttigh et al., *Time-of-Flight and Structured Light Depth Cameras*,
DOI 10.1007/978-3-319-30973-6_7

Fine registration is generally performed by the Iterative Closest Points (ICP) algorithm, independently introduced by [7] and [10], or its variants. An extensive presentation of ICP in all its components and possible variations of the base algorithm is given in [37].

The second step, global registration, limits the effects of error propagation due to a sequence of pairwise registration errors and redistributes the registration error among the single pairwise alignments. Earlier solutions for this problem are given in [3, 4, 14, 31, 32]. In particular, the algorithm of [4], also implemented in commercial products, generally provides good quality global alignments but only after several iterations. As the total number of 3D views increases, the total iteration number and the computation time and memory allocation per single iteration increase accordingly. The global registration technique of [32] reaches the same registration quality of [4] with improved memory allocation and performance, and it was successfully used for the global registration of large 3D models of statues [25].

Since the 3D point cloud obtained from the registration of a sequence of 3D views has an extremely redundant and non uniform 3D point distribution, it is customary to turn it into a surface description based on low-order polygons, such as triangles, which can be converted to NURBS-based mathematical descriptions of the 3D surface for mechanical applications oriented to CAD models. Although various approaches directly adopting surface elements have been proposed [45], standard solutions to this issue such as [13] and [46] instead use volumetric approaches based on marching cubes [26]. We recall that volumetric methods ensure the generation of a manifold surface, allowing one to control the surface sampling density and to conveniently handle surface averaging. On the other hand, since they deal with volumetric entities, time and storage requirements are much higher than those of methods using surface elements [45]. Moreover, although the topology delivered by volumetric methods is consistent, it may not necessarily be the desired one, since volumetric entities are intrinsically unsuited for modeling thin surfaces.

Low cost systems capable of automatically building a 3D model without any manual intervention represent one of the major and longest sought after targets of research in 3D modeling. The need for both inexpensive acquisition instruments and automatic 3D modeling tools received special attention in cultural heritage applications like digitally recording and reconstructing archaeological artifacts [1, 40].

Compared to laser scanners, although consumer depth cameras have a lower accuracy, they are less expensive and can acquire a large number of views at video rate. The quantity of close views collected in a very short time by consumer depth cameras often overcomes the need of computationally demanding post-processing techniques, like *hole-filling* and *mesh-completion*, which are mandatory when merging a few high quality views acquired by laser scanners.

Even though all the standard 3D modeling pipeline methods can still be directly applied to consumer depth camera data, their peculiar characteristics suggest a substantial rethinking of the standard algorithms. Indeed 3D reconstruction from consumer depth camera data involves the registration and fusion of a large number of noisy views, unlike the small number of accurate views needed for 3D reconstruction by laser scanners.

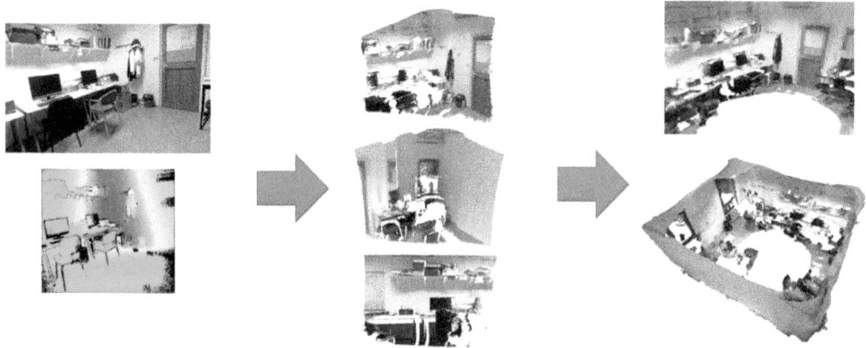

Fig. 7.1 3D scene reconstruction from Kinect™ v2 data: the color and depth data streams from the Kinect™ v2 are used to produce a set of 3D views, incrementally combined together to give a full 3D reconstruction of the scene

Methods for 3D reconstruction based on consumer depth cameras proliferated in the last few years, leading to systems that are inexpensive, robust, and capable of producing fairly accurate 3D models, such as the ones presented in Fig. 7.1.

7.1 3D Reconstruction from Depth Camera Data

The standard 3D reconstruction approach by consumer depth cameras consists of moving them around the object or scene to be acquired. Each frame of the acquired depth and color video corresponds to a textured 3D view, equivalent to a point cloud with color information. Several textured 3D views must be acquired and combined to obtain a first 3D model of the scene, to be refined by iteratively adding the missing parts until the full geometry of the scene is reconstructed. As with the standard 3D modeling case, the two fundamental issues are how to compute the relative position between the newly acquired point cloud and the previously acquired ones (i.e., how to register the 3D views) and how to integrate the registered 3D point clouds into a single representation without redundant information (i.e., how to fuse the 3D views).

Most algorithms for 3D scene reconstruction from depth camera data follow the basic pipeline shown in Fig. 7.2, accounting for the noisy nature of the input data, the spatial proximity of the views, and the eventual presence of color information.

The first block of the pipeline computes 3D points from the acquired depth maps, unless the input data are already provided as 3D point clouds, using the camera's calibration information. This step is often preceded by pre-processing the depth maps or the single view point clouds to reduce noise and remove outliers. It is then necessary to refer all the point clouds to a common coordinate reference system, which, as already stated, is a task typically solved in two stages. The first step, the coarse alignment, is computed by selecting a set of corresponding points (the correspondences can be found using various feature extraction algorithms)

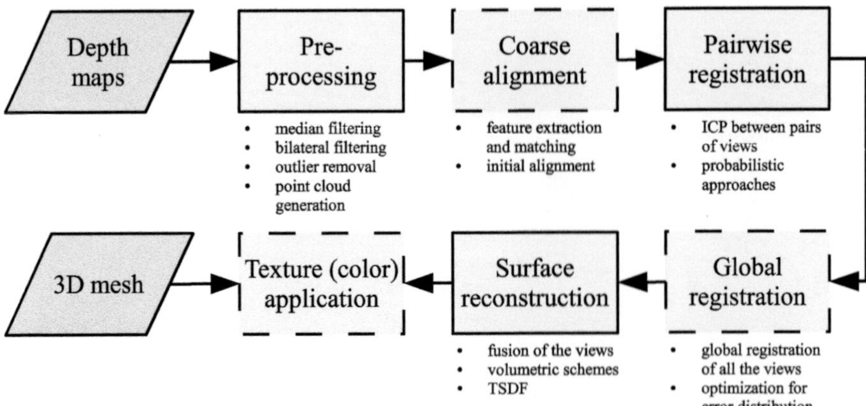

Fig. 7.2 3D reconstruction pipeline for depth camera data. *Dashed boxes* correspond to optional stages not performed by all the reconstruction procedures

and by obtaining the corresponding roto-translation aligning them (e.g., by the Horn algorithm). The second step, refining the previous alignment, is typically performed with the ICP algorithm or its variations [37]. For 3D views corresponding to spatially distant acquisition positions, as in the case of the standard 3D modeling pipeline, the coarse alignment is necessary in order to avoid local minima. When consumer depth cameras are employed, however, especially if the frame rate is high, the subsequent 3D views to be aligned are similar since they correspond to close acquisition positions, and the coarse alignment step may be avoided. With consumer depth cameras, due to the high number of noisy 3D views to be aligned, the global registration step assumes greater relevance compared to traditional modeling situations, in order to obtain more accurate results.

The fusion of different point clouds into a single surface involves two main tasks, i.e., the removal of redundant points and surface reconstruction from the point cloud. Such tasks can be executed either sequentially or simultaneously, e.g., by volumetric approaches commonly employed in this step. Finally, for a textured 3D reconstruction, color information associated with the different views must be fused into a single color representation.

Summarizing, typical 3D registration methods encompass the following main steps:

- pre-processing the acquired depth maps, to reduce the noise and remove outliers;
- rough alignment, typically computed from feature extraction and matching schemes to get a better starting point for the fine registration algorithm (not mandatory in all the approaches);
- pairwise registration, where each view is registered with the previous one (or with some subset of previously acquired ones), typically using the ICP algorithm or its variations;

- global optimization to avoid the accumulation of the pairwise registration error (optional in some approaches);
- fusion of the various views into a single point cloud, from which a mesh structure is built;
- combination of the color information from the various views to be added to the reconstructed model.

7.2 Pre-processing of the Views

The pre-processing step has two main targets: noise reduction and outlier removal (e.g., flying pixels or acquisition artifacts). A simple low-pass filter applied to the acquired depth maps reduces measurement noise, but may also introduce critical artifacts near edges or geometry discontinuities, typically creating virtual 3D points between foreground objects and the background. As reported in Chap. 3, median filters are better suited to the noise characteristics of ToF sensors (e.g., Kinect™ v2, MESA or PMD depth cameras) and effectively reduce their noise level. For this reason, simple 3×3 or 5×5 median filters are often directly implemented in the ToF camera firmware or driver to return already pre-filtered depth maps.

Many state of the art approaches prefer instead to use a bilateral filter [43] for noise removal, since it reduces noise while simultaneously preserving edges, thus avoiding the creation of flying pixels due to the low pass smoothing. The standard bilateral filter already returns good results, though some of its variations can be more effective. Let us first notice that the weights used by the bilateral filter depend on both the spatial distance and range value difference between the samples. This filter can be either applied on the depth data alone, or used with an associated color image for range dependent weights, to exploit high resolution color data for a precise description of edges, as already discussed in Sect. 5.2.1. Also notice that as stated in Chap. 2, structured light sensors have noise levels and accuracy strongly dependent on their distance from the target (e.g., the quantization error increases quadratically with distance). For this reason, an extended version of the bilateral filter with parameters varying according to the depth values at each location has been proposed in [9]. Figure 7.3 shows an example of this approach applied to a Kinect™ v1 frame.

Note that the filtering step can be performed either in the depth map domain, considering depth maps like images, or in the 3D point cloud domain. The first approach typically leads to faster and simpler implementations and allows one to use the implicit point connectivity given by the depth map sampling grid. The second method is typically slower and more difficult to implement, but better suited to the 3D nature of the input data. It also leverages the effective 3D distances between the samples with better results.

The second issue is the presence of completely misplaced pixels due to edge artifacts (i.e., mixed pixels due to the capture of objects at different distances), saturation, multipath, and many other problems described in Chaps. 2 and 3.

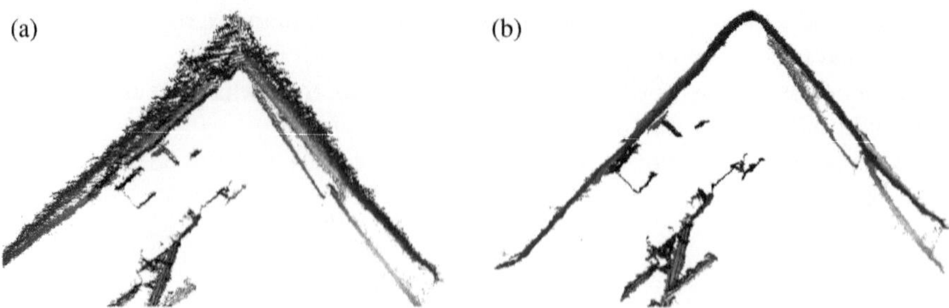

Fig. 7.3 Example of bilateral filtering from [9]. View of a room corner from the *top*: (**a**) before filtering; (**b**) after modified bilateral filtering

Although median filtering may reduce these artifacts, better results are obtained by ad hoc filtering schemes. A common solution is *the Statistical Outlier Removal* filter [38], which assumes that the samples are modeled by Gaussian distributions. It computes the mean μ and the standard deviation σ of the nearest neighbor distances and removes samples with a distance from their neighbors larger than $\mu \pm \alpha\sigma$. The parameter α depends on the size of the considered neighborhood, e.g., a neighborhood of 30 samples with $\alpha = 1$, reported in [38], leads to the removal of about 1 % of the samples of the considered datasets.

Finally, since ToF cameras are characterized by low resolution, it is also possible to apply a super-resolution scheme in the pre-processing step, e.g., any of the methods presented in Chap. 5. In [12] the set of acquired frames is first divided into blocks of close views and then a high resolution depth map is computed from each block of low resolution views using a variation of the approach presented in [39]. The high resolution depth maps are then used in the 3D reconstruction algorithm.

7.3 Rough Pairwise Registration

As previously stated, the first registration step in standard 3D reconstruction pipelines for laser scanners or depth camera data is a rough pairwise alignment computed from a set of correspondences. The extraction of corresponding 3D points is a delicate task which can either be performed manually or automatically using suitable 3D feature extraction and description algorithms [20, 22, 48]. Once a set of at least three corresponding 3D points on the two views has been determined, the corresponding roto-translation between them can be computed by solving the absolute orientation problem, using Horn's algorithm or by solving the Orthogonal Procrustes Problem. Even though three points are sufficient to solve the absolute orientation problem since there are six degrees of freedom, a larger number of points is often used to obtain more stable results. This initial step, motivated by needing a good starting point for the ICP algorithm, is not present in many 3D reconstruction approaches for consumer depth cameras [9, 21], since the spatial proximity of the acquired views ensures good starting points on its own for the ICP algorithm, due to the high acquisition frame rate. However, this is a critical assumption not enforced by all the approaches. Indeed, the proximity between the views depends on the acquisition frame rate and the speed at which the depth camera is moved. In practice, the full frame rate of the depth camera cannot be used for real time operation due to computation time limitations. For this reason, some 3D registration schemes still enforce a rough alignment based on the matching of feature points extracted from color or depth data.

7.4 Fine Pairwise Registration

Once an initial estimate of the relative positions between two subsequent frames has been determined, either by a coarse alignment as described in the previous section, or because the two views are very close each other, the fine registration between the two views is typically obtained by the ICP algorithm [7] or by some of its variations (see Fig. 7.4).

The interesting characteristic of the ICP algorithm, outlined in Algorithm 1, is its ability to compute the roto-translation between a pair of 3D views without knowing the point correspondences between them. The basic idea behind the ICP algorithm is simple: first, each point in the source view S is assumed to correspond to the closest point on the target view T. Second, the orientation problem is solved using this assumption. Finally, a new set of correspondences is computed using the roto-translated views. The procedure is iterated until it converges to a (local) minimum [7].

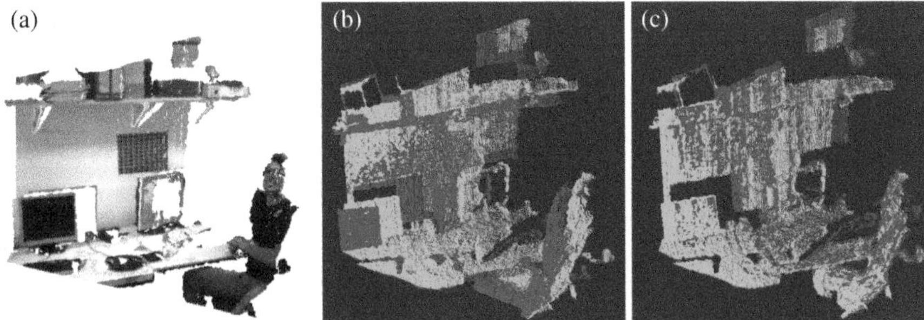

Fig. 7.4 Example of 2 view registration with the ICP algorithm: (**a**) colored representation of the source view S; (**b**) source view S and target view T before applying the ICP algorithm; (**c**) S and T after the application of the algorithm. The reference view S corresponds to the *green* points and the target view T to the *purple* ones

Algorithm 1 ICP registration procedure

1: Select a subset S' of the 3D points in the source point cloud S that will be used for the registration
2: For each 3D point in S' find the closest point in the target point cloud T
3: Estimate the roto-translation between the two views using a mean squared error cost function minimizing the distance between matched points in the two views computed in the previous step.
4: Roto-translate S using the obtained transformation
5: Go back to step 2 and iterate until convergence.

Some critical issues of the ICP algorithm are pointed out next:

- the time required for the computation of the distance between each pair of points in the two 3D views is quadratic with respect to the points cardinality. For this reason, efficient data structures like KD-trees and optimized implementations of ICP are often adopted. It is also possible to replace the search for the closest point with approximations, e.g., by constructing the surface corresponding to one of the clouds and projecting the other view's points on the surface. However, the most common way of dealing with quadratic time with respect to point cardinality is using a reduced set of points. Selecting the points near corners, edges and high curvature regions gives better results than random subsampling. Different strategies for selecting points to be used in registration are presented in [37].
- Not all points have a matching point in the other view. This issue is typically solved by using a threshold excluding samples without close neighbors in the other view. The setting of this threshold is critical: if the threshold is too strict many correspondences can be discarded, especially during the first iterations and in the presence of noisy data. Conversely, if the threshold is too large, the algorithm is forced to match points without a valid correspondence in the other view. In another common approach, called *projective data association*, the points of one view are associated with the surface corresponding to the other view. Notice that this method requires the construction of a 3D surface from the point cloud.
- The algorithm can be easily trapped in a local minimum if the starting point is far from the solution. This is the reason why a rough alignment is used to initialize the algorithm.
- Large planar surfaces can slide over each other (see Fig. 7.5). If a large number of samples are collected in planar areas and the data are noisy, the edge information associated with such regions may not suffice to correctly drive the registration. This issue can be overcome by intelligently selecting points to be used for the registration, or by extending the distance function in order to consider the associated color values. An example of the improvement offered by texture information in properly registering the data over flat surfaces is shown in Fig. 7.6.

Several variations of the ICP algorithm have been proposed in the literature [37] to overcome the above and other issues. By using consumer depth cameras at high

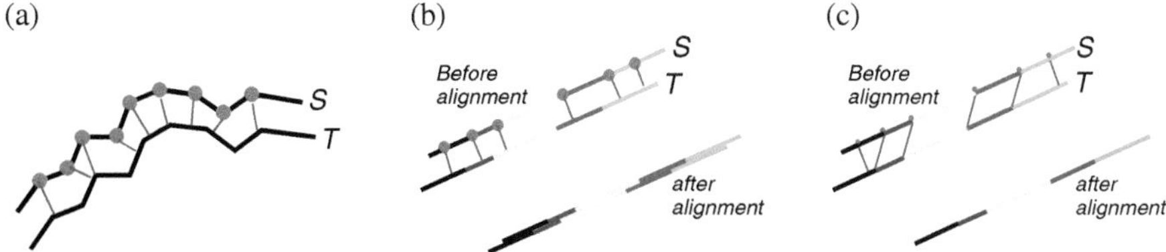

Fig. 7.5 Examples of the alignment of a target point cloud T with an already registered source view S: (**a**) alignment of two surfaces containing enough geometry information to constrain the registration; (**b**) alignment of two textured planar surfaces with geometry information only; (**c**) alignment of two textured planar surfaces with both color and geometry constraints

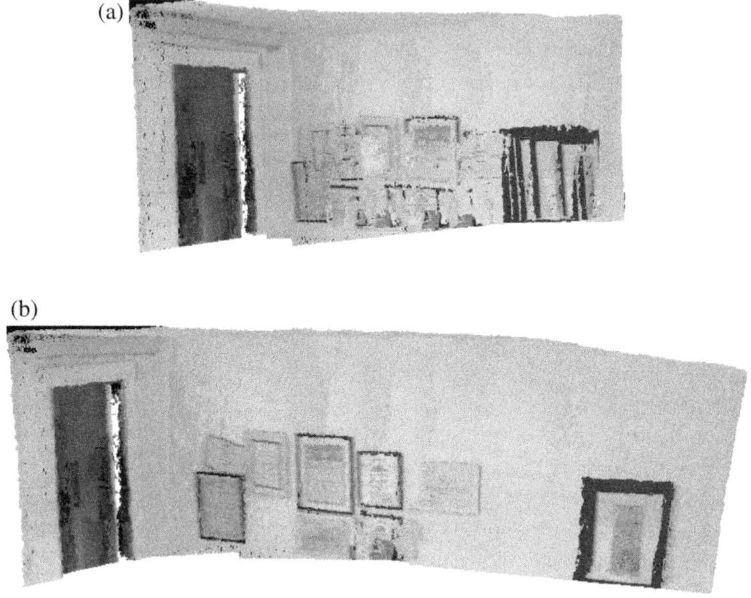

Fig. 7.6 Example of the reconstruction of a planar scene: (**a**) Reconstruction with geometry-based distance; (**b**) Reconstruction with color and geometry-based distance

frame rates (i.e., using very close views) the risk of local minima can be reduced but not completely eliminated. Most consumer depth cameras also have an associated color camera that can provide useful information to constrain registration on large planar patches. In general, with depth camera data, the point cloud subsampling is critical because of the high noise levels and the presence of various artifacts. Furthermore, even though edge or corner points are in principle the best features to constrain registration, they can be easily misplaced by noise when using consumer depth cameras data. In the case of structured light depth cameras, noise may be due to jagged edges, while in ToF depth cameras it may be due to multipath or mixed pixels. For the above reasons, a proper pre-processing stage is critical for accurate

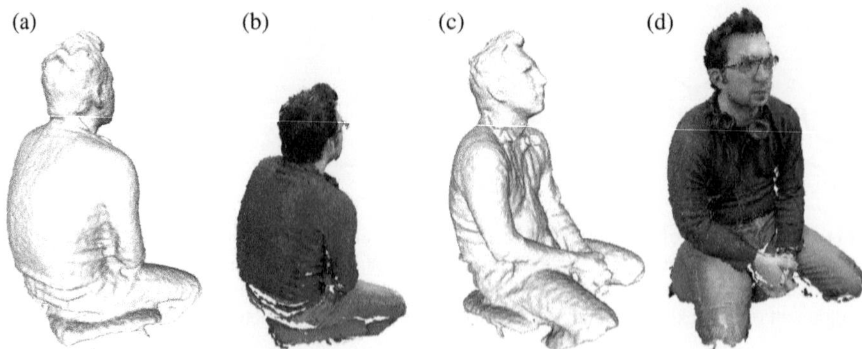

Fig. 7.7 Automatic 3D reconstruction of a sitting person from 800 frames, using the approach of [9]

registration. An approach for the automatic selection of samples to be used in the ICP procedure targeted to depth cameras has been presented in [8, 9]. It selects the points that correspond to high variance in both the color and depth domain. An example of 3D reconstruction performed by this method is shown in Fig. 7.7.

We observe that although the ICP algorithm is the most popular registration tool, it is not the only possible solution for the pairwise registration problem. For instance, a probabilistic global non-rigid scan alignment approach is proposed in [11], where the non-rigid component is used to account for the systematic biases introduced by the sensor acquisition.

A final issue of interest concerns the selection of the pairs of views to align, namely, one can iteratively register each view to the subsequent one, the simplest and most common approach, or register each view with the whole 3D point cloud built from the previous ones, as done for example by the KinectFusion [21, 28] method. Another possible approach is to first build a graph representing the adjacency information between the various views and then use it to select the alignments to be performed.

7.5 Global Registration

The scene geometry estimate, obtained by sequentially registering pairs of views, is often corrupted by error accumulation. This issue is solved by a global optimization step that updates the previously computed registrations in order to redistribute and balance the pairwise registration errors. Many approaches have been developed for this task, from simple heuristics to very complex optimization schemes. A key observation is that if the chain of views contains a loop, the accumulation error becomes quite evident when the loop is closed, i.e., when the first and last views of the loop are registered together. Following this rationale, simple heuristic approaches (e.g., [42]) build a graph of the connections between the views, whose

Fig. 7.8 Reconstruction of a research lab interior, from 800 frames, with the approach of [9]

nodes are associated with the different views and the edges represent their possible registrations, and find all the loops contained in the graph (or the ones longer than a pre-defined threshold). Then the error in the loop closing, i.e., the difference between the registration of the first and last views computed by the ICP along the registration chain and their relative positions, is redistributed among all the pairwise registrations in the loop. Different heuristics, such as the distance between the viewpoints, the difference in orientation, the estimated error on each registration, and several others, can be used to decide how to redistribute the error. Figure 7.8 shows how this technique avoids the error accumulation in the reconstruction of large scale environments.

More complex approaches, either from the 3D scanning field or from Simultaneous Localization and Mapping (SLAM), described in Sect. 7.8, instead perform a global optimization on all the acquired views. For example, the method of [12] optimizes a global energy function across all views, computed on a subset of the points to make the problem tractable, using an Expectation-Maximization (EM) algorithm.

7.6 Fusion of the Registered Views

The registration steps seen above lead to a point cloud representing the shape of the acquired object or scene, or better yet to a set of aligned and partially overlapping point clouds. However, since one is usually interested in a representation of the surface or volume of the acquired objects, the simplest solution is to obtain the surface from the set of aligned point clouds using any of the methods developed in the 3D reconstruction field. There are various approaches for this task, based either on the point cloud representation or volumetric integration schemes. The volumetric representation is the most commonly used. In particular, it is adopted by the KinectFusion approach [21, 28], which proved to be very effective, and paved the way to many extensions and modifications of the original idea. This method will be described in detail in Sect. 7.6.1. The point cloud representation can also be directly used to fuse the various views, e.g., the approach of [23] uses a point cloud representation where each point is associated a stability value. Each time a new view is added, its points are projectively associated with other points already present in the cloud. If a correspondence is found, the points are merged and the corresponding stability value is increased, otherwise, a new point with a low stability value is introduced.

7.6.1 KinectFusion

The KinectFusion [21, 28] reconstruction pipeline solves the registration and fusion problems in a single step, and integrates the data of each newly added view within a volumetric representation. This approach is described next because it has very good performance and has become very popular, inspiring a number of variants. Note that it is just one of the possible volumetric methods. As the name suggests, the approach is targeted to Kinect data, but in principle it can be applied to data from any depth camera.

The KinectFusion method, outlined in Fig. 7.9, is targeted to real-time 3D reconstruction and requires the execution of four main steps each time a new view is added:

1. conversion of the depth map to a 3D point cloud;
2. camera tracking;
3. volumetric integration;
4. ray casting.

The first two steps are standard: each newly acquired depth map is converted to a 3D point cloud and then aligned with the scene model, represented by a point cloud made of all previously acquired 3D views registered by the ICP algorithm, as discussed in the previous sections. However KinectFusion introduces a few interesting innovations in these steps.

Firstly KinectFusion uses a highly optimized GPU implementation exploiting projective data association, i.e., the previously introduced idea of projecting the points from a view to the surface corresponding to other views. This implementation allows one to run the ICP algorithm in real time on the complete set of samples of the new view without any of the previously discussed subsampling schemes.

Moreover the alignment is performed between the current 3D view and a virtual 3D view obtained by ray casting the volume representing the scene model built from

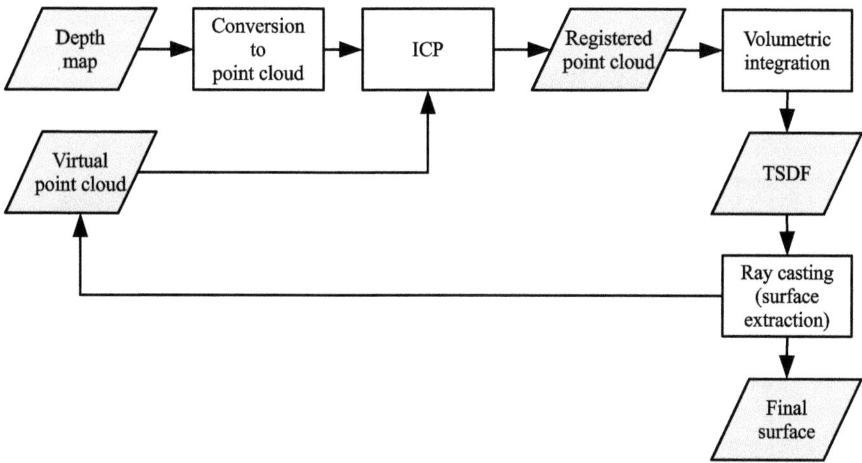

Fig. 7.9 Pipeline of the KinectFusion approach

all the previous acquisitions. This virtual 3D view of the scene model is typically a less noisy and more accurate representation allowing to obtain better alignment of the newly acquired view with the scene model. This represents a critical point since the KinectFusion approach does not rely on a global registration step to refine noisy alignments, therefore, the alignment error can propagate to the final result.

The next step is the volumetric integration algorithm used to accumulate the samples of each added view to the 3D scene model on the fly. The scene model is subdivided into a 3D grid of voxels and the volumetric representation consists of a cumulative weighted Signed Distance Function (SDF) representing the distance of the voxel from the acquired 3D point. In order to compute this distance, assuming the sensor pose is known, each voxel center v, with distance $d(\mathbf{v})$ from the camera viewpoint, is projected to the image plane and the closest sample x is selected. The distance $d(\mathbf{x})$ associated with sample $\mathbf{x}$ in the current view is then compared with $d(\mathbf{v})$. The SDF associated with voxel v is the difference between the two distances

$$SDF(\mathbf{v}) = d(\mathbf{v}) - d(\mathbf{x}) \tag{7.1}$$

as exemplified in Fig. 7.10. In particular, in order to mitigate the effects of noise and make the aggregation more robust, KinectFusion implements a Truncated Signed Distance Function (TSDF) by removing SDF measures greater than a fixed threshold. Each time a new view is added, the TSDF at each voxel location is updated by using a weighted average of all the values belonging to each voxel. Finally, the position of the implicit surface is obtained by extracting the zero-crossing of the TSDF. Note how the KinectFusion algorithm stores the complete volumetric representation at full resolution and computes the TSDF for each voxel. This is suited to a simple parallel implementation on the GPU, but it increases the memory requirements and does not allow one to obtain a high accuracy and large

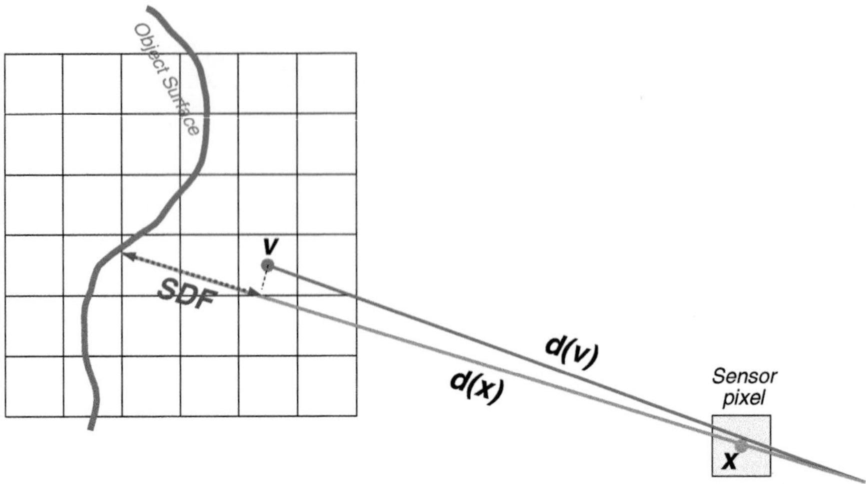

Fig. 7.10 Computation of the signed distance function (SDF)

reconstruction volume at the same time. With 1 GB of GPU memory, one can create a 640^3 cube, thus obtaining a 1.6 mm resolution on a volume of 1 m^3. Notice that for a given number of voxels, the resolution scales with the acquired volume size.

Some variations of the original KinectFusion approach have been proposed in order to solve this issue. In [35] the reconstruction volume is periodically updated by translating and rotating it according to the depth camera movements. The TSDF values are remapped to the new volume each time this operation is performed. This provision allows one to handle much larger scenes. A similar idea is also used in [47]. In [30], the explicit realization of the voxel structure is avoided by using a hashing data structure.

Finally, the KinectFusion system includes also a highly efficient ray casting technique that traverses the voxels and finds the zero-crossing, thus extracting the position of the implicit surface and the surface normals at each location. This is useful not only for visualization purposes, but also for improving registration accuracy. In fact, as previously noted, at each step a virtual view is rendered from the point of view of the target view T and is used to perform the registration in place of T itself. Since the virtual view contains information from all the previous views and is less noisy, its use leads to better accuracy and precision. Examples of 3D reconstructions with this method are shown in Fig. 7.11.

7.7 Reconstruction of Dynamic Scenes

All the presented approaches are based on the assumption that the scene to be reconstructed is static, i.e., the displacement of each single scene point between different views is due only to camera motion. The pose estimation approaches of

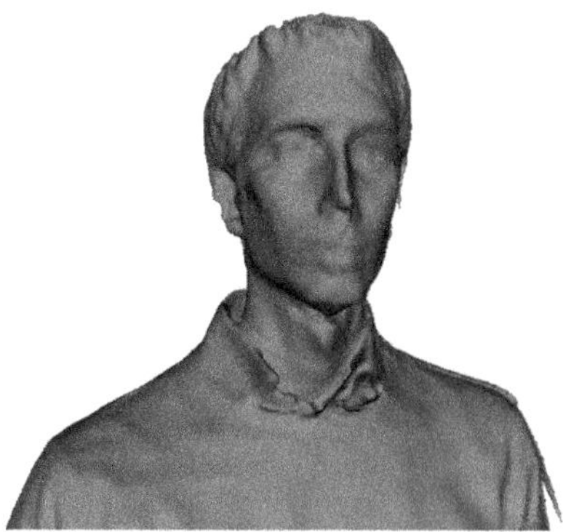
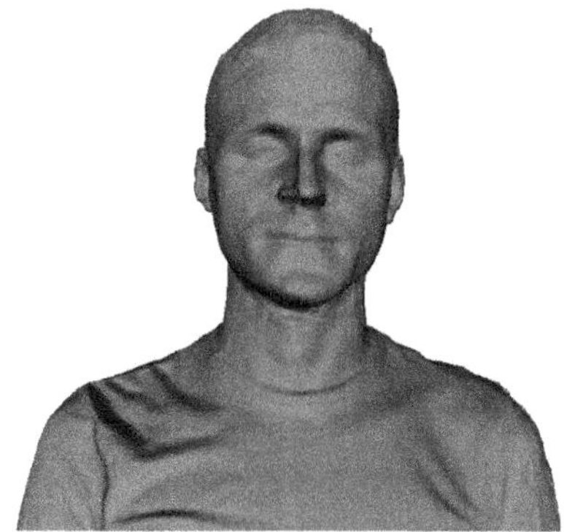

Fig. 7.11 Examples of 3D reconstructions of people from Kinect™ v2 data using the KinectFusion implementation from Microsoft™

Chap. 8, on the other hand, assume the use of a fixed camera or a set of fixed cameras, and that the point displacement is only due to the person's movements. The most challenging case is the 3D reconstruction of dynamic scenes with a hand-held sensor, since it is necessary to simultaneously estimate the camera motion and the motion of the objects in the scene. In this case, there is an intrinsic ambiguity in the problem, since the estimated motion of each point in the scene can be due to either a movement of the observed point or to a movement of the camera, and the employed algorithm must disambiguate between the two cases.

A possible simplification is adopting a pre-defined template to represent the object's motion, such as the skeleton model used in Chap. 8 to represent human body movements. The generic case where a priori models cannot be considered for the acquired scene is very difficult to handle; until recently, no approaches were available for this task. The recent method of [23] is able to recognize moving objects in the scene, under the rationale that moving points do not have a valid ICP correspondence even if they are close to stable points in the model (stable points are those observed in several views). A region growing segmentation scheme is adopted to segment the scene into static and dynamic points, then only static points are used for computing the registration between views. Unfortunately, this method allows one to recognize the moving objects but not to properly register them across multiple frames.

The DynamicFusion [29] method is a very interesting extension of the Kinect-Fusion scheme, designed to capture dynamic scenes in real time by a moving depth camera. The key idea is to estimate the scene geometry and a volumetric warping map that encodes the motion information at the same time, and instantly maps the acquired 3D view to a fixed canonical scene model. The algorithm has three main steps performed each time a new frame is acquired:

1. estimation of the parameters of the volumetric warping map relating the current view to the canonical frame;

2. fusion of the current depth map into the canonical scene model using the estimated warping map and the KinectFusion reconstruction approach;
3. update of the warping map to capture the newly added geometry.

The volumetric warp function is conceptually a roto-translation applied to each single voxel in order to model its motion and to map its current position to the corresponding position in the canonical scene model (i.e., the reference frame corresponding to the initial state). This representation, defined by six parameters per voxel, would require about ten million parameters for a typical 256^3 voxel space, and is not practically feasible. DynamicFusion instead adopts a method called *dual-quaternion blending* (DQB) that exploits a set of deformation nodes. The transformation associated with each voxel is a weighted average of the transformations defined by the close nodes. Each voxel center is transformed according to the warping map and the transformed voxel center is then used within the volumetric integration scheme of KinectFusion, i.e., the approach of Sect. 7.6.1 is used for this first step after the warping. The only difference is that the weights used to average the TSDF values in this case also account for the reliability of the estimated warping map. Then, the warping map is updated by computing the set of parameters minimizing an energy function made by two terms. The first term is determined by a dense ICP registration between the current frame and the canonical scene model. The second one is a regularization term, enforcing smooth motion and rigid deformations among the related deformation nodes arranged in a graph structure representing their relationships. The last step also updates the structure of the warping map, by introducing new deformation nodes and by updating the graph used for the regularization term. Figure 7.12 shows an example of the canonical scene model and some acquired frames along with the relative warped canonical scene model.

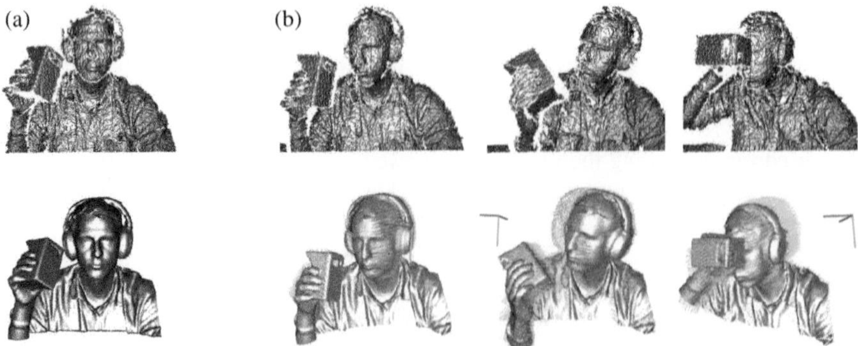

Fig. 7.12 Raw depth data and 3D reconstruction with the DynamicFusion [29] approach: (**a**) reference raw depth map and canonical scene model; (**b**) raw depth maps for different frames and corresponding warped canonical scene model. (*Courtesy of the authors of [29]*)

7.8 SLAM with Depth Camera Data

Simultaneous Localization and Mapping (SLAM) is a classic problem of robotics and other fields, concerning the simultaneous estimation of the scene structure and of the trajectory of a moving camera within a framework, similar to the structure from motion techniques mentioned in Chap. 1. SLAM has been widely studied in the past, although approaches based on features extracted from either standard or panoramic video streams are typically of limited effectiveness.

Recently this problem has been revisited by using RGB-D data from consumer depth cameras. A major advantage of depth data is that they allow one to directly obtain the 3D structure of the scene with a higher accuracy than that obtained by data from a standard camera, and furthermore they simplify the camera pose estimation problem. Standard SLAM methods extract relevant features from a video stream in order to estimate the camera motion and reconstruct a 3D representation of the framed environment. If the video stream is replaced by a depth map stream generated by a consumer depth camera, the technical issues faced by SLAM, which in this case will be called RGB-D SLAM in order to distinguish it from SLAM based on standard cameras, become very similar to those of 3D reconstruction with hand-held depth cameras. Indeed, all the methods presented in the previous sections are perfectly suited to estimate the camera pose and deliver a 3D reconstruction of the explored environment.

Even though in principle 3D reconstruction methods can be directly used for RGB-D SLAM, some ad hoc provisions are necessary because the objectives of 3D reconstruction do not exactly coincide with those of SLAM. In particular, typical SLAM settings involve extended motion within a large environment. The required algorithms, with respect to 3D registration, must be able to handle larger scenes with two major differences: the error drift in subsequent registrations must be carefully handled and the 3D models needed for SLAM purposes can be more sparse and less accurate than those needed for 3D reconstruction.

Two main families of methods have been considered in order to solve RGB-D SLAM. The first is based on the extraction of a set of relevant features from each frame with the associated descriptors. In this case, RGB-D SLAM is performed on the basis of a limited number of robust features rather than on a dense 3D point cloud, in order to reduce the computational and memory requirements necessary to handle larger size scenes. The features to be tracked in RGB-D SLAM are typically extracted from the color information using well known descriptors, e.g., SIFT [27], SURF [2], FAST [34], ORB [36] and many others, rather than from depth information. Indeed, in spite of the conceptual equivalence, color data turn out to be easier to recognize and track. After extracting the feature points from the color frames, it is possible to associate the corresponding 3D locations to each feature point by using depth and calibration information. Finally, from the corresponding 3D points in two different frames, it is possible to estimate the roto-translation between the camera positions, as already seen in Sect. 7.3. Given this, the reliability of the features becomes critical, as typically there are several incorrect correspondences and points with unreliable depth values. Robust estimation schemes, such as the RANSAC framework [16], can be used to deal with this issue. For instance, [15] uses RANSAC with an adaptive threshold for inlier selection, but different schemes have also been proposed. Since SLAM approaches deal with a larger number of views and can tolerate less accurate registrations, the error drift issue is even more critical than in 3D registration. Therefore, some global optimization scheme is needed. The approach of [15] uses an environment measurement model to estimate the reliability of each frame-to-frame alignment. This information is then used within a graph optimization framework that allows one to find the optimal solution for the global alignment of all the frames.

The second family of methods is instead based on the registration of dense point clouds using the ICP algorithm, similar to the approaches presented for 3D reconstruction in the previous sections. Notice that a major difference between RGB-D SLAM and 3D reconstruction concerns the cardinality of the number of views in 3D registration. For this reason, the global optimization step in RGB-D SLAM is particularly critical and some proposed solutions are based on loop closure [24] or graph optimization [18].

The families of methods described above can be also combined to generate more accurate solutions, e.g., by using the ICP algorithm after the feature-based RANSAC procedure, as done in [18]. Furthermore, since the starting point is already close to the optimal solution, the computation time is reduced as well as the risk of falling into local minima. Another possibility is to estimate the correspondences between each pixel in the newly acquired view and the reconstructed 3D scene using machine learning techniques, as done in [41].

7.9 Conclusions and Further Reading

Obtaining a 3D model of a framed scene from a set of pictures or a video is still a challenging task, requiring dedicated hardware and skilled people. Consumer depth cameras have paved the way for 3D reconstruction to become a commodity available to end users. However, while data acquisition is very simple, its processing to obtain reliable 3D models remains an open research problem and one of the key challenges in depth camera applications. In this chapter, several 3D reconstruction approaches explicitly targeted to depth cameras are presented.

Automatic camera localization and mapping and 3D scene reconstruction are strictly related problems, as they face similar technical issues. The same methods, after proper adjustments, can be used to solve both of them. SLAM from visual data remains a challenging problem (see [17] for a recent review) which may continue to inspire ideas for RGB-D SLAM.

References

1. M. Andreetto, N. Brusco, G.M. Cortelazzo, Automatic 3d modeling of textured cultural heritage objects. IEEE Trans. Image Process **13**(3), 354–369 (2004)
2. H. Bay, T. Tuytelaars, L. Van Gool, SURF: speeded up robust features, in *Proceedings of IEEE European Conference on Computer Vision* (Springer, Heidelberg, 2006), pp. 404–417
3. R. Benjemaa, F. Schmitt, Fast global registration of 3d sampled surfaces using a multi-z-buffer technique. Image Vis. Comput. **17**(2), 113–123 (1999)
4. R. Bergevin, M. Soucy, H. Gagnon, D. Laurendeau, Towards a general multi-view registration technique. IEEE Trans. Pattern Anal. Mach. Intell. **18**(5), 540–547 (1996)
5. F. Bernardini, H.E. Rushmeier, The 3d model acquisition pipeline. Comput. Graph. Forum **21**(2), 149–172 (2002)
6. P.J. Besl, Active, optical range imaging sensors. Mach. Vis. Appl. **1**(2), 127–152 (1988)
7. P.J. Besl, N.D. McKay, A method for registration of 3-D shapes. IEEE Trans. Pattern Anal. Mach. Intell. **14**(2), 239–256 (1992)
8. E. Cappelletto, P. Zanuttigh, G.M. Cortelazzo, Handheld scanning with 3d cameras, in *Proceedings of IEEE International Workshop on Multimedia Signal Processing* (Pula, 2013), pp. 367–372
9. E. Cappelletto, P. Zanuttigh, G.M. Cortelazzo, 3D scanning of cultural heritage with consumer depth cameras. Multimedia Tools Appl. **75**(7), 3631–3654 (2016)
10. Y. Chen, G. Medioni, Object modeling by registration of multiple range images, in *Proceedings of IEEE International Conference on Robotics and Automation* (Sacramento, 1991), pp. 2724–2729
11. Y. Cui, S. Schuon, D. Chan, S. Thrun, C. Theobalt, 3D shape scanning with a time-of-flight camera, in *Proceedings of IEEE Conference on Computer Vision and Pattern Recognition* (2010), pp. 1173–1180
12. Y. Cui, S. Schuon, S. Thrun, D. Stricker, C. Theobalt, Algorithms for 3d shape scanning with a depth camera. IEEE Trans. Pattern Anal. Mach. Intell. **35**(5), 1039–1050 (2013)
13. B. Curless, M. Levoy, A volumetric method for building complex models from range images, in *Proceedings of ACM SIGGRAPH* (New York, 1996), pp. 303–312
14. D.W. Eggert, A.W. Fitzgibbon, R.B. Fisher, Simultaneous registration of multiple range views for use in reverse engineering of cad models. Comput. Vis. Image Underst. **69**(3), 253–272 (1998)
15. F. Endres, J. Hess, N. Engelhard, J. Sturm, D. Cremers, W. Burgard, An evaluation of the RGB-D slam system, in *Proceedings of IEEE International Conference on Robotics and Automation* (2012), pp. 1691–1696
16. M.A. Fischler, R.C. Bolles, Random sample consensus: a paradigm for model fitting with applications to image analysis and automated cartography. Read. Comput. Vis. Issues Probl. Princ. Paradig. **1**, 726–740 (1987)
17. J. Fuentes-Pacheco, J. Ruiz-Ascencio, J.M. Rendón-Mancha, Visual simultaneous localization and mapping: a survey. Artif. Intell. Rev. **43**(1), 55–81 (2015)
18. P. Henry, M. Krainin, E. Herbst, X. Ren, D. Fox, RGB-D mapping: using kinect-style depth cameras for dense 3D modeling of indoor environments. Int. J. Robot. Res. **31**(5), 647–663 (2012)
19. B.K.P. Horn, Closed-form solution of absolute orientation using unit quaternions. J. Opt. Soc. Am. **4**, 629–642 (1987)
20. D.F. Huber, M. Hebert, Fully automatic registration of multiple 3D data sets. Image Vis. Comput. **21**(7), 637–650 (2003)
21. S. Izadi, D. Kim, O. Hilliges, D. Molyneaux, R. Newcombe, P. Kohli, J. Shotton, S. Hodges, D. Freeman, A. Davison, A. Fitzgibbon, KinectFusion: real-time 3D reconstruction and interaction using a moving depth camera, in *Proceedings of ACM Symposium on User Interface Software and Technology* (2011)

22. A.E. Johnson, M. Hebert, Using spin images for efficient object recognition in cluttered 3D scenes. IEEE Trans. Pattern Anal. Mach. Intell. **21**(5), 433–449 (1999)
23. M. Keller, D. Lefloch, M. Lambers, S. Izadi, T. Weyrich, A. Kolb, Real-time 3D reconstruction in dynamic scenes using point-based fusion, in *Proceedings of IEEE International Conference on 3D Vision* (Washington, 2013), pp. 1–8
24. C. Kerl, J. Sturm, D. Cremers, Dense visual SLAM for RGB-D cameras, in *Proceedings of IEEE/RSJ International Conference on Intelligent Robots and Systems* (Tokyo, 2013), pp. 2100–2106
25. M. Levoy, K. Pulli, B. Curless, S. Rusinkiewicz, D. Koller, L. Pereira, M. Ginzton, S. Anderson, J. Davis, J. Ginsberg, J. Shade, D. Fulk, The digital michelangelo project: 3D scanning of large statues, in *Proceedings of ACM SIGGRAPH* (New York, 2000), pp. 131–144
26. W.E. Lorensen, H.E. Cline, Marching cubes: a high resolution 3D surface construction algorithm, in *Proceedings of ACM SIGGRAPH* (New York, 1987), pp. 163–169
27. D.G. Lowe, Distinctive image features from scale-invariant keypoints. Int. J. Comput. Vis. **60**(2), 91–110 (2004)
28. R.A. Newcombe, S. Izadi, O. Hilliges, D. Molyneaux, D. Kim, A.J. Davison, P. Kohli, J. Shotton, S. Hodges, A. Fitzgibbon, Kinectfusion: real-time dense surface mapping and tracking, in *Proceedings of IEEE International Symposium on Mixed and Augmented Reality* (2011)
29. R. Newcombe, D. Fox, S. Seitz, Dynamicfusion: reconstruction and tracking of non-rigid scenes in real-time, in *Proceedings of IEEE Conference on Computer Vision and Pattern Recognition* (2015)
30. M. Nießner, M. Zollhöfer, S. Izadi, M. Stamminger, Real-time 3D reconstruction at scale using voxel hashing. ACM Trans. Graph. **32**(6), 169 (2013)
31. K. Pulli, *Surface Reconstruction and Display from Range and Color Data.* PhD thesis, University of Washington (1997)
32. K. Pulli, Multiview registration for large data sets, in *Proceedings of IEEE International Conference on 3-D Digital Imaging and Modeling* (1999), pp. 160–168
33. M. Rioux, Laser range finder based on synchronized scanners. Appl. Opt. **23**(21), 3837–3844 (1984)
34. E. Rosten, T. Drummond, Fusing points and lines for high performance tracking, in *Proceedings of IEEE International Conference on Computer Vision* (Beijing, 2005), pp. 1508–1515
35. H. Roth, M. Vona, Moving volume kinectfusion, in *Proceedings of British Machine Vision Conference* (2012), pp. 1–11
36. E. Rublee, V. Rabaud, K. Konolige, G. Bradski, ORB: an efficient alternative to sift or surf, in *Proceedings of IEEE International Conference on Computer Vision* (Barcelona, 2011), pp. 2564–2571
37. S. Rusinkiewicz, M. Levoy, Efficient variants of the ICP algorithm, in *Proceedings of IEEE International Conference on 3-D Digital Imaging and Modeling* (Quebec City, 2001), pp. 145–152
38. R.B. Rusu, Z.C. Marton, N. Blodow, Mi. Dolha, M. Beetz, Towards 3D point cloud based object maps for household environments. Robot. Auton. Syst. **56**(11), 927–941 (2008)
39. S. Schuon, C. Theobalt, J. Davis, S. Thrun, Lidarboost: depth superresolution for tof 3D shape scanning, in *Proceedings of IEEE Conference on Computer Vision and Pattern Recognition* (2009), pp. 343–350
40. R. Scopigno, P. Pingi, C. Rocchini, P. Cignoni, C. Montani, 3D scanning and rendering cultural heritage artifacts on a low budget, in *Proceedings of European Workshop on High Performance Graphics Systems and Applications* (2000), pp. 16–17
41. J. Shotton, B. Glocker, C. Zach, S. Izadi, A. Criminisi, A. Fitzgibbon, Scene coordinate regression forests for camera relocalization in RGB-D images, in *Proceedings of IEEE Computer Vision and Pattern Recognition* (Washington, 2013)

42. J. Sprickerhof, A. Nüchter, K. Lingemann, J. Hertzberg, An explicit loop closing technique for 6D SLAM, in *Proceedings of European Conference on Mobile Robots* (2009)
43. C. Tomasi, R. Manduchi, Bilateral filtering for gray and color images, in *Proceedings of IEEE International Conference on Computer Vision* (1998), pp. 839–846
44. M. Trobina, Error model of a coded-light range sensor. Technical report, Communication Technology Laboratory Image Science Group, ETH-Zentrum, Zurich (1995)
45. G. Turk, M. Levoy, Zippered polygon meshes from range images, in *Proceedings of ACM SIGGRAPH* (New York, 1994), pp. 311–318
46. M.D. Wheeler, Y. Sato, K. Ikeuchi, Consensus surfaces for modeling 3D objects from multiple range images, in *Proceedings of IEEE International Conference on Computer Vision* (Bombay, 1998), pp. 917–924
47. T. Whelan, M. Kaess, M. Fallon, H. Johannsson, J. Leonard, J. McDonald, Kintinuous: spatially extended kinectfusion, in *Proceedings of RSS Workshop, RGB-D: Advanced Reasoning with Depth Cameras* (2012)
48. D. Zhang, M. Hebert, Harmonic maps and their applications in surface matching, in *Proceedings of IEEE Conference on Computer Vision and Pattern Recognition* (Fort Collins, 1999)

Chapter 8
Human Pose Estimation and Tracking

The estimation of the movements and posture of human beings is one of the key applications of consumer depth cameras. It motivated the development of the Kinect^{TM} v1 and v2, and favored the diffusion of ToF and structured light technologies from the industrial and research fields to the mass market. The appeal of human pose estimation and tracking is due to its vast range of applications solving daily life tasks. Console games using the body or the hands as controller were the first commercial application of consumer depth cameras, and the skeletal tracking approach introduced with Kinect^{TM} v1 represents the first reliable and efficient solution to the pose estimation problem in a home environment. Human-computer interaction is another intriguing field, as the various hand configurations and body movements are often exploited to convey non-verbal information, either by explicitly associating gestures to specific meanings or more implicitly by augmenting speech information. Besides people interaction, human pose can also have a fundamental role in many situations requiring the manipulation of an object or the possibility of controlling a machine by performing intuitive (natural) movements, e.g., in the robotics field. Historically, computer animation has been one of the first and more active areas successfully exploiting human pose data derived from motion capture. Complex movements performed by a human actor can be tracked and recorded in order to be used either in real-time or in a second time, to drive the movements of some computer-generated character or avatar (motion retargeting). Finally, many other applications exploit information from human pose estimation and tracking, e.g., video surveillance and control, posture and movement analysis in medical applications and data compression through the use of representations more compact than full 3D point clouds.

Various solutions have been proposed for human pose estimation and tracking task (Fig. 8.1). Marker-based systems are able to acquire reliable information about body or hand posture but they are expensive and invasive, therefore their usage is confined to highly controlled industrial or medical environments. Colored gloves and special suits equipped with reflective or LED lights markers require delicate

P. Zanuttigh et al., *Time-of-Flight and Structured Light Depth Cameras*,
DOI 10.1007/978-3-319-30973-6_8

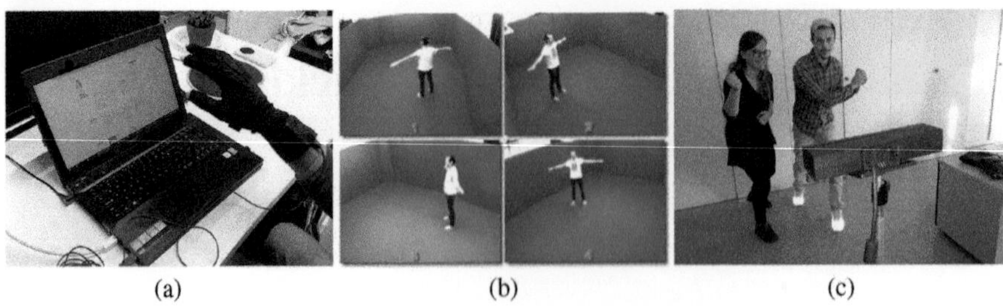

(a) (b) (c)

Fig. 8.1 Examples of pose estimation setups: (**a**) motion capture glove; (**b**) markerless motion capture setup; (**c**) typical Kinect^{TM} setting

and complex calibration procedures and may significantly restrict body or hand movements, making them unsuitable for most commercial applications. For these reasons vision-based approaches, due to their non-invasivity, received great attention in the last years. Several methods have been proposed to estimate the body pose from a single image or video stream [1, 21, 37] but they never led to completely satisfactory performance. Better results can be obtained by setups made by a large number of calibrated cameras and a cooperative background [37], but they require large dedicated rooms and complex calibration procedures besides being typically very expensive.

Even though the geometric 3D information embedded in depth data can solve some of the problems of systems based on a single color camera, such approach requires to properly solve a number of issues. For example, the depth data provided by many depth cameras are affected by a considerable amount of noise and artifacts. Moreover, single-view approaches based on 3D geometry often present a large number of self-occlusions, resulting in missing data. Deploying multiple depth cameras (whenever the mutual interference is negligible) reduces missing data at the expenses of more complex calibration procedures. Self-occlusions probably represent one of the most difficult problems within single view pose recovery, and as well known even multi-view acquisition setups may not be able to resolve all the occlusions. Indeed, complex articulated objects like the human body or hand can assume a high number of configurations with self-occlusions, thus making the design of human pose recovery algorithms rather challenging especially when the detection of smaller parts is required as an intermediate step.

The hand pose estimation and tracking is even more challenging than its full body counterpart, since the number of pose configurations characterizing the hand is usually much higher. Color-based segmentation and recognition techniques can improve the discrimination of the various body parts, but their effectiveness depends on the clothes textures and colors, thus being unsuited to hand pose estimation since the skin color is rather uniform in the whole hand. In any case, depth data allow to capture body parts either hard or impossible to detect from color information only, e.g., an arm pointing in the camera direction may be impossible to detect by silhouette-based methods but is clearly visible from depth data.

As detailed in Sect. 8.1, many approaches model the human body or hand as articulated objects and represent their pose by a three-dimensional kinematic graph, with vertices and edges corresponding to their joints and segments respectively. The pose is then represented by a specific arrangement of vertices in the 3D space, or more generally by a specific set of values for the pose parameters (or configuration) controlling the model. Considering, for example, that a generic hand kinematic model has typically about 27 degrees of freedom, it is evident that the parameter search space for the hand model has a rather high dimensionality. The huge number of degrees of freedom (DoF) characterizing the body and hand pose, makes pose estimation rather challenging for its high dimensionality. Although the pose estimation search space can be considerably restricted by preventing the evaluation of unfeasible configurations, this provision may not always make the problem more tractable since some of the required constraints may not be injected in simpler models. Avoiding to look for the complete pose recovery and focusing, instead, on partial pose estimates is another way to reduce the search space which simplifies the problem solution at the expenses of a reduced model representational power. To this aim, one may only track the more relevant joints or body parts, e.g., the upper part of the body instead of the whole body, or one may simplify the models, e.g., by modeling the fingers by kinematic chains made by just two segments.

Human movements can also be "too fast" for being captured by current depth cameras, with frame rates ranging from 30 to 60 Hz. In this case, the assumption of small target pose changes between subsequent frames made by several tracking approaches may not be realistic.

Despite the vast literature about body and hand pose estimation and tracking, several unsolved challenges make this problem rather intriguing as will be seen in this chapter articulated in three sections.

Section 8.1 describes the most popular body and hand models for pose and tracking purposes. Sections 8.2 and 8.3 respectively address the pose estimation and tracking problems, and overview the main depth-based approaches based on the body and hand models treated in Sect. 8.1. Although the two problems are often considered equivalent, in this chapter they are separately treated for clarity's sake, within the convention that pose estimation will be assumed to be performed only on single frames and pose tracking on multiple frames.

8.1 Human Body Models

Human body models play a fundamental role in many pose tracking algorithms, since they represent the main mechanism to generate pose hypotheses. In fact, a very common pose solution approach is the evaluation of each hypothesis against depth camera observations followed by the selection of the most likely one. Sections 8.1.2 and 8.1.3 describe pure skeleton models, skeleton models augmented with primitive shapes and skeleton models augmented with surface meshes, which are popular 3D graphical human models often employed to represent the structure and motion

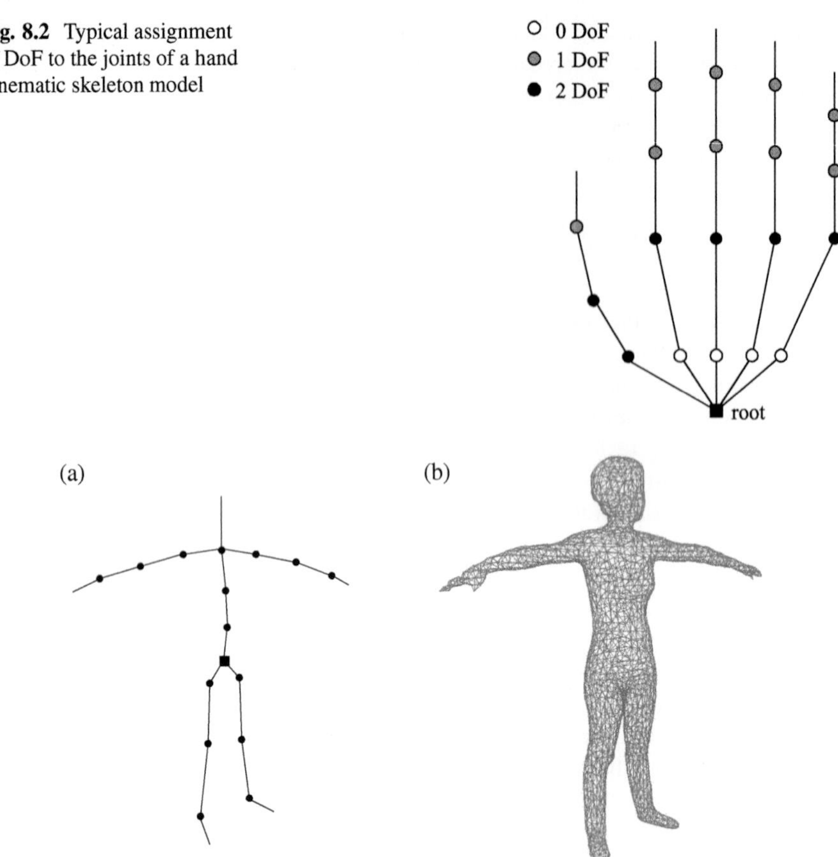

Fig. 8.2 Typical assignment of DoF to the joints of a hand kinematic skeleton model

Fig. 8.3 Two possible models for the human body at dressing pose configuration: (*left*) a simple kinematic skeleton model; (*right*) a skin mesh model. The pose with straight legs and arms pointing sideways (*T-pose*) is typically used as dressing pose

of a human hand or body, both seen as articulated objects (Fig. 8.3). A few basic concepts about the pose and the motion of kinematic chains and generic articulated objects are briefly discussed beforehand in Sect. 8.1.1 for a better understanding of the subsequent material. It is worth noting that there exist many types of graphical models besides the ones considered in this chapter, which can be successfully applied in order to represent the human body or some of its parts.

According to a first possible basic division, all graphical models can be classified into 2D and 3D representations. *2D models* represent the structure and motion of the human body in the camera image plane, with the idea of recovering the body pose by directly searching in the projection space. 2D models are particularly useful when the target movement is parallel to the camera image plane, as in the case of gait analysis. A simple example are the *cardboard* models introduced in [25], where each body part is represented by quadrilateral patches and special constraints are enforced between the corners of the different patches. Another example of planar models are the *prismatic* models [39], which represent the human body as a set of 2D chains of trapezoidal elements. The pictorial structures introduced in [14] are another example of 2D graphical models where each individual body part is represented by suitable templates and the spatial arrangement of all the parts is constrained by spring-like constraints between them. *3D models* represent an effective alternative solution, and allow to directly take advantage of depth information when available. Besides the *skeleton-based* models, presented in Sect. 8.1.2, many other 3D graphical models have been proposed in the literature, including *loose-limbed* models where the human body is represented as a set of rigid components connected by elastic links. An example of such models is given in [53], where the position of each single part with respect to the others is described by a conditional probability distribution.

8.1.1 Articulated Objects

Non-deformable objects are often represented by the three-dimensional rigid-object model, subject only to *rigid transformations*. According to this model, the distance between each pair of object points does not change over time and all the object points solidly undergo the same roto-translation. By denoting with O-3D the object reference coordinate system, it is natural to represent a *pose configuration* with the rigid transformation $(\mathbf{R}, \mathbf{t})$ mapping O-3D to the world coordinate system W-3D so that for any object point P holds

$$\mathbf{P}_W = \mathbf{R}\mathbf{P}_O + \mathbf{t} \tag{8.1}$$

where $\mathbf{P}_O$ and $\mathbf{P}_W$ are the coordinates of object point P with respect to the O-3D and W-3D reference systems. As it can be noticed, the pose of the object is completely determined once the rotation matrix $\mathbf{R}$ and translation vector $\mathbf{t}$ are known. Since any rotation $\mathbf{R}$ around the origin of the W-3D world reference system can be equivalently expressed as the non-commutative product of three elementary rotation matrices $\mathbf{R}_x(\alpha_x)$, $\mathbf{R}_y(\alpha_y)$, $\mathbf{R}_z(\alpha_z)$ around axes x, y, z by angles α_x, α_y and α_z respectively, (8.1) can be reformulated as

$$\mathbf{P}_W = \mathbf{R}_x(\alpha_x)\mathbf{R}_y(\alpha_y)\mathbf{R}_z(\alpha_z)\mathbf{P}_O + \mathbf{t} \tag{8.2}$$

and the pose of the rigid object can thus be represented in a more compact way by

$$\theta = \{\mathbf{t}, \alpha_x, \alpha_y, \alpha_z\}. \tag{8.3}$$

Notice that three angle values describe the full rotation. From now on, for simplicity's sake, the symbol θ will be equivalently used to denote both a specific pose configuration and the set of parameters actually needed to describe it.

A motion model can be defined in a similar way. In particular, for any transformation of a rigid object from an initial pose θ_1 to a final pose θ_2, there exist suitable rotation axes φ_1, φ_2, φ_3, rotation angles α_1, α_2, α_3 and a translation vector $\mathbf{t}$ such that

$$\mathbf{P}_{W,2} = (r_{\varphi_1}(\alpha_1) \circ r_{\varphi_2}(\alpha_2) \circ r_{\varphi_3}(\alpha_3))(\mathbf{P}_{W,1}) + \mathbf{t} \tag{8.4}$$

where $r_\varphi(\alpha) : \mathbb{R}^3 \to \mathbb{R}^3$ denotes the generic rotation around axis φ by an angle α while the world coordinates of an object point prior to the motion are denoted by $\mathbf{P}_{W,1}$ and after motion by $\mathbf{P}_{W,2}$. Notice that, different from (8.2), the rotation axes in (8.4) are arbitrary, i.e., the rotations in (8.4) may not be around the origin, hence the composition of rotation functions in (8.4) cannot in general be replaced by a multiplication of elementary rotation matrices [20].

Once the rotation axes have been chosen, the set of motion parameters $\mathbf{t}$, α_1, α_2, α_3 defining the object transformation from pose θ_1 to pose θ_2 can be considered as an alternative representation of the object final pose θ_2 with respect to its initial pose θ_1. Therefore, given some known rotation axes φ_1, φ_2, φ_3, one can define a motion function f_m taking as input both the world coordinates of an object point P denoted as $\mathbf{P}_W$, some pose configuration with respect to the object initial pose denoted as $\theta = \{\mathbf{t}, \alpha_1, \alpha_2, \alpha_3\}$, and returning as output the world coordinates of P when the object pose is θ, as

$$f_m(\theta, \mathbf{P}_W) = (r_{\varphi_1}(\alpha_1) \circ r_{\varphi_2}(\alpha_2) \circ r_{\varphi_3}(\alpha_3))(\mathbf{P}_W) + \mathbf{t}. \tag{8.5}$$

An *articulated object* can be thought as a collection of rigid objects called *components* whose relative orientations and positions are bounded by special constraints called *joints*. Both the human hand and human body can be naturally modeled as articulated objects, where the joints play the role of the various skeletal connections and the components are associated with specific hand or body parts, such as limb or finger segments. Many human models proposed so far in the literature have in common the fact that they can be easily split into a number of simpler structures called *kinematic chains*. A kinematic chain is a special kind of articulated object consisting of an ordered sequence of rigid components where only consecutive components are connected by a joint. Usually, the first sequence component is called *root* and the last *end effector*. A hierarchy is implicitly imposed to the chain components in such a way that the orientation and position of the first chain components influence the pose of all the subsequent components. Each chain joint may enforce different kinds of constraints, leaving at most three DoF between the poses of two consecutive components. A more natural way of looking at a joint is as a 3D point around which two consecutive components are allowed to rotate. Notice that no displacement is allowed between consecutive components. Depending on the joint type, connected components can either freely rotate around a joint, or can only rotate by limited angular ranges. Two components may also not be allowed to rotate in some specific dimension, further limiting to one or two instead of three the degrees of freedom associated with their connecting joint.

A simple characterization of the pose and motion of a kinematic chain will be given next from the concepts previously introduced for rigid objects. It will be straightforward extending these results to articulated objects made by collections of multiple kinematic chains, such as the skeleton models described in Sects. 8.1.2 and 8.1.3. For simplicity's sake, from now on we will assume a simple constrained kinematic chain made by $N + 1$ components and N joints, where each joint J^j for $j = 1, \ldots, N$ is subject to one DoF only. Under these assumptions, there is a single rotation axis φ_j passing through the jth joint. Any possible chain configuration can be obtained by *forward kinematics*: starting from an initial configuration, the root is moved first by a rigid transformation and then each chain component is iteratively rotated around the associated joint axis. As it can be readily seen, this process may change the rotation axes positions with respect to the world reference system, since any rotation around a joint affects the positions of all subsequent chain joints up to the Nth. As for rigid objects, the pose of a kinematic chain with respect to some initial configuration and known joint rotation axes $\varphi_1, \ldots, \varphi_N$ can be described in a compact way as

$$\theta = \{\mathbf{t}, \alpha_x, \alpha_y, \alpha_z, \alpha_1, \ldots, \alpha_N\} \tag{8.6}$$

where $\mathbf{t}$ denotes the root displacement with respect to the world coordinate system origin, α_x, α_y, α_z denote the root joint rotation around the world axes, and α_j is the joint rotation angle around axis φ_j, for $j = 1, \ldots, N$. Following this rationale, one can derive for kinematic chains a motion function analogous to (8.5) as

$$f_m(\theta, \mathbf{P}_W) = (r_x(\alpha_x) \circ r_y(\alpha_y) \circ r_z(\alpha_z) \circ r_{\varphi_1}(\alpha_1) \circ \cdots \circ r_{\varphi_N}(\alpha_N))(\mathbf{P}_W) + \mathbf{t} \tag{8.7}$$

where $\mathbf{P}_W$ are the world coordinates with respect to the W-3D system of an object point P belonging to the (N+1)th chain component.

8.1.2 Kinematic Skeleton Models

Since both the human hand and human body can be naturally conceived as articulated objects, skeleton models are often chosen to describe their structure and motion in a straightforward way. Their simplicity, however, comes with a coarser representation of the hand or body surface. Skeleton models are often referred to as kinematic models, in that they efficiently model the motion of the body while neglecting its appearance. As will be seen in the following, pure skeleton models can be augmented with primitive shapes or surface meshes to account also for the body shape appearance. The idea is to represent the body by modeling its inner skeletal structure by a collection of line segments connected each other at their end points. For this reason skeleton models are also called *stick figures*.

More formally, skeleton models are represented by undirected graphs whose vertices are points in the 3D space. Each edge is associated with a bone or body segment, while the graph vertices are in correspondence with the joints connecting the bones. Generally, instead of a generic graph, one prefers to use a tree-like structure in order to obtain a more convenient description of the body as a collection of kinematic chains sharing a common root. According to this representation, the pose of a human body can be defined as the pose of all the kinematic chains composing its skeleton model, and can be described by a set of real-valued parameters θ. A compact representation is provided by the set of all the joint angles, considered with respect to some suitable rotation axes passing through the skeleton joints, and by the rotation and translation of the root with respect to the world coordinate system.

As for kinematic chains, each joint is associated with as many DoF as the number of directions in which it can move. Different from loose-limbed models, no displacement is allowed between linked segments. Even though most skeleton models share the same global structure, the number of joints and the number of DoF associated with each joint can vary depending on the required level of detail. Very simple models for the upper-body use just 10 DoF (3 DoF for each shoulder and 2 DoF for each elbow), while more complex skeletons may use up to 50 DoF [1, 4]. A common skeleton model for the human hand is the one proposed in [32], made by 27 bones and 19 joints (see Fig. 8.2). In this case, the entire skeleton is composed by a set of five kinematic chains. Many other skeleton models have been proposed for the human hand which are just slight variations of this model. There are models [2, 31] where the trapezio-metacarpal joint accounts for 3 DoF, in order to better describe the complex configurations taken by the thumb with respect to the palm. Other models allow the four carpo-metacarpal joints to move for better describing palm deformations. In general, the higher is the number of DoF permitted at the various joints, the larger is the set of poses the model can describe. On the other hand, this entails an increase in the search space dimensionality, and makes the problem harder to solve. Anyway, even for simple models parameterized by few variables, the space of possible poses may still be too large to be explored in a reasonable time. For this reason, additional constraints are often enforced to forbid unfeasible poses, thus limiting the search space. Static constraints may be enforced to limit the range of joint angles, and dynamic constraints are sometimes used to account for specific relationships between different model parameters. An example of dynamic constraint, supported by bio-mechanical studies, is the one equating the angle between the distal and middle phalanx of a finger to two thirds of the angle between the middle and proximal phalanx of the same finger. Other constraints can be enforced to account for the fact that different body parts cannot occupy the same space volume, since they are not penetrable.

8.1.3 Augmented Skeleton Models

As already observed in Sect. 8.1.2, both primitive shapes and surface meshes can be attached to the skeleton segments or joints in order to obtain more realistic descriptions of the hand or the body shape.

3D skeleton models can be augmented with elementary solids such as spheres, cylinders or cones. One of the earliest works making use of this type of graphical models is the one of [44, 45], employing cylinders and hemispheres to model the shape of phalanges and finger tips respectively. Spheres and truncated cones are, instead, used in [18] to approximate the surface of a human hand. Similarly, [54] uses various kinds of quadrics such as cylinders, truncated cones, half-ellipsoid and hemispheres. Models for body pose tracking may use spheres [41], truncated cones [12], cylinders [24, 46, 51] or tapered super-quadrics [11, 18, 27].

Even though primitive shapes provide a rather unrealistic representation of the hand or body shape, they can be still considered a valid choice since they require only a few parameters to encode their position and size, and one can exploit standard properties from projective geometry to implement fast projection into the camera plane, such as in [54]. Shape parameters accounting for the proportions of the various body parts can be initialized with values learned from large datasets of individuals. Otherwise ad hoc procedures can be implemented in order to automatically estimate the values of these parameters for the specific person to be tracked.

Skinning techniques can be leveraged to represent the human hand or body surface at a finer level of detail by means of polygonal meshes. This more realistic representation comes at the expenses of an increase of the computation needed to evaluate its consistency with respect to the observations, usually performed by comparing the features obtained by rendering the 3D shape attached to the skeleton model with the features observed in the input data. Nevertheless, several projection operations can be avoided when 3D observation features are available, making this solution quite attractive for tracking applications based on depth data.

Skinning techniques can be divided into two main groups, namely *rigid skinning* and *deformable skinning* methods. Among the latter, *Linear Blend Skinning* is the most popular one due to its simplicity. More complex and nonlinear solutions model skin deformations in a more realistic way at the expenses of a considerable increase of the computational demands [33]. Both rigid and linear blend skinning take advantage of the underlying skeleton model to compute the position of mesh vertices according to the pose configuration assumed by the model. In *rigid skinning* each mesh vertex V is bound exactly to one single segment or joint J of the skeleton. Let $\mathbf{V}_S$ denote the position of V with respect to some skin coordinate system, usually determined in a pre-processing phase by registering the skeleton at some specific dressing pose to the corresponding mesh surface, e.g., obtained by accurate 3D laser scans. The world coordinates $\mathbf{V}_W$ of vertex V when the skeleton is at some pose θ with respect to its dressing pose can be computed as

$$\mathbf{V}_W = (M_J^{\theta} \circ (M_J^D)^{-1} \circ M_V^D)(\mathbf{V}_S) \tag{8.8}$$

where M_V^D is the transformation converting the skin coordinates of V to its corresponding world coordinates, M_J^D is the transformation expressing the local coordinate system of the joint J at dressing pose with respect to the world coordinate system, and M_J^{θ} is the transformation accounting for the motion of J when the skeleton goes from the dressing pose to pose θ. Notice that the transformation M_J^θ can be computed following the same rationale of (8.7).

Linear blend skinning is different from rigid skinning in that each mesh vertex can be tied to more than one joint. Let $J^1, \ldots, J^N$ be a set of binding joints for vertex V and let $w_1, \ldots, w_N$ be a set of corresponding weights. The idea is to consider each binding joint separately and apply rigid skinning accordingly, obtaining a set of new candidate positions. The final position of vertex V is computed as a linear combination of the candidate positions according to the given weights. More formally, by the notation of (8.8), the world coordinates $\mathbf{V}_W$ of vertex V, when the skeleton is at some pose θ with respect to its dressing pose, can be expressed as

$$\mathbf{V}_W = (\sum_{j=1}^{N} w_j M_{Jj}^\theta \circ (M_{Jj}^D)^{-1} \circ M_V^D)(\mathbf{V}_S). \tag{8.9}$$

Examples of human models using linear blend skinning can be found in [4, 5, 7, 13, 26], while further algorithmic details can be found in [33, 38].

8.2 Human Pose Estimation

Visual pose estimation refers to the automatic inference from visual data of either a complete or a partial description of the pose of a given target relative to a single time instant. This section presents some examples of the broad family of algorithms devised for this task, focusing on methods recovering the pose of a human hand or body from a single depth map. Following the notation of Sect. 8.1, where θ denotes a set of parameters describing a specific hand or human body pose configuration drawn from the set Θ of all the possible poses, pose estimation can be formulated as the search for the most likely configuration in Θ conditioned to the observation relative to a single time instant. Even though the following relationships hold for any type of observation, each observation will correspond to a single depth map Z for the rest of the section for simplicity. In this case the pose estimation problem can be formally stated as

$$\hat{\theta} = \underset{\theta \in \Theta}{\operatorname{argmax}}\, P(\theta | Z) \tag{8.10}$$

where $P(\cdot)$ is the conditional probability of pose θ given observation Z. It is worth noting that, since no information about the past is available, the search for the pose best explaining the observation potentially extends to the whole space of pose configurations Θ.

This section is focused on *discriminative* algorithms, which are often the preferred choice for pose recovering from single frames. Discriminative algorithms directly model the conditional probability distribution of (8.10), or in other words, model the boundaries of the observations space between the different pose configurations. In a deterministic framework, discriminative algorithms reduce pose estimation to the problem of determining a suitable function mapping the observations (in our case, depth maps) to the pose configuration taken by the observed body or hand.

Learning-based algorithms are a popular option to solve the pose estimation problem, in that they give a way to automatically learn such mapping from the training set. The major drawback of these methods is that they need a large dataset for an adequate generalization, though, on the other hand, a ground truth for each training sample may not be available for large datasets. Although a synthetic dataset generated by rendering a 3D parametric model of the body or hand surface can be used in place or real data, the synthetic data may not accurately approximate real observations. One of the major benefits with learning approaches is that most computation is performed offline, a desirable characteristic for real-time or super real-time applications.

Section 8.2.1 will present an example of learning-based approaches using Random Forests. Section 8.2.2 will present a family of discriminative methods solving pose estimation by a non-learning approach, reducing the problem to a database lookup. Section 8.2.3 discusses the detection of points of interest as an example of partial-pose estimate, by an algorithm specifically targeted to depth data applications.

8.2.1 Learning Based Approaches and the Kinect™ pose Estimation Algorithm

When Microsoft started the development of the tracking system for the Kinect™ v1 in 2008, most approaches available at the time had critical limitations such as the need to wear ad hoc markers, very high computational requirements or the fact that they strongly relied on tracking information from previous frames (many approaches typically lost the tracking after a few critical frames). All these issues prevented the usage of such systems in typical home environments as needed for gaming applications.

The key idea of the pose estimation scheme used for the Kinect™ is to formulate the problem as a *labeling* problem where the acquired depth samples are associated with the various body parts [19, 49, 50]. The rationale behind this choice is that the efficiency of marker-based approaches is due to their capability to associate the different body parts to the corresponding marker, and the same result can be obtained in a marker-less environment by employing *natural markers* [15] produced by the body part recognition scheme. This association simplifies the computation of the position of the various joints from the depth samples of the corresponding body part, and avoids typical issues of tracking-based approaches, such as the need for a pose initialization or the inability to recover when the tracking is lost.

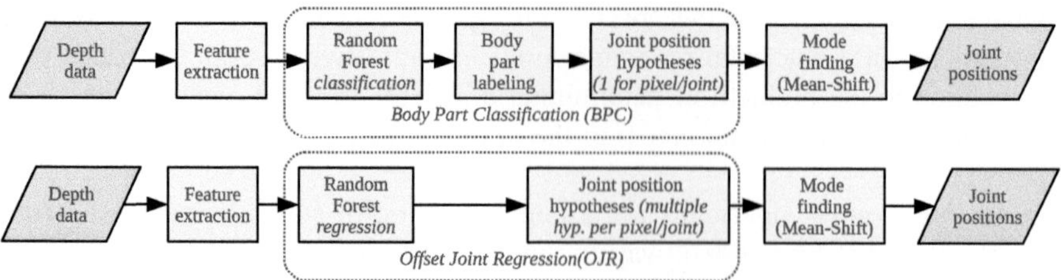

Fig. 8.4 Pipeline of the body pose estimation approach of [50]. The figure shows the basic steps of the two different schemes proposed in [50], i.e., BPC and OJR. Notice that the key difference is the fact that OJR directly estimates the joint positions by a Regression Forest without explicit body part labeling

The basic pipeline of the approach developed by Microsoft, shown in Fig. 8.4, is rather common: a set of relevant features is extracted from the depth data, a machine learning algorithm is used for the classification of the various body parts and finally the joint positions and orientations are estimated. However, the method outlined in Fig. 8.4 includes several key contributions in order to overcome the most critical issues.

The first step in Fig. 8.4 is feature extraction from depth data. The features are extracted on the basis of the comparison between the depth of two samples surrounding the considered sample p with coordinates $\mathbf{p} = [u, v]^T$. More precisely, a feature value $f(\mathbf{p}, \mathbf{d}^1, \mathbf{d}^2)$ from a set of possible couples of displacements $(\mathbf{d}^1, \mathbf{d}^2) = ([d_u^1, d_v^1]^T, [d_u^2, d_v^2]^T)$ is computed as

$$f(\mathbf{p}, \mathbf{d}^1, \mathbf{d}^2) = z\left(\mathbf{p} + \frac{\mathbf{d}^1}{z(\mathbf{p})}\right) - z\left(\mathbf{p} + \frac{\mathbf{d}^2}{z(\mathbf{p})}\right) \tag{8.11}$$

where $z(\mathbf{p})$ is the depth value at $\mathbf{p}$ and $\mathbf{d}^1$ and $\mathbf{d}^2$ are the displacements with respect to $\mathbf{p}$ randomly selected within a rectangular window of fixed size [50]. In particular $\mathbf{d}^2$ is set to 0 with probability $1/2$, i.e., half of the features are computed by comparing the depth of two random points in the window and half by comparing the depth of the considered point $\mathbf{p}$ only with one of the two random points (Fig. 8.5). Notice that the displacement in (8.11) is normalized by the depth value at the considered location in order to make the approach invariant with respect to the distance of the acquired person from depth camera. The samples belonging to the background or falling outside the acquisition range receive a high constant value from (8.11). It is worth noting that each feature value is computed extremely fast from 3 pixels only. The features, although being invariant to the person position in 3D space, are not invariant to perspective distortion and rotation (although the assumption that the depth camera is placed horizontally and the person is standing facing the depth camera is quite reasonable for the considered application).

According to the pipeline of Fig. 8.4 the features are then fed to a classifier, based on Random Forests [23], which assigns each sample to one of the body parts. The

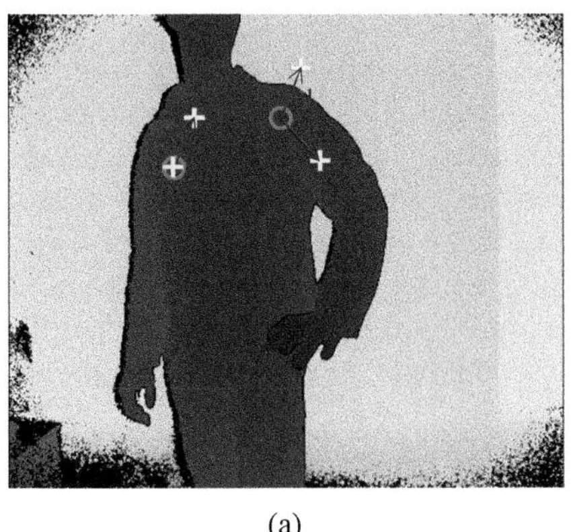

(a)

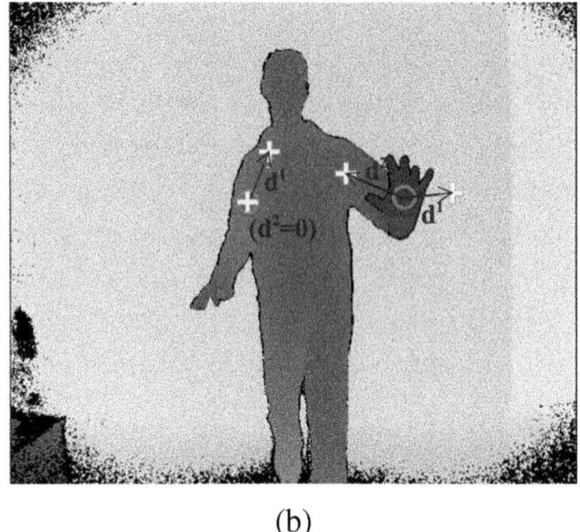

(b)

Fig. 8.5 Computation of features $f(\mathbf{p}^i, \mathbf{d}^1, \mathbf{d}^2)$ of (8.11). Notice that half of the features are computed by comparing two random points (denoted by *yellow crosses*) close to the considered location (denoted by the *red circle*) and half by comparing the considered location with a random point, i.e., with $\mathbf{d}^2 = 0$

body is partitioned in 31 parts associated with the skeletal joints to be estimated. The classifier is made by a set of decision trees where each split node is characterized by a simple thresholding on a feature selecting one of the two branches, i.e.,

$$f(\mathbf{p}^i, \mathbf{d}^1, \mathbf{d}^2) < \tau_n. \tag{8.12}$$

The threshold values τ_n and the set of features used at each node of the various trees are selected by the training procedure.

The training set is made by 2000 depth samples $\mathbf{p}^i, i = 1, \ldots, 2000$ randomly selected from each depth map, that is, the scheme adopts the standard training procedure for Random Forests where a random set of splitting features and threshold values are selected first. Then a sample set of the training data is divided in two subsets according to (8.12) and the decision criteria leading to the highest information gain (according to Shannon's entropy) is selected for each node of the tree. Notice that the proper pose selection is very difficult due to the large number of possible human body poses, and proper training of the classification forest requires millions of different depth maps with ground truth data. Since the real world acquisition and labeling of all these data would be impracticable, the training set was built only from synthetic data. On the other hand, synthetic data do not capture all the variability of real human motion. In order to cope with this problem the researchers of Microsoft acquired a huge set of motion capture data and used this information to render about a million of different synthetic depth maps of human poses with the associated ground truth labeling.

Two different approaches exploiting Random Forests are proposed in [50]. The first, called *Body Part Classification* (BPC), uses a random forest classifier in order to predict in two steps the most likely body part label assignment for each pixel. In the first step the set of randomly selected samples is assigned to the corresponding body part. In the second step this information is used to estimate the joint positions. More specifically, a probability distribution $P_l(k)$ representing the likelihood of each possible assignment (i.e., the probability that pixel p^i is associated with the kth body part b^k) is associated with each leaf l of each tree of the random forest. For each pixel p^i the various trees in the forest are descended until reaching a set of leaves $\mathcal{L}_i$ of cardinality $|\mathcal{L}_i|$, each one corresponding to a certain probability distribution $P_l(k)$. The various probability distributions are finally averaged in order to obtain the estimated probability distribution $P_i(k)$ corresponding to the considered sample p^i

$$P_i(k) = P(k|\mathbf{p}^i) = \frac{\sum\limits_{l \in \mathcal{L}_i} P_l(k)}{|\mathcal{L}_i|}. \tag{8.13}$$

The next step of the processing pipeline of Fig. 8.4 maps these assignments from the body parts to the body joint positions of the model, namely each body joint J^j is associated with a body part b^k trough a mapping function $k = B(j)$. The body joint position is computed by a final aggregation step. Each pixel p^i is back-projected to the corresponding 3D location $\mathbf{P}^i$ and shifted by a predefined amount along the z-axis direction in order to account that the joint lies inside the body while the observed samples are on the surface, i.e.,

$$\hat{\mathbf{P}}^{\mathbf{i}} = \mathbf{P}^i + [0, 0, \delta_j]^T. \tag{8.14}$$

Each pixel p^i then casts a vote for the placement of the jth joint at location $\hat{\mathbf{P}}^i$ with a weight w_j^i based on the product between the probability distribution $P_i(k)$ and the squared depth of p^i, i.e.,

$$w_j^i = P_i(B(j))z(p^i)^2. \tag{8.15}$$

Notice that the $z(p^i)^2$ component accounts for the fact that farther body parts appear smaller due to perspective projection and makes the approach depth invariant. Each pixel gives a single contribution to the position of each joint, even if in practice the probability distributions are 0-valued for many joint/pixel pairs.

An alternative approach, also presented in [19], is the *Offset Joint Regression* (OJR) method, where the joint positions are directly estimated without the intermediate representation given by the body part assignments. In this case the Random Forest is a regression forest providing continuous predictions. Each leaf contains a distribution representing the relative offset from each pixel position in 3D space to the coordinates of each joint and, different from the BPC approach, each pixel can provide multiple contributions to every body joint. The distribution of the possible offsets at each leaf node is very difficult to model and cannot be captured by simple approximations like Gaussian distributions. For this reason, a clustering scheme has been used in [19] in order to represent the offsets by a small set of 3D vectors.

Although the final position of each joint could be simply determined by computing the weighted average by weights w_i^j of the location estimates, this solution proved to be unreliable due to the presence of many outliers. Better results instead can be obtained by a local mode finding approach based on Mean Shift [10], clustering votes corresponding to the same 3D position within a preassigned tolerance. A confidence given by the sum of the corresponding weights w_j^i is assigned to each cluster found by Mean Shift (that corresponds to a candidate joint position). Among the various candidate positions, the one with the highest confidence value is selected and the position of its centroid is considered as the final joint position. This approach is used both for the BPC and OJR schemes.

The just described pose estimation scheme is very reliable, and in real-life environments allows one to recognize even complex partially occluded poses, as shown in the examples of Fig. 8.6. This approach has been extended in various

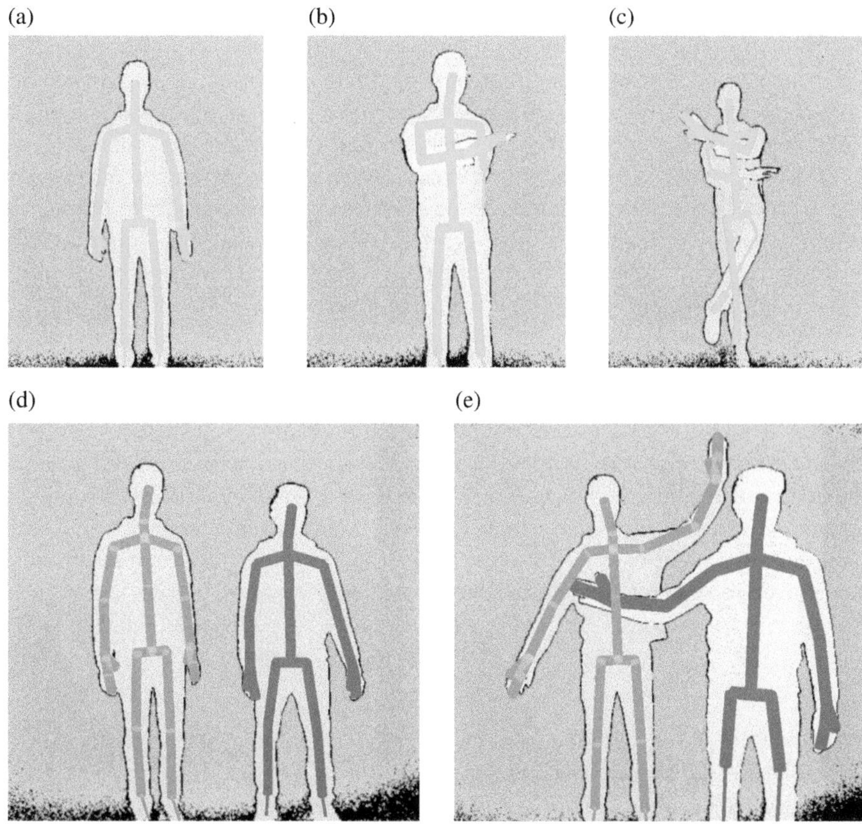

Fig. 8.6 Examples of body joint position estimation by the approach of [50]. Notice how this approach produces correct results not only in simple configuration (**a**, **d**), but also in presence of self-occlusions (**b**, **c**) or occlusions from other users (**e**) provided the occlusions extensions are not too large

recent works, for instance [48] uses a conditional regression forest model in order to include the relationships between the output variables through a global latent variable encoding additional information, like the torso orientation or the height of the acquired person.

The approach of [50] has also been adapted to the context of hand pose estimation. Even though the two problems are conceptually similar, the higher number of self-occlusions makes hand pose estimation a more challenging problem. Various extensions of the previously described body pose estimation methods have been proposed to solve the specific issues of the hand pose estimation [28, 29, 55].

8.2.2 *Example-Based Approaches*

Example-based methods reduce pose estimation to the search for the template best fitting the observations, selected from a database of previously collected and labeled templates (examples) representing the target in various pose configurations. The labels are associated with the different pose configurations θ. A common choice to represent each template is by a synthetically generated mesh approximating the full-body surface, or alternatively one may use real or synthetic depth maps.

These approaches, in order to be effective, require the template set stored in the database to be adequately representative of all the pose configurations θ assumed in the considered application. For this reason, and for the high dimensionality characterizing the space of body or hand poses, a very large number of templates is often needed, making storage resources a major issue for this kind of methods.

Given an input depth map or point cloud, the idea behind this family of methods is to retrieve from the database all the templates with distance from the observation lower than a preset threshold. The distance is measured by a similarity function f_s crucial in the design of any example-based pose estimation algorithm. Ideally, f_s should assign a high similarity score (or equivalently a low distance value) to an input-template pair if and only if their associated pose configurations are similar. Denoting with $A_1, \ldots, A_N$ the templates stored in the database, and with $\theta_1, \ldots, \theta_N$ their corresponding pose labels, a common solution of the pose estimate problem for a given input depth map Z can be computed as follows

$$\hat{\theta} = \theta_i \quad s.t. \quad i = \underset{1 \leq i \leq N}{\operatorname{argmax}} f_s(A_i, Z). \tag{8.16}$$

Note how in (8.16) one may use a point cloud $\mathscr{P}$ in place of the depth map Z, depending on the chosen data representation.

In general, the algorithms of this family after retrieving the subset of all templates with distance from the input lower than a given threshold, chose as final pose either the *nearest* pose to the observation, or a weighted combination of the extracted pose candidates to account for all the pose candidate hypotheses. There are hybrid approaches combining example-based pose estimation methods with generative methods in order to achieve better performance [56]. In particular, for further refinement purposes, the best pose configuration returned by (8.16), or the poses corresponding to the templates most similar to the input, can be fed to local optimization algorithms (e.g., those presented in Sect. 8.3.1) as reliable pose hypotheses for their initialization.

In spite of its apparent simplicity, this general approach exhibits a number of critical issues that must be properly tackled for practical solutions. One of the major drawbacks characterizing example-based pose estimation methods for monocular systems is the dependency of the input data from the specific camera view point. In particular, the scores associated by the similarity function to observations of the same target at identical poses but from different view points can vary considerably. Simple similarity functions between 3D point clouds compare only the point-to-point distances between the template mesh vertices and the input point cloud points, after a prior *registration* (e.g., by ICP). However, since the templates generally describe the full body or hand surface, while the input point clouds often describes only the partial surface framed from the considered camera view point, a naive pair-wise point-to-point comparison typically returns poor similarity estimates. A possible solution, proposed in [56], is to generate from each template mesh A_i in the database four point clouds $\mathscr{P}_{i,1}$, $\mathscr{P}_{i,2}$, $\mathscr{P}_{1,3}$, $\mathscr{P}_{i,4}$ corresponding to four different view points. More specifically, four local coordinate systems are defined with origin at the template mesh centroid and axes derived from the three principal components obtained by applying PCA to the mesh vertices. For each coordinate system, one axis is computed from the component with largest eigenvalue, fixing its positive direction accordingly to the template up-direction. The other two axes are computed from the remaining components, by choosing their positive directions with respect to one of the four possible configurations. The four point clouds are generated first by transforming the template mesh into each local coordinate system, then by rendering the result to a corresponding depth map and finally by projecting each depth map pixel back to the 3D space. In order to compare an input point cloud $\mathscr{P}$ with the pose templates $\mathscr{P}_{i,j}$, for $1 \leq i \leq N$, $1 \leq j \leq 4$, $\mathscr{P}$ is firstly transformed into a local coordinate system with origin at the cloud centroid, and axes computed from the three principal components of $\mathscr{P}$, by fixing the positive direction of the first component accordingly to the camera up-direction, and by fixing at random the positive direction of the remaining two components. Simple similarity functions are then applied to compare the transformed input point cloud against the template point clouds $\mathscr{P}_{i,j}$, for $1 \leq i \leq N$, $1 \leq j \leq 4$, without any concern about view dependencies.

Another drawback is related to the time required for each input-template comparison. Similarity metrics comparing an input depth map or point cloud to a given template on a per-pixel or per-point basis respectively are usually too expensive for real time applications. A common trick to overcome this problem is to embed both inputs and templates to a common low-dimensional space, and perform the matching in it. Denoting by $\mathbb{R}^D$ the space of inputs and templates, and by $h : \mathbb{R}^D \longrightarrow \mathbb{R}^K$ an embedding of $\mathbb{R}^D$ into $\mathbb{R}^K$ such that $K \ll D$, problem (8.16) can be restated as

$$\hat{\theta} = \theta_i \quad s.t. \quad i = \underset{1 \le i \le N}{\text{argmax}} \, \|h(A_i) - h(Z)\|_2^2. \tag{8.17}$$

For an embedding which, for each input or template point cloud, returns the three eigenvalues obtained by applying PCA to the point cloud, see [56].

Finally, ad hoc efficient retrieval methods should be used to speed up the search for the best-fitting templates in the database, which is usually done by minimizing the number of input-template comparisons. An efficient indexing of the templates based on hashing functions is proposed in [47].

8.2.3 Point of Interest Detection

Human pose estimation from visual data related to the same time instant, as previously noted, is affected by a number of issues due to self-occlusions, noisy samples, and the great variability characterizing the appearance and shape of individuals. Although some of these difficulties can be tackled by specific pose estimation methods, the problem of the high dimensionality of the space of human pose configurations still remains a major obstacle.

Luckily, complete pose recovery is not always required. Indeed, several applications only require a partial knowledge of the human body or hand pose configuration, with a consequent reduction of the number of considered pose parameters. Applications based on human pose estimation in indoor environments, for example, often limit their focus to the upper body only, as the lower part is likely to be occluded by the scene objects and does not carry relevant information. Pointing systems are another example for which only a rough estimate of the position and orientation of one or both hands is usually needed. *Goal-oriented* pose estimation algorithms allow to safely neglect all the unnecessary pose parameters, thus reducing the dimensions of the pose configurations space. Some methods, even when the final task is a complete recovery of the full body or hand pose configuration, introduce an intermediate step where a partial pose estimate is computed by simple and fast algorithms. This estimate is then refined by more accurate algorithms delivering a full pose estimate. In other cases, the partial knowledge of the human pose at a given time can be exploited as additional information by robust tracking algorithms in order to help detection and avoid drifting away from the right track. The detection of *points of interest*, returning information about their locations and orientations, is an effective approach for partial pose estimation.

The method of [43] is of special interest, since it tries to directly take advantage of the three-dimensional nature of depth data. It represents a valid alternative to many naive solutions simply applying to the depth data variations of standard algorithms originally developed for color information. The key observation is that human points of interests, such as those corresponding to the head, hands or feet can be located by maximizing their mutual *geodesic* distance on the body surface mesh. Furthermore, geodesic distances can be considered almost invariant with respect to changes in the

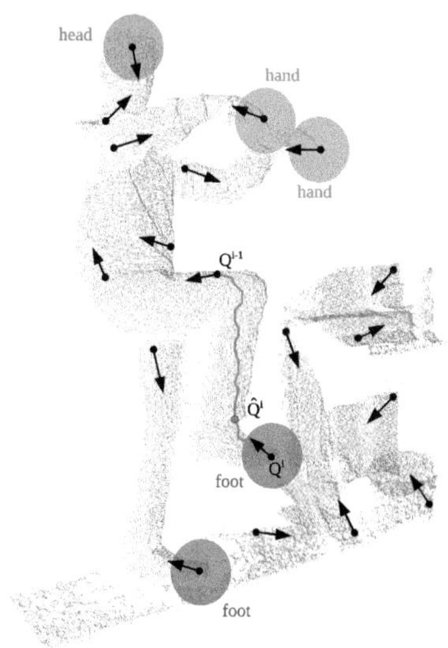

Fig. 8.7 Points of interest by the method of [43] based on surface geodesics. The geodesic path from the (i-1)th point of interest Q^{i-1}, to the point of interest Q^i is shown in *red*. The detected body parts, namely the head, hands and feet, are shown with *colored disks*

body pose configuration. Notice that the latter property may not hold whenever the body surface is segmented in more than one connected component, e.g., because of self-occlusions. However, since only a few components are typically used, this difficulty can be overcome by a small effort. The method of [43] considers the 3D undirected graph induced by the body mesh and, starting from its geodesic centroid, iteratively finds a set of points of interest by selecting at each step the mesh point maximizing the geodesic distance from all the previously found points of interest by Dijkstra's algorithm. Dijkstra's algorithm firstly computes all the shortest paths between every current point of interest (i.e., every point of interest found up to the current step) to each vertex of the body surface mesh. A new point of interest is then selected by choosing the longest of these paths. For each point of interest Q, the shortest path leading to it during the search phase is traced back until some maximum distance threshold is reached, leading to the estimate of point $\hat{Q}$. The orientation associated with the point of interest is thus given by vector $\hat{\mathbf{Q}} - \mathbf{Q}$. The body part detection for each point of interest Q is obtained by feeding a fixed size patch centered at Q to a properly trained supervised classifier (Fig. 8.7).

Algorithms for the points of interest detection are well suited to applications targeted to specific tasks, e.g., pointing systems. Alternatively, if the points of interest are identified by labels, they can be easily matched against the vertices of a synthetic human model. Correspondence-based local optimization methods such as those presented in Sect. 8.3.1 can then be applied to recover a full-pose estimate. In a similar way, it is possible to compute fingertip locations and finger orientations. The complete hand pose can be estimated by inverse kinematics starting from this information. Methods for detecting points of interest can also be deployed to compute high-level features to use as input data for training purposes, in order to improve the accuracy of gesture or activity recognition, as will be seen in Chap. 9.

8.3 Human Pose Tracking

The goal of pose tracking is the recovery of the pose configuration of a given target over time. Different from the single frame pose estimation methods, considered in Sect. 8.2, pose tracking needs a sequence of frames acquired over an interval of time instants $n = 1, \ldots, T, \ldots, T_f$, where T is the current frame instant and T_f the total length of the considered sequence. It is typical to initialize the pose estimate for the current frame at instant $n = T$ with the pose recovered from the previous frame at instant $n = T - 1$, therefore the pose estimate at instant T ends up to indirectly depend from all the previous poses up to the first instant $n = 0$.

Pose tracking can be modeled as a MAP optimization problem over the set of all possible pose sequences $\theta_{1:T} = \{\theta_1, \ldots, \theta_T\}$ of length T, with $\theta_n \in \Theta$ denoting the body pose at instant $n = 1, \ldots, T$ drawn from the pose set Θ, conditioned to the observation sequence $Z_{1:T} = \{Z_1, \ldots, Z_T\}$ with Z_n denoting the depth map acquired at instant n, that is

$$\hat{\theta}_{1:T} = \underset{\theta_{1:T}}{\operatorname{argmax}} P(\theta_{1:T}|Z_{1:T}). \tag{8.18}$$

Following the same convention adopted in the previous sections, θ_n will be used to refer both to some specific human hand or body pose configuration and to the set of parameters needed to describe it, disregarding the chosen data representation. Note how, analogously to Sect. 8.2, the pose observation Z_n at instant n is assumed to be a single depth map for simplicity's sake, although (8.18) holds for any type of data observation.

Different from pose estimation algorithms, most pose tracking methods follow a *generative* approach, better suited to directly exploit the pose information inherited from past estimates and the prior knowledge about the target structure and dynamics. The key idea behind generative approaches can be formalized within a MAP framework similar to the one presented in Chap. 5, i.e., as

$$\hat{\theta}_{1:T} = \underset{\theta_{1:T}}{\operatorname{argmax}} P(Z_{1:T}|\theta_{1:T})P(\theta_{1:T}) \tag{8.19}$$

where the first term denotes the likelihood of all the observations until instant T about the sequence of poses $\theta_{1:T}$, estimated by the selected observation or measurement model used to generate pose hypotheses and compare them against observations. Often, the observation likelihood is modeled by an error function evaluating the dissimilarity between the observations and the tracked pose according to a selected metric (e.g., the Euclidean distance), under the rationale that the lower the computed error the higher is the observation likelihood. The second term, often simply referred to as *prior*, encodes the shared knowledge, namely the available information about the current pose in absence of observations. The prior can be used to explicitly model knowledge about unfeasible or unlikely pose configurations when considering specific motion or space-time locality constraints. Expert knowledge can have an important role in the design of a realistic prior.

Generative approaches differ from the discriminative ones in the fact that they allow to model the joint probability distribution of poses and observations, by estimates of the likelihood and prior as in (8.19). Conversely, discriminative approaches aim to directly derive the posterior probability distribution without attempting to reconstruct the corresponding joint probability distribution. For the aforementioned reasons, generative algorithms are also called *model-based* algorithms.

The solution of problem (8.19) is not trivial, mainly because of the high dimensionality of the search space which makes infeasible the naive evaluation of all the pose sequences $\theta_{1:T}$. The use of graphical human models such those presented in Sects. 8.1.2 and 8.1.3, for example, requires the optimization of about 10–50 real-valued parameters for each frame. The problem remains intractable even limiting the search to sub-optimal solutions, and further assumptions such as the statistical independence of the observations and the Markovian nature of the pose transition process are needed in order to reduce the problem complexity.

The first assumption assesses that each observation Z_n at instant n only depends from the target pose θ_n that generated it and not from the previous observations, hence the observation model in (8.19) can be rewritten as

$$P(Z_{1:T}|\theta_{1:T}) = \prod_{n=1}^{T} P(Z_n|\theta_n). \tag{8.20}$$

The second assumption allows one to model the target pose evolution as a first-order Markov process. In this case, by the Markov property and the Bayes' rule the pose conditional probability with respect to previous poses can be written as

$$P(\theta_T|\theta_{T-1},\ldots,\theta_1) = P(\theta_T|\theta_{T-1}). \tag{8.21}$$

Namely, the pose transition probability from instant $n = T - 1$ to instant $n = T$ is assumed to depend only from the pose at instant $n = T - 1$ and not from the complete pose evolution from instant $n = 1$. It is worth noting that pose tracking by the above formulation is also used as an intermediate step in several approaches addressing the broader problem of *dynamic gesture recognition* treated in Chap. 9, which exploits the complete pose evolution from instant $n = 1$ instead of the single tracked pose at instant $n = T$.

The solution of (8.19) based on assumptions (8.20) and (8.21) leads to two macro-families of approaches: *single-hypothesis* and *multiple-hypotheses* methods. Single-hypothesis approaches assume that a reliable pose configuration estimate at instant $n = T - 1$ is *always* available. In this case the tracking problem can focus on the pose at instant $n = T$ and look for the most probable pose θ_T conditioned to its observation Z_T and to the previous pose estimate θ_{T-1}, i.e., as

$$\hat{\theta}_T = \underset{\theta_T \in \Theta}{\operatorname{argmax}}\, P(Z_T|\theta_T)P(\theta_T|\theta_{T-1}). \tag{8.22}$$

Equation (8.22) is typically solved by the *optimization-based* approaches introduced in Sect. 8.3.1.

Multiple-hypotheses based approaches, instead, are adopted when the pose estimate at instant $n = T - 1$ is not reliable due to a number of factors such as noisy data, inter and self occlusions or fast movements, and cope with the estimate uncertainty by considering also the past observations and pose estimates. In this case the posterior probability distribution of θ_T conditioned to the observations $Z_{1:T}$ can be better formulated as

$$\hat{\theta}_T = \underset{\theta_T \in \Theta}{\operatorname{argmax}} P(\theta_T | Z_{1:T}). \tag{8.23}$$

Probability $P(\theta_T | Z_{1:T})$ in (8.23), also known as the *filtering distribution*, can be derived by marginalization from the full posterior distribution $P(\theta_{1:T} | Z_{1:T})$. Equation (8.23) is typically solved by the *filtering-based* approaches introduced in Sect. 8.3.3.

8.3.1 Optimization-Based Approaches

This section presents a large family of methods modeling human hand and body tracking as an optimization problem to be solved by standard optimization techniques, classified into single-hypothesis and multiple-hypotheses methods.

Single-hypothesis methods at each time retain only the best pose hypothesis and use it as a starting point for the recovery of a new pose in the next time step. Hence the optimization at each time step searches for the best pose in a neighborhood of the previously estimated pose. This provision in practice avoids the search for a global optimum over the whole pose configuration space Θ, which would be computationally prohibitive in most of cases. Its main drawback is that the performance at each step is highly dependent on the quality of the pose estimate computed at the previous step. A poor estimate will heavily affect present and subsequent results.

The methods based on multiple hypotheses, instead, presented in the next section, try to overcome this issue by retaining more than one hypothesis at each time step.

Optimization-based approaches for pose tracking reduce the problem to the minimization of a suitable error function E, measuring the mismatch between a given pose hypothesis θ for the target at current time T and the set of current observations, i.e.,

$$\hat{\theta}_T = \underset{\theta \in \Theta}{\operatorname{argmin}} E(\theta). \tag{8.24}$$

A probabilistic interpretation of (8.24) is readily given by looking at the error function E as a model for the likelihood of observations in (8.22). The higher the mismatch between model and observations, the smaller is the likelihood. A transition prior, instead, is implicitly induced by the locally biased search strategy, by which the poses closer to θ_{T-1} are more likely to be encountered and selected as best estimates. Clearly the design of a suitable error function and the choice of a convenient optimization strategy deserve special care.

The desired error function has to measure the dissimilarity between observations and model predictions effectively, and to be fast to compute. Graphical models such as those presented in Sect. 8.1.3 are used to evaluate the error function in (8.24) at any specific pose hypothesis θ. More precisely, the error function is evaluated by measuring the discrepancy between the observation features extracted from the input data and the features predicted by rendering the body or hand model at a given pose hypothesis θ. The error function is evaluated on a given set of N correspondences between the observation features, taken from the depth map or the 3D point cloud of the scene depending on the adopted input representation and the points of the graphical model (e.g., skin vertices if skinning models are employed), as

$$E(\theta) = \sum_{i=1}^{N} r_i(\theta) \tag{8.25}$$

where each term r_i, called the ith residual, accounts for the dissimilarity evaluated at the ith correspondence pair, for $i = 1, \ldots, N$. When the observations at a given time come from a 3D point cloud, the set of N correspondences takes the form $\{(Q^1, P^1), \ldots, (Q^N, P^N)\}$, where P^i is a point of the cloud representing an observation feature and Q^i is a point of the 3D graphical model, for $i = 1, \ldots, N$ (Fig. 8.8). By denoting with $\mathbf{P}^i_W$ the coordinates of P^i with respect to the world coordinate system, and with $\mathbf{Q}^i_S$ the coordinates of Q^i with respect to some local coordinate system associated for instance to the model underlying the skeleton components, one can define a prediction function $f_p : \mathbb{R}^D \times \mathbb{R}^3 \longrightarrow \mathbb{R}^3$ such that $\tilde{\mathbf{Q}}_W = f_p(\theta, \mathbf{Q}_S)$, which associates a set of pose parameters $\theta \in \mathbb{R}^D$ and the 3D local coordinates $\mathbf{Q}_S$ to the world coordinates $\tilde{\mathbf{Q}}_W$ of the same point (when the model is at pose θ). Notice that, by using skin models, f_p can be directly computed by (8.8) or (8.9). The function f_p allows one to straightforwardly define the error function as

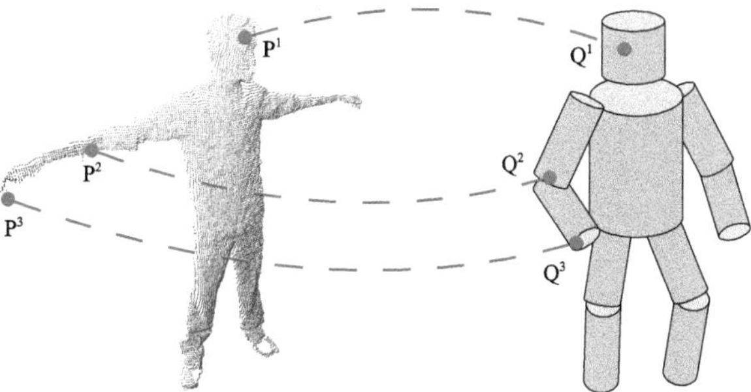

Fig. 8.8 Correspondences between points of a cylinder-based human model [30] (*right*) and points of an input point cloud (*left*)

$$E(\theta) = \sum_{i=1}^{N} \|\tilde{\mathbf{Q}}_W^i - \mathbf{P}_W^i\|_2^2 \tag{8.26}$$

where $\tilde{\mathbf{Q}}_W^i = f_p(\theta, \mathbf{Q}_S^i)$.

8.3.1.1 Local Optimization Methods

Standard local numerical optimization methods such as gradient descent can be directly applied to the solution of (8.24) with error function defined as in (8.26) once a set of correspondences between observations features and model points has been determined. In other words, the pose configuration θ_T of the target at current time step $n = T$ is recovered by looking for a pose minimizing the error function E in a neighborhood of θ_{T-1}, by one of the methods described next.

Stochastic Gradient Descent (SGD) can be effectively employed to locally optimize the error function (8.25) if each residual is a differentiable function with respect to the pose parameters θ. Stochastic gradient descent follows the same rationale as batch gradient descent, except that the direction taken at each iteration is computed from the gradient associated with a few residuals only. It differs from standard gradient descent methods in that at each iteration, it randomly selects a different subset of residuals according to a suitable probability distribution. The updating rule is

$$\theta^{i+1} = \theta^i - \lambda \nabla E_s(\theta^i) \tag{8.27}$$

where $E_s(\theta^i) = \sum_{j \in S} r_j^2(\theta^i)$ is an approximation of the error function computed only on a random subset $S \subset \{1, \ldots, N\}$ of the residuals. Despite its simplicity, stochastic gradient descent benefits of an increased robustness with respect to spurious minima, quite frequent with noisy depth data. Using stochastic subsampling, these minima are likely to change at each iteration, thus preventing the algorithm from getting stuck in one of them. Deterministic subsampling may also be applied to traditional algorithms in order to hasten the computation. While the use of few samples simplifies the computation, it also introduces spurious minima drifting the search towards erroneous poses. This scenario is likely to worsen when only lacking data are available due to noise or self-occlusions, e.g., fingers or limbs orthogonal to the camera plane. Unfortunately, there is no way to easily solve this situation and discriminate spurious minima from the minima corresponding to proper pose hypotheses. A major drawback of stochastic gradient descent is its slow convergence, which may become rather tricky on multidimensional plateaux, where the gradient tends to vanish, or when passing through long and narrow valleys.

The Gauss-Newton method is a well known variation of the Newton-Rhapson method for nonlinear least squares problems, in which the objective function takes the form of (8.25) and the residuals are possibly nonlinear functions. Newton methods move step by step in the search space Θ by locally approximating E by a quadratic function. With this assumption, the step direction decreasing the value of the objective function can be easily determined by straightforward convex analysis. Denoting with θ^i the pose computed at the ith iteration, the error function can be approximated by a quadratic function in a neighborhood of θ^i by Taylor expansion

$$E(\theta) \approx \tilde{E}(\theta) = E(\theta^i) + \Delta_\theta^T \nabla E(\theta^i) + \frac{1}{2}\Delta_\theta^T \mathbf{H}(\theta^i)\Delta_\theta \tag{8.28}$$

where $\Delta_\theta = \theta - \theta^i$, and $\nabla E(\theta^i)$ and $\mathbf{H}(\theta^i)$ are the gradient and the Hessian matrix of the error function evaluated at θ^i. The step to be taken in order to reach the minimum of the approximated error function $\tilde{E}$ is computed by setting its gradient equal to zero, therefore obtaining

$$\hat{\Delta}_\theta = -\mathbf{H}(\theta^i)^{-1}\nabla E(\theta^i). \tag{8.29}$$

It can be shown that the gradient term can be written as $\nabla E(\theta^i) = 2\mathbf{J}(\theta^i)r(\theta^i)$ where $r(\theta^i) = [r_1(\theta^i), \ldots, r_N(\theta^i)]^T$ and $\mathbf{J}(\theta^i)$ is the Jacobian matrix of the first derivatives of E evaluated at θ^i, while the Hessian matrix can be approximated as $\mathbf{H}(\theta^i) \approx 2\mathbf{J}(\theta^i)^T\mathbf{J}(\theta^i)$. The Gauss-Newton updating rule can be written as

$$\theta^{i+1} = \theta^i - (\mathbf{J}(\theta^i)^T\mathbf{J}(\theta^i))^{-1}\mathbf{J}(\theta^i)^T r(\theta^i). \tag{8.30}$$

It is important to notice that, in general, the better the residual functions can be approximated by linear functions the faster is the algorithm convergence. Indeed, if the residuals are linear function with respect to θ, then the objective function E is a quadratic function and the Newton-Rhapson assumption is exactly satisfied. For an error function such as the one of (8.26), it is crucial to assume that small movements between subsequent time steps, so that the residuals associated with the various correspondence pairs can be reasonably approximated by linear functions. Rapid movements, which are quite common for human hands, represent a serious issue for this kind of methods.

Gauss-Newton method rely on the assumption that the error function E can be locally approximated by a quadratic function. The size of the neighborhood where this approximation can be considered acceptable may greatly vary in the space of pose configurations Θ. For this reason, the step size should be tuned accordingly. The Levenberg-Marquardt method automatically penalizes large variations in the value of the pose parameters from one iteration to the other, enforcing small steps when necessary. This is done by adding a scaled identity matrix to the Hessian matrix approximation appearing in (8.30), thus obtaining

$$\theta^{i+1} = \theta^i - (\mathbf{J}(\theta^i)^T\mathbf{J}(\theta^i) + \lambda\mathbf{I})^{-1}\mathbf{J}(\theta^i)^T r(\theta^i) \tag{8.31}$$

where the scaling factor λ is conveniently adapted. Observe that when λ is large the method behaves similar to gradient descent.

Performing the descent by a constant step size as in (8.27) may cause slow convergence with some particular shapes of the error function. The key idea to overcome these situations is to allow the step size to change both with respect to time between subsequent iterations, and with respect to each pose parameter. Various policies have been proposed in order to increase the step size for each pose parameter whenever the gradient keeps its sign constant with respect to that parameter for a certain number of iterations and to decrease it when the sign alternates. The Stochastic Meta-Descent (SMD) proposed by Bray in [6] works in a different way, namely by applying a meta-level descent strategy to update, at each iteration i, a column vector a^i accounting for the descent step size in each parameter direction. The rule for computing a new set of pose parameters is the same of gradient descent, i.e., it is

$$\theta^{i+1} = \theta^i - a^i \cdot \nabla E(\theta^i) \tag{8.32}$$

where $\cdot$ denotes the Hadamard product. The step vector a^i is instead updated accordingly to

$$\ln a^i = \ln a^{i-1} - \mu \nabla E(\theta^i) \cdot v^i \tag{8.33}$$

$$v^{i+1} = \lambda v^i + a^i \cdot (\nabla E(\theta^i) - \lambda \mathbf{H}(\lambda^i) v^i) \tag{8.34}$$

where μ denotes the scalar meta-step size, λ is another scalar, v^i is a column vector whose elements encode the influence of each step size on the corresponding parameter value. Notice that by increasing $0 \leq \lambda \leq 1$ the system can account for the influence of the step sizes on the parameter values over longer time intervals. Approximations are typically applied both to reduce (8.34) to the form $a^i = a^{i-1} \max\{0.5, 1 + \mu v^i \nabla E(\theta^i)\}$ in order to fasten the computation and reduce the evaluation of the Hessian matrix to a much cheaper multiplication of Jacobians. The main advantage over traditional gradient descent methods is that the automatic step size adjustment compensates for small gradients when encountering large plateaux and for oscillating gradients when passing through long and narrow valleys.

8.3.2 ICP and Ray Casting Approaches

All the methods discussed so far assume the availability of a set of correspondences between points of the observed features and graphical model. Approaches based on the Iterative Closest Points (ICP) algorithm and on ray casting combine a strategy for obtaining the correspondences with local optimization methods, thus providing an effective solution to the pose tracking problem.

The ICP algorithm, described in Sect. 7.4, allows one to find the roto-translation that aligns a 2D or 3D source point cloud to a reference one. At each iteration, the correspondences are computed by associating each point of the source cloud with the closet point of the reference cloud. A rigid transformation minimizing the sum of squared distances between corresponding points is then applied to the source point cloud. The process is repeated by updating the correspondences and the roto-translation of the target point cloud until the measured mismatch between the two clouds is less than a given threshold, or until some other stopping condition is met. ICP-like approaches for human pose tracking use a similar strategy to register a 3D skin mesh model to the input point cloud. These approaches can be naturally implemented whenever depth data are available, since depth measurements can be easily converted to 3D points by camera calibration. The correspondences are computed as in ICP, i.e., each skin vertex Q, is associated with the corresponding input point P so that the Euclidean distance $\|\tilde{\mathbf{Q}}_W - \mathbf{P}_W\|_2$ is minimized, where $\tilde{\mathbf{Q}}_W = f_p(\theta_t, \mathbf{Q}_S)$ as in (8.26). A new pose θ_{t+1} is then computed by applying one of the local optimization methods presented in the previous section. Notice that the alignment process is inherently different with respect to the one applied in the ICP, since the pose parameters are jointly optimized and the resulting transformation is no more a rigid transformation. In particular, given the exact correspondences, the original ICP is able to optimally register the source point cloud to the reference point cloud in just one iteration, while many iterations may be needed to register a skin mesh model to an input point cloud.

Another approach not requiring the availability of a correspondence set, is given by rendering the 3D human model in order to generate a synthetic depth map $\tilde{Z}_\theta$ for each pose hypothesis θ, and by comparing it with the input depth map Z_t. The rendering is performed assuming an hypothetical camera placed at a fixed position from the model and by casting rays from the camera to the model. An error function taking into account the discrepancies between the two depth maps can be written as

$$E(\theta) = \sum_i \left|\tilde{Z}(p^i) - Z(p^i)\right|^2 \tag{8.35}$$

where the difference between the synthetic and input depth maps is measured on a per-pixel basis. Different from ICP, *ray casting* allows for a natural modeling of the observed scene capable to capture all the information provided by each single depth pixel. Indeed, ray casting approaches can take into account not only information about the presence of a surface at a given distance for each depth pixel, but also negative information about the absence of any solid object along the ray between the surface and the camera. Despite ray casting models are usually far more powerful than the ones based on the ICP, they are less frequently used, since the error function (8.35) is quite hard to optimize because of the complex dependencies between the pose parameters and the depth samples. On the contrary, simple gradient-based optimization methods can successfully minimize the ICP error function. An approach combining the benefits of both the ICP and ray casting is presented in [17].

8.3.3 Filtering Approaches

A human body can be considered as a dynamic system evolving in time according to a discrete-time state-space model. Let θ_n be the body pose configuration vector at instant n and, with little abuse of notation, let Z_n be the vector collecting the depth values measured by the camera at instant n. The relationships between the body poses and the corresponding observations can be modeled by the following system

$$\theta_n = f(\theta_{n-1}) + V_n \tag{8.36}$$

$$Z_n = g(\theta_n) + W_n \tag{8.37}$$

where θ_n is the *state* vector of the system at time n, Z_n is the *measurement* vector at time n, V_n and W_n are *noise* vectors and f and g are two functions. Equations (8.36) and (8.37) describe a generally nonlinear and non-Gaussian system whose state θ_n is *hidden*, namely, internal to the system and thus not known. The hidden state θ_n can only be estimated at each instant from the measurements Z_n and the physical laws ruling the system evolution. In particular, (8.36) enforces the fact that the state estimate at instant n depends both on the underlying physical process and an uncertainty factor modeled by noise vector V_n, while (8.37) describes the measurement process relating the state value to the measurements Z_n. It is worth noting that state and measurements do not necessarily refer to the same physical quantities, indeed the state represents internal variables only required for estimation purposes.

The goal of body or hand tracking is the estimate, at each instant n, of the most likely state or pose configuration $\theta_n \in \Theta$ assumed by the system, where Θ is the space of all possible poses. When employing a skinning model as presented in Sect. 8.1.3, θ_n typically accounts for the relative orientations and translations of the body joints with respect to their ancestors in the kinematic chain. State θ_n can also possibly include the speed of the joints movements.

State equations (8.36) and (8.37) can be formalized in the Bayesian setting of Sect. 8.3 as the MAP problem of (8.23). According to optimal filtering theory, (8.23) can be recursively solved by iterating two phases, namely a *prediction* step

$$P(\theta_T|Z_{1:T-1}) = \sum_{\theta_{T-1}\in\Theta} P(\theta_T|\theta_{T-1})P(\theta_{T-1}|Z_{1:T-1}) \tag{8.38}$$

where $P(\theta_{T-1}|Z_{1:T-1})$ is computed by recursion at instant $n = T-1$, and an *update* step

$$P(\theta_T|Z_{1:T}) \propto P(Z_T|\theta_T)P(\theta_T|Z_{1:T-1}) \tag{8.39}$$

where the pose estimate is corrected by measurements Z_T.

8.3.3.1 Pose Tracking with Particle Filter

Although the optimal filtering problem is well defined, (8.38) and (8.39) cannot be solved analytically, as the underlying probability distribution generally remains undisclosed. By assuming the system to be a first-order Markov random process, as in Sect. 8.3, only $P(\theta_T|\theta_{T-1})$ and the measurement model $P(Z_T|\theta_T)$ are known from (8.36) and (8.37) respectively.

For these reasons, several body and hand tracking methods in literature based on filtering employ Monte Carlo methods to approximate the unknown posterior $P(\theta_{T-1}|Z_{1:T-1})$ thus making tractable the solution of (8.38) and (8.39), at the expenses of its optimality. Sequential Monte Carlo Approximation, also known as *particle filtering*, and its variations are the most common approaches. Particle filtering (see [3] for a tutorial on this topic) is based on *importance sampling*, namely on describing certain properties of a *target* distribution (e.g., the expectation) by exploiting some known and easily manageable probability distribution, called the *importance distribution*. For instance, the expected pose configuration according to a target probability distribution P can be approximated by a suitable importance distribution Q. Assuming $P > 0$ whenever $Q \neq 0$, it can be easily proved that

$$\mathbb{E}_P[\theta] = \sum_{\theta\in\Theta} \theta P(\theta) = \sum_{\theta\in\Theta} \theta \frac{P(\theta)}{Q(\theta)} Q(\theta) = \mathbb{E}_Q[\theta w(\theta)]. \tag{8.40}$$

Equation (8.40) formalizes the fact that a properly chosen importance distribution Q and a suitable weighting function $w(\theta)$ give the same expectation of the target probability distribution P. The P over Q ratio is named *importance ratio* and the weights compensate for the discrepancies between the target and importance distributions. The selection of the importance distribution is a crucial design step that affects the method effectiveness.

Particle filtering computes the full posterior probability distribution $P(\theta_{0:T}|Z_{1:T-1})$ by iteratively approximating $P(\theta_{0:T-1}|Z_{1:T-1})$ at instant $n = T - 1$ by drawing samples from a suitable importance distribution $Q(\theta_{0:T-1}|Z_{1:T-1})$. Notice that, with a small abuse of notation, the index of θ starts from 0 in order to underline that the algorithm needs a proper initialization. The samples are taken according to a set of N *particles* $\theta^1_{0:T-1}, \ldots, \theta^N_{0:T-1}$ and corresponding weights $w^1_{T-1}, \ldots, w^N_{T-1}$, that is

$$P(\theta_{0:T-1}|Z_{1:T-1}) \approx \sum_{i=1}^{N} w^i_{T-1} \delta^i_{\theta_{0:T-1}} \tag{8.41}$$

where $\delta^i_{\theta_{0:T-1}}$ is the Dirac's delta function centered at value $\theta^i_{0:T-1}$ and the weights w^i_{T-1} are in direct proportionality with the importance ratio between $P(\theta_{0:T-1}|Z_{1:T-1})$ and $Q(\theta_{0:T-1}|Z_{1:T-1})$. Assume that the importance distribution at time $n = T$ can be factored as

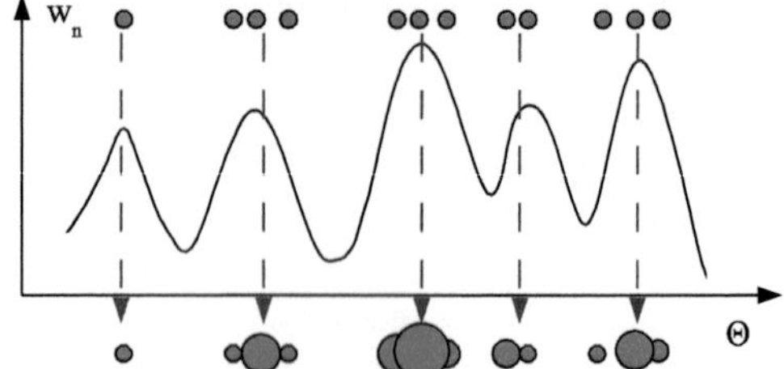

Fig. 8.9 Pictorial illustration of particle filtering. The particle weights are represented by circles of size directly proportional to the weight values

$$Q(\theta_{0:T}|Z_{1:T}) = Q(\theta_T|\theta_{0:T-1}, Z_{1:T})Q(\theta_{0:T-1}|Z_{1:T-1}) \tag{8.42}$$

where $Q(\theta_{0:T-1}|Z_{1:T-1})$ has been computed recursively at instant $n = T-1$. Each particle $\theta^i_{0:T-1}$ at instant $n = T-1$ is updated just by adding θ^i_T sampled from $Q(\theta_T|\theta_{0:T-1}, Z_{1:T})$ at instant $n = T$. Although the full derivation is beyond the scope of this book, let us note how, assuming $Q(\theta_T|\theta_{0:T-1}, Z_{1:T}) = Q(\theta_T|\theta_{T-1}, Z_T)$ (first order Markov process), the weight update can be recursively computed as

$$\theta^i_T \sim Q(\theta_T|\theta^i_{T-1}, Z_T) \tag{8.43}$$

$$w^i_T \propto w^i_{T-1}\frac{P(Z_T|\theta^i_T)P(\theta^i_T|\theta^i_{T-1})}{Q(\theta^i_T|\theta^i_{T-1}, Z_T)} \tag{8.44}$$

where $P(Z_T|\theta^i_T)$ models the similarity of the ith particle solution candidate θ^i_T with the measurement Z_T and $P(\theta^i_T|\theta^i_{T-1})$ models the probability of the ith particle moving from state θ_{T-1} at instant $n = T-1$ to state θ_T at instant $n = T$. A pictorial example of particle filtering is shown in Fig. 8.9.

It is worth noting that the standard particle filtering method is affected by a few problems that without a proper solution would make the algorithm rather inefficient. The first problem is *degeneracy*: after few iterations, only a very small number of particles maintain significant weights, while the other particles have small weights. Degeneracy is typically measured by an estimate of the so-called *effective sample size*, that is

$$N_{eff} = \frac{1}{\sum_{i=1}^{N}(w^i_T)^2} \tag{8.45}$$

where small N_{eff} values imply a large weights variance, hence a most likely degeneracy. Most tracking methods in literature aim at lowering the risk of degeneracy by *resampling* the particle set from the discrete approximation of (8.41) whenever N_{eff} drops below a preset threshold T_S. The particle weights are then all reset to $1/N$.

Since the sampling is performed with replacement, the particles with higher weights are more likely to be sampled multiple times, while the less significant ones are hardly sampled. In this way only the most significant particles are likely to be retained, while the others are discarded. Unfortunately, this solution is plagued by another problem named *sample impoverishment*: as the particles with higher weights are likely to be sampled multiple times, the number of *distinct* particles can be considerably reduced thus making the discrete approximation of (8.41) defined only by a few particles. In the worst case, the approximation can collapse into a single particle describing the whole distribution.

A possible solution to sample impoverishment comes from a variation of particle filtering based on *sample importance re-sampling*, which consists in selecting the state transition distribution $P(\theta_T|\theta_{T-1}^i)$ as the importance distribution $Q(\theta_T|\theta_{T-1}^i, Z_T)$ and by forcing particle re-sampling at each iteration. In this case, (8.43) and (8.44) can be rewritten as

$$\begin{aligned} \theta_T^i &\sim P(\theta_T|\theta_{T-1}^i) \\ w_T^i &\propto P(Z_T|\theta_T^i). \end{aligned} \tag{8.46}$$

The drawbacks of sample importance filtering are the higher risk of sample impoverishment, as the re-sampling is performed at each step, and the fact the weights are no longer directly proportional to the similarity of the estimated state with the measurements.

Another major problem of particle filtering due to the discrete approximation of (8.41) is the sampling of identical samples which, instead, does not happen for a continuous posterior $P(\theta_T|Z_{1:T})$. Such problem can be solved by another variation of particle filtering named *regularized particle filter*: instead of discretizing $P(\theta_T|Z_{1:T})$ by mean of N Dirac's delta functions as in (8.41), the posterior is approximated by a weighted combination of kernel density estimate functions centered at the various θ_T^i, resulting in

$$P(\theta_T|Z_{1:T}) \approx \sum_{i=1}^{N} w_T^i f_K(\theta_T - \theta_T^i) \tag{8.47}$$

where f_K denotes the selected kernel function. Particle re-sampling in this case is only applied when required, as in the sequential importance sampling version of the particle filter algorithm.

Finally, note how the particle filter can also be jointly used with gradient descent approaches both to leverage the advantages of the two families of methods for body and hand tracking and to compensate for their weaknesses. Optimization methods based on gradient descent lead to the best fitting of the body or hand model but are likely to fail in presence of occlusions as they may converge to local minima requiring a complete re-initialization, while particle filtering methods are designed to deal with multiple hypotheses but require a large number of particles. The method of [7], for example, combines the Stochastic Meta-Descent (SMD) algorithm with particle filtering to optimize an error function similar to (8.26). In particular this method, named *Smart Particle Filtering* (SPF), after propagating the particles for the next instant optimizes each particle with SMD leading to a new set of states representing optimized multiple pose hypotheses estimates. This provision allows a more robust tracking of the hand pose with a sensible reduction of the number of particles with respect to standard particle filtering.

8.3.3.2 Pose Tracking with Kalman Filter

If the system defined by (8.36) and (8.37) is *linear* and the noise described by V_n and W_n is *Gaussian*, as in the case of several first or second order dynamic systems used to model mechanical processes, one may prove that the posterior $P(\theta_n|Z_{1:n})$ is Gaussian as well and, in this case, the optimal filtering problem of (8.38) and (8.39) can be rewritten without any approximation as

$$\theta_n = \mathbf{F}_n \theta_{n-1} + V_n \tag{8.48}$$

$$Z_n = \mathbf{G}_n \theta_n + W_n \tag{8.49}$$

where θ_n and Z_n denote the current system state and measurement vector respectively, $\mathbf{F}_n$ is the state transition matrix from system state θ_{n-1} to θ_n, $V_n \sim \mathcal{N}(0, \boldsymbol{\Sigma}_n)$ is the state estimation process noise vector drawn from a multivariate normal distribution with zero mean and covariance matrix $\boldsymbol{\Sigma}_n$, $\mathbf{G}_n$ is the measurement model matrix and $W_n \sim \mathcal{N}(0, \boldsymbol{\Gamma}_n)$ is the measurement process noise vector, assumed to be Gaussian white noise with zero mean and covariance matrix $\boldsymbol{\Gamma}_n$. Note how the generic nonlinear functions f and g of (8.36) and (8.37) have been replaced by linear maps $\mathbf{F}_n$ and $\mathbf{G}_n$ respectively. Equations (8.48) and (8.49) form a *Kalman filter*. The Kalman filter is here used to model autonomous systems, that is systems whose evolution does not depend on any input. However, in the more general formulation, a term accounting for possible inputs is included in (8.48).

The prediction and update steps of (8.38) and (8.39) can be performed by the following steps:

1. Predict the state estimate

$$\hat{\theta}_{n|n-1} = \mathbf{F}_n \hat{\theta}_{n-1|n-1} \tag{8.50}$$

   where $\hat{\theta}_{n|n-1}$, the state estimate at instant n given measurements up to, and inclusive of, time $n-1$, is computed from $\hat{\theta}_{n-1|n-1}$ the state estimate at previous time $n-1$.
2. Predict the covariance estimate

$$\mathbf{M}_{n|n-1} = \mathbf{F}_n \mathbf{M}_{n-1|n-1} \mathbf{F}_n^T + \boldsymbol{\Sigma}_n \tag{8.51}$$

   where $\mathbf{M}_{n|n-1}$, the covariance matrix estimate at instant n, is computed from the estimated covariance matrix $\mathbf{M}_{n-1|n-1}$ at instant $n-1$.
3. Compute the innovation or measurement residual

$$\tilde{Z}_n = Z_n - \mathbf{G}_n \hat{\theta}_{n|n-1} \tag{8.52}$$

   where $\tilde{Z}_n$ denotes the difference between the measurements Z_n at instant n and the observation model $\mathbf{G}_n \hat{\theta}_{n|n-1}$ applied to the estimated state $\hat{\theta}_{n|n-1}$ at instant n.

4. Compute the innovation or measurement residual covariance

$$\mathbf{R}_n = \mathbf{G}_n \mathbf{M}_{n|n-1} \mathbf{G}_n^T + \boldsymbol{\Gamma}_n \tag{8.53}$$

where $\mathbf{R}_n$ denotes the measurement residual covariance matrix at instant n.
5. Compute the optimal Kalman gain

$$\mathbf{K}_n = \mathbf{M}_{n|n-1} \mathbf{G}_n^T \mathbf{R}_n^{-1} \tag{8.54}$$

where $\mathbf{K}_n$ denotes the optimal Kalman gain matrix at instant n.
6. Update the state estimate

$$\hat{\theta}_{n|n} = \hat{\theta}_{n|n-1} + \mathbf{K}_n \tilde{Z}_n \tag{8.55}$$

where $\hat{\theta}_{n|n}$ denotes the updated state estimate at instant n corrected by new measurements Z_n.
7. Update the covariance estimate

$$\mathbf{M}_{n|n} = (\mathbf{I} - \mathbf{K}_n \mathbf{G}_n) \mathbf{M}_{n|n-1} \tag{8.56}$$

where $\mathbf{M}_{n|n}$ denotes the updated covariance matrix estimate at instant n and $\mathbf{I}$ denotes the identity matrix.

In connection with human pose tracking, Kalman filtering is often used to track *saliency points*, e.g., the fingertip positions in the Euclidean space [16] or the hand center [42], and for *trajectory smoothing*, where trajectory means the "history" of the states θ_n, assumed by the system at instants $n = 0, \ldots, T$. The pose configuration state θ_n is typically not fully observable and can only be estimated from the measurements Z_n which, along with the state transition process of (8.48), are affected by measurement noise.

Trajectory smoothing by Kalman filtering has many practical applications in pose estimation from depth data. It can also be used in applications simpler than full pose estimation, for example consider a virtual 3D mouse realized by tracking the palm center or the index fingertip trajectory in the Euclidean space from a sequence of depth maps acquired by a depth camera (i.e., this application uses the position of a single joint). The estimated hand trajectory does not generally appear smooth due to inaccurate estimates of the hand center position either because of the process noise or because of inaccurate modeling of the system dynamics. This system can be modeled as a first-order autonomous dynamic linear system with partially observable state accounting for the mouse position and speed, and observations only considering the measured hand center position. Kalman filtering can be applied to refine the estimated trajectory and obtain a fluid mouse motion.

8.3.4 Approaches Based on Markov Random Fields

Let us recall that Sect. 8.3.1 presented several approaches solving the body or hand tracking by a sequence of optimization problems, targeting suitable objective functions quantifying the discrepancies between the measurements and the estimated pose at each instant. Such methods, however, often employ complex parametric models representing the *full* pose, instead of partitioning the original problem into smaller (and easier) sub-problems to be solved by a *divide and conquer* strategy, and do not fully exploit the hierarchical relationships between the various body parts of interest. Recall also that global optimization algorithms often try to avoid exploring regions of the state space associated with unfeasible pose configurations, due to violations of anatomical constraints, by injecting them directly in the objective function as *penalty* terms.

Consider the body or hand pose configuration θ at a given time, and assume it can be described by the absolute position of the chain root and by the relative orientation θ^k of each body part b^k, for $k = 1, \ldots, N$, with respect to the previous joint in the kinematic chain (see Sect. 8.1). Assume also, for brevity of treatment, that the body or hand surface can be modeled as in Sect. 8.1.3 by a set of primitive solids, e.g., cylinders [9] or "ear-plugs-like" shapes [22] approximating the limbs and thorax in the case of body pose or the fingers in the case of hand pose. Finally, assume that the primitive solids are scaled to tailor the tracked person's dimensions, e.g., by regression on a training set. In this context, the body or hand pose is represented by the relative orientations of the single components of a parametric articulated rigid model approximating the deformable body surface. In particular note how the actual phalanx surfaces can be approximated without any substantial information loss for tracking purposes by 3D cylinders, due to the limited deformation phalanges undergo when bending and to the reduced measurement accuracy of low-cost depth cameras.

With these premises, the original global pose tracking task can be split into independent *local* tracking tasks of each segment, enforcing the anatomical constraints in a further step aggregating each single body part pose estimate in order to discard the local configurations leading to unfeasible global pose estimates. The uncertainty introduced in the pose estimation process by the observation inaccuracies makes this problem suitable to be solved by probabilistic approaches. In particular, the hierarchical body and hand structure can be represented by a *Markov Random Field (MRF)* [9, 22, 35, 52] where each body part b^k is associated with a random variable θ^k representing the part local orientation or its absolute position in the Euclidean space. Call Z^k the random variable accounting for the depth measurements of b^k. Denoting by $\theta = \{\theta^1, \ldots, \theta^N\}$ a random variable associated with the body or hand full-pose and by $Z = \{Z^1, \ldots, Z^N\}$ its global measurements, the MAP problem of (5.11) can be reformulated as [52]

$$P(\theta|Z) \propto P(Z|\theta)P(\theta) = \prod_{k=1}^{N} P(Z^k|\theta^k)P(\theta^k) \prod_{(k,h)\in\mathscr{C}} P(\theta^k, \theta^h) \tag{8.57}$$

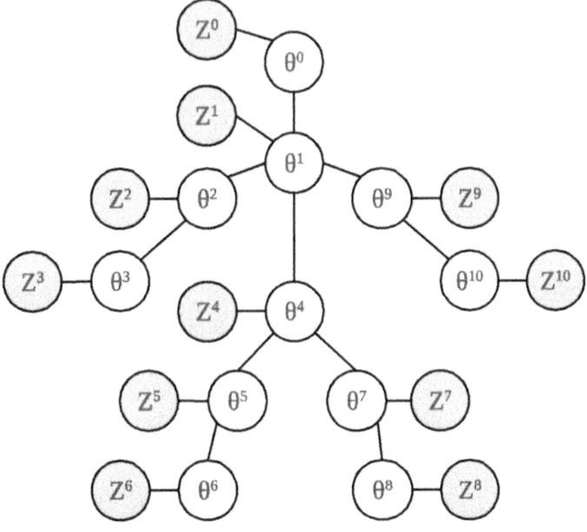

Fig. 8.10 Example of MRF modeling a body kinematic chain

with $\mathscr{C}$ denoting the MRF realizing θ and (k, h) denoting the edges in the graph of $\mathscr{C}$ linking body parts b^k and b^h bounded by precedence relationships in the kinematic chain and by anatomical constraints. Note how, different from the MRF $\mathscr{X}$ of (5.11), $\mathscr{C}$ in this case is no longer a 2D regular lattice but rather a *tree* (Fig. 8.10).

The inference model given by

$$\hat{\theta} = \underset{\theta \in \mathscr{C}}{\operatorname{argmax}} \prod_{k=1}^{N} P(Z^k|\theta^k) P(\theta^k) \prod_{(k,h) \in \mathscr{C}} P(\theta^k, \theta^h) \tag{8.58}$$

can be solved by belief propagation [34], as in this case loopy belief propagation is not required since the graph of Fig. 8.10 is acyclic. If follows that the method is guaranteed to converge in a finite number of steps. The outcome of belief propagation is a more accurate estimate of each segment roto-translation with respect to its parent in the kinematic chain. Due to the selection of the previously described geometric modeling of the segment surfaces, the model likelihood $P(Z^k|\theta^k)$ can be better reformulated as

$$P(Z^k|\theta^k) = P(Z^k|\tilde{Z}^k) \tag{8.59}$$

with Z^k denoting a depth patch extracted from a "real" (acquired) depth map Z, centered at the expected segment position $\mathbf{p}^k$ in the image plane of the depth camera (e.g., by projecting its 3D position $\mathbf{P}^k$ on the image plane), and $\tilde{Z}^j$ denoting a *synthetic depth patch* generated by the rendering of the body or hand model and centered at the neighborhood of the expected position $\mathbf{P}^k$ of the kth body part.

As seen in (5.15), the observation model $P(Z^k|\theta^k)$ plays a fundamental role in the actual pose estimate since it can account for multiple clues like pixel colors or depth [35]. In this case a straightforward approach to model $P(Z^k|\tilde{Z}^k)$ would be aggregating the pixel-wise dissimilarities between the acquired and synthetic depth patches (e.g., by the sum of squared differences (SSD) or the sum of absolute differences (SAD)). However this approach is not used since such metrics are likely to be corrupted by depth measurement noise. For this reason, methods like [40] propose to evaluate the *proximity* of the two patches instead of computing the pixel-wise distances from the patches superposition. Chamfer and Haussdorff [8] distances are two proximity metrics often used for this purpose.

Once selected an appropriate patch similarity metric $d(Z^k, \tilde{Z}^k)$, the model likelihood $P(Z^k|\tilde{Z}^k)$ can be expressed as

$$P(Z^k|\tilde{Z}^k) = \rho \exp\left(-\frac{d^2(Z^k, \tilde{Z}^k)}{\sigma_{Z^k}^2}\right) \tag{8.60}$$

where σ_{Z^k} is a term proportional to the accuracy of the acquired depth data and ρ a proper scaling factor. Methods like [35] compute, instead, probability distribution images by mean of histogram binning techniques applied to each body (or hand) part depth map, color image or edge map and, assuming each pixel statistically independent from the others, estimate the overall image likelihood as

$$P(I^k|S(\tilde{I}^k)) = \prod_{p \in S(\tilde{I})} P_f(p) \prod_{p \in I^k \setminus S(\tilde{I}^k))} P_b(p) \tag{8.61}$$

where I^k denotes a generic image patch of part b^k, $S(\tilde{I}^k)$ denotes the silhouette of the rendered part b^k of the body or hand model and $P_f(p)$, $P_b(p)$ respectively denote the probability that pixel p of patch I^k belongs to the foreground or to the background. The rationale is that the silhouette of the most likely part pose will correspond to the matching of the foreground pixels of the observed body part (detected by mean of histogram techniques) with the part silhouette of the synthetic patch obtained by rendering the oriented part model.

The probability $P(\theta^k, \theta^h)$ of (8.58) models, instead, various anatomical constraints related, for example, to the *proximity* or the relative orientation of two body parts (e.g., the thumb phalanges are expected to have a rather fixed offset or an arm and the related forearm may not form an angle wider than 180° [22]). The probability $P(\theta^k, \theta^h)$ may also model more complex situations like body part occlusions [52] (Fig. 8.11a) or temporal constraints between the same body parts in consecutive frames [9] (Fig. 8.11b).

8.4 Conclusions and Further Reading

Human pose estimation and tracking is a challenging task with many applications in different fields from gaming to research and industry. Depth data can give a significant contribution to the solution of this problem, as demonstrated by the rich literature addressing this topic [36]. This chapter shows how depth data allow to overcome several limitations and ambiguities of earlier vision-based methods like varying lighting conditions or self-occlusions, without relying on marker-based

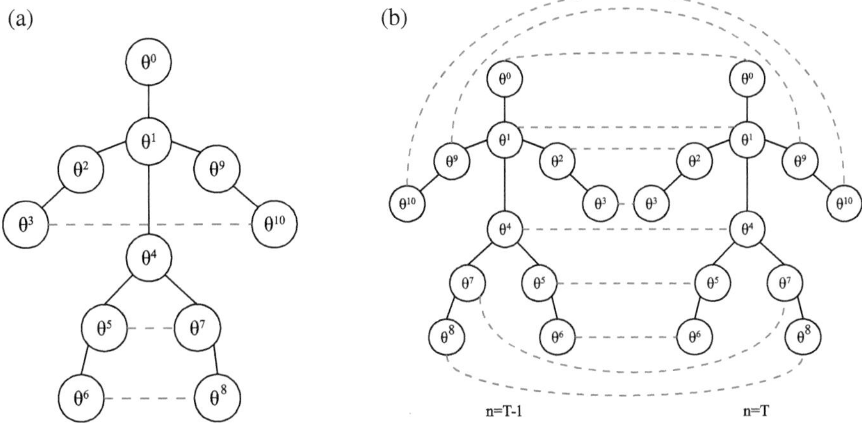

Fig. 8.11 Example of complex constraints that can be handled by MRF: (**a**) body part occlusions; (**b**) temporal constraints. Note how the anatomical constraints are represented by *continuous lines* while the occlusions and temporal constraints are represented by *dashed lines* (in *red*)

systems or on other aiding tools and equipment which have the disadvantage of constraining the movements and forcing to operate in highly controlled environments.

The chapter is organized in three main parts. The first one introduces the theory behind pose estimation and tracking within a probabilistic framework, giving an overview on the most used body and hand models and describes their hierarchical structure and kinematic.

The human pose estimation and tracking, although being instances of the same task, are separately treated in the second and third part of this chapter to highlight how they are generally solved by different families of approaches (generative and discriminative) working on the same type of data. In particular, the second part shows how the human pose in the Euclidean space can be effectively recovered directly from depth data provided by low-cost depth cameras. The third part shows, instead, how pose tracking mostly relies on intermediate body representations (models) and how this task can be either effectively solved in a global fashion or by a sequence of independent sub-tracking problems enforcing the constraints between the different body parts in a subsequent step.

References

1. A. Agarwal, B. Triggs, Recovering 3D human pose from monocular images. IEEE Trans. Pattern Anal. Mach. Intell. **28**(1), 44–58 (2006)
2. I. Albrecht, J. Haber, H.P. Seidel, Construction and animation of anatomically based human hand models, in *Proceedings of ACM SIGGRAPH* (Aire-la-Ville, 2003), pp. 98–109
3. M.S. Arulampalam, S. Maskell, N. Gordon, T. Clapp, A tutorial on particle filters for online nonlinear/non-Gaussian Bayesian tracking. IEEE Trans. Signal Process. **50**(2), 174–188 (2002)

4. C. Barrón, I.A. Kakadiaris, Estimating anthropometry and pose from a single uncalibrated image. Comput. Vis. Image Underst. **81**(3), 269–284 (2001)
5. A. Bottino, A. Laurentini, A silhouette based technique for the reconstruction of human movement. Comput. Vis. Image Underst. **83**(1), 79–95 (2001)
6. M. Bray, E. Koller-Meier, P. Muller, L. Van Gool, N.N. Schraudolph, 3D hand tracking by rapid stochastic gradient descent using a skinning model, in *Proceedings of IEEE European Conference on Visual Media Production* (2004), pp. 59–68
7. M. Bray, E. Koller-Meier, L. Van Gool, Smart particle filtering for 3D hand tracking, in *Proceedings of IEEE International Conference on Automatic Face and Gesture Recognition* (Washington, 2004), pp. 675–680
8. P. Breuer, C. Eckes, S. Muller, Hand gesture recognition with a novel ir time-of-flight range camera: a pilot study, in *Proceedings of International Conference on Computer Vision/Computer Graphics Collaboration Techniques* (Springer, Berlin, 2007), pp. 247–260
9. N.G. Cho, A.L. Yuille, S.W. Lee, Adaptive occlusion state estimation for human pose tracking under self-occlusions. Pattern Recogn. **46**(3), 649–661 (2013)
10. D. Comaniciu, P. Meer, Mean shift: a robust approach toward feature space analysis. IEEE Trans. Pattern Anal. Mach. Intell. **22**, 603–619 (2002)
11. Q. Delamarre, O. Faugeras, 3D articulated models and multiview tracking with physical forces. Comput. Vis. Image Underst. **81**(3), 328–357 (2001)
12. J. Deutscher, A. Blake, I. Reid, Articulated body motion capture by annealed particle filtering, in *Proceedings of IEEE Conference on Computer Vision and Pattern Recognition* (2000), pp. 126–133
13. G. Dewaele, F. Devernay, R. Horaud, Hand motion from 3D point trajectories and a smooth surface model, in *Proceedings of IEEE European Conference on Computer Vision* (Springer, Berlin/Heidelberg, 2004), pp. 495–507
14. M.A. Fischler, R.A. Elschlager, The representation and matching of pictorial structures. IEEE Trans. Comput. **C-22**(1), 67–92 (1973)
15. A. Fossati, J. Gall, H. Grabner, X. Ren, K. Konolige, *Consumer Depth Cameras for Computer Vision: Research Topics and Applications* (Springer, London, 2012)
16. V. Frati, D. Prattichizzo, Using kinect for hand tracking and rendering in wearable haptics, in *Proceedings of IEEE World Haptics Conference* (2011), pp. 317–321
17. V. Ganapathi, C. Plagemann, D. Koller, S. Thrun, Real-time human pose tracking from range data, in *Proceedings of IEEE European Conference on Computer Vision* (Springer, Berlin/Heidelberg, 2012), pp. 738–751
18. D.M. Gavrila, L.S. Davis, 3-D model-based tracking of humans in action: a multi-view approach, in *Proceedings of IEEE Conference on Computer Vision and Pattern Recognition* (1996), pp. 73–80
19. R. Girshick, J. Shotton, P. Kohli, A. Criminisi, A. Fitzgibbon, Efficient regression of general-activity human poses from depth images, in *Proceedings of IEEE International Conference on Computer Vision* (Washington, 2011), pp. 415–422
20. D. Grest, J. Woetzel, R. Koch, Nonlinear body pose estimation from depth images, in *Proceedings of DAGM Conference on Pattern Recognition* (Springer, Berlin/Heidelberg, 2005), pp. 285–292
21. P. Guan, A. Weiss, A.O. Balan, M.J. Black, Estimating human shape and pose from a single image, in *Proceedings of IEEE International Conference on Computer Vision (2009)*, pp. 1381–1388
22. H. Hamer, K. Schindler, E. Koller-Meier, L.J. Van Gool, Tracking a hand manipulating an object, in *Proceedings of IEEE International Conference on Computer Vision* (Kyoto, 2009), pp. 1475–1482
23. T.K. Ho, Random decision forests, in *Proceedings of International Conference on Document Analysis and Recognition* (1995), pp. 278–282
24. D. Hogg, Model-based vision: a program to see a walking person. Image Vis. Comput. **1**(1), 5–20 (1983)

25. S.X. Ju, M.J. Black, Y. Yacoob, Cardboard people: a parameterized model of articulated image motion, in *Proceedings of IEEE International Conference on Automatic Face and Gesture Recognition* (1996), pp. 38–44
26. I.A. Kakadiaris, D. Metaxas, Three-dimensional human body model acquisition from multiple views. Int. J. Comput. Vis. **30**(3), 191–218 (1998)
27. R. Kehl, L. Van Gool, Markerless tracking of complex human motions from multiple views. Comput. Vis. Image Underst. **104**(2), 190–209 (2006)
28. C. Keskin, F. Kıraç, Y.E. Kara, L. Akarun, Real time hand pose estimation using depth sensors, in *Proceedings of IEEE International Conference on Computer Vision Workshops* (2011), pp. 1228–1234
29. C. Keskin, F. Kıraç, Y.E. Kara, L. Akarun, Hand pose estimation and hand shape classification using multi-layered randomized decision forests, in *Proceedings of IEEE European Conference on Computer Vision* (2012)
30. S. Knoop, S. Vacek, R. Dillmann, Sensor fusion for 3D human body tracking with an articulated 3D body model, in *Proceedings of IEEE International Conference on Robotics and Automation* (Orlando, 2006), pp. 1686–1691
31. J.J. Kuch, T.S. Huang, Human computer interaction via the human hand: a hand model, in *Proceedings of Asilomar Conference on Signals, Systems and Computers* (1994), pp. 1252–1256
32. J. Lee, T.L. Kunii, Constraint-based hand animation, in *Models and Techniques in Computer Animation*, ed. by N.M. Thalmann, D. Thalmann. Computer Animation Series (Springer, Tokyo, 1993), pp. 110–127
33. J.P. Lewis, M. Cordner, N. Fong, Pose space deformation: a unified approach to shape interpolation and skeleton-driven deformation, in *Proceedings of ACM SIGGRAPH* (New York, 2000), pp. 165–172
34. S.Z. Li, *Markov Random Field Modeling in Image Analysis*, 3rd edn. (Springer, New York, 2009)
35. T. Liu, W. Liang, X. Wu, L. Chen, Tracking articulated hand underlying graphical model with depth cue, in *Proceedings of IEEE International Congress on Image and Signal Processing* (Washington, 2008), pp. 249–253
36. Z. Liu, J. Zhu, J. Bu, C. Chen, A survey of human pose estimation: the body parts parsing based methods. J. Vis. Commun. Image Represent. **32**, 10–19 (2015)
37. T.B. Moeslund, E. Granum, A survey of computer vision-based human motion capture. Comput. Vis. Image Underst. **81**(3), 231–268 (2001)
38. A. Mohr, M. Gleicher, Building efficient, accurate character skins from examples, in *Proceedings of ACM SIGGRAPH* (New York, 2003), pp. 562–568
39. D.D. Morris, J. Rehg, Singularity analysis for articulated object tracking, in *Proceedings of IEEE Conference on Computer Vision and Pattern Recognition* (1998), pp. 289–296
40. I. Oikonomidis, N. Kyriazis, A. Argyros, Efficient model-based 3D tracking of hand articulations using kinect, in *Proceedings of British Machine Vision Conference* (BMVA, Dundee, 2011), pp. 101.1–101.11
41. J. O'Rourke, N.I. Badler, Model-based image analysis of human motion using constraint propagation. IEEE Trans. Pattern Anal. Mach. Intell. **2**(6), 522–536 (1980)
42. S. Park, S. Yu, J. Kim, S. Kim, S. Lee, 3D hand tracking using kalman filter in depth space. EURASIP J. Adv. Signal Process. **2012**(1) (2012)
43. C. Plagemann, V. Ganapathi, D. Koller, S. Thrun, Real-time identification and localization of body parts from depth images, in *Proceedings of IEEE International Conference on Robotics and Automation* (2010), pp. 3108–3113
44. J. Rehg, *Visual Analysis of High DOF Articulated Objects with Application to Hand Tracking*. PhD thesis, Robotics Institute, Carnegie Mellon University, Pittsburgh (1995)
45. J.M. Rehg, T. Kanade, Digiteyes: vision-based hand tracking for human-computer interaction, in *Proceedings of IEEE Workshop on Motion of Non-rigid and Articulated Objects* (1994), pp. 16–22

46. K. Rohr, Towards model-based recognition of human movements in image sequences. CVGIP Image Underst. **59**(1), 94–115 (1994)
47. G. Shakhnarovich, P. Viola, T. Darrell, Fast pose estimation with parameter-sensitive hashing, in *Proceedings of IEEE International Conference on Computer Vision* (Washington, 2003), p. 750
48. J. Shotton, Conditional regression forests for human pose estimation, in *Proceedings of IEEE Conference on Computer Vision and Pattern Recognition* (Washington, 2012), pp. 3394–3401
49. J. Shotton, A. Fitzgibbon, M. Cook, T. Sharp, M. Finocchio, R. Moore, A. Kipman, A. Blake, Real-time human pose recognition in parts from single depth images, in *Proceedings of IEEE Conference on Computer Vision and Pattern Recognition* (Washington, 2011), pp. 1297–1304
50. J. Shotton, R. Girshick, A. Fitzgibbon, T. Sharp, M. Cook, M. Finocchio, R. Moore, P. Kohli, A. Criminisi, A. Kipman, Efficient human pose estimation from single depth images. IEEE Trans. Pattern Anal. Mach. Intell. **35**(12), 2821–2840 (2013)
51. H. Sidenbladh, M. J. Black, D. J. Fleet, Stochastic tracking of 3D human figures using 2D image motion, in *Proceedings of IEEE European Conference on Computer Vision* (Springer, London, 2000), pp. 702–718
52. L. Sigal, Human pose estimation, in *Computer Vision*, ed. by K. Ikeuchi (Springer, New York, 2014), pp. 362–370
53. L. Sigal, M. Isard, B.H. Sigelman, M.J. Black, Attractive people: assembling loose-limbed models using non-parametric belief propagation, in *Proceedings of Conference on Neural Information Processing Systems* (Cambridge, 2003), pp. 1539–1546
54. B. Stenger, P.R.S. Mendona, R. Cipolla, Model-based 3D tracking of an articulated hand, in *Proceedings of IEEE Conference on Computer Vision and Pattern Recognition* (2001)
55. D. Tang, T.H. Yu, T.K. Kim, Real-time articulated hand pose estimation using semi-supervised transductive regression forests, in *Proceedings of IEEE International Conference on Computer Vision* (2013), pp. 3224–3231
56. M. Ye, X. Wang, R. Yang, L. Ren, M. Pollefeys, Accurate 3D pose estimation from a single depth image, in *Proceedings of IEEE International Conference on Computer Vision* (2011), pp. 731–738

Chapter 9
Gesture Recognition

Gesture recognition, namely the task of automatically recognizing people's gestures, always received great attention among the researchers due to the number of possible applications related to it. In particular, the rapid development of 3D applications and technologies has created the need of novel human-machine interfaces, offering to the users the possibility of interacting with machines in easier and more natural ways than those of traditional input devices, such as keyboard and mouse.

Let us introduce the terminology for a first possible classification of body and hand gestures based on their dynamism and their granularity, where the latter denotes the capability of gestures to be decomposed in atomic movements:

- **Static gestures**: are often characterized by the shape or the pose assumed by the body or the hand at a given instant, e.g., a gesture from the American Sign Language alphabet (Fig. 9.1a).
- **Dynamic gestures**: represent continuous and atomic movements, e.g., raising an arm (Fig. 9.1b). In the case of hand gestures, they are often characterized by the trajectory followed by the hand's center throughout the whole input sequence [10, 50, 51], or by its speed [75].
- **Actions**: are dynamic body gestures made by a hierarchical sequence of atomic movements, e.g., drinking from a glass (Fig. 9.1c). They are usually characterized by body pose variations throughout the whole input sequence.
- **Activities**: are dynamic body gestures made by a sequence of actions, not necessarily bounded by precedence constraints, and describe complex motion sequences like playing soccer (Fig. 9.1d).

Most gesture recognition methods share a common pipeline, depicted in Fig. 9.2. First, the body or its parts of interest, like hands, are identified in the framed scene and segmented from the background. Then, relevant features are extracted from the segmented data and eventually the performed gesture is identified from a set

P. Zanuttigh et al., *Time-of-Flight and Structured Light Depth Cameras*,
DOI 10.1007/978-3-319-30973-6_9

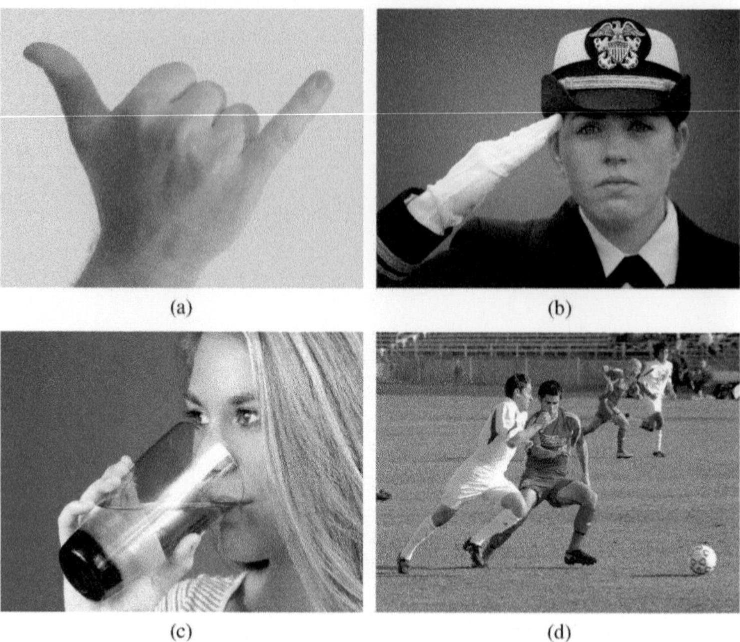

Fig. 9.1 Example of different kinds of gestures: (**a**) static gesture, (**b**) dynamic gesture, (**c**) action and (**d**) activity

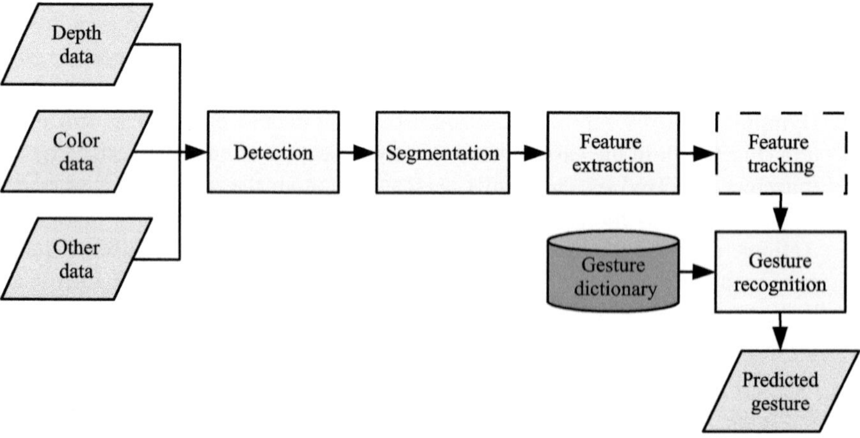

Fig. 9.2 Pipeline of a generic gesture recognition algorithm

of predefined gestures, possibly exploiting suitable machine learning techniques. In the case of non-static gestures, the general pipeline also includes tracking features among multiple frames.

The earlier gesture recognition approaches, mainly based on the processing of color pictures or videos only, suffered from the typical issues characterizing such type of data. Indeed, inter and self occlusions, different skin colors among different users, and unstable illumination conditions, often made the gesture recognition problem intractable. Other approaches solved the previous issues by making the user wear reflective markers, gloves equipped with sensors, or hold tools or devices designed to help the localization of body and hands, thus compromising the user experience.

The recent introduction in the mass market of the consumer depth cameras paved the way to innovative gesture recognition approaches exploiting the geometry of the framed scene. In particular, as already seen in Chap. 8, body and hand detection and segmentation problems are effectively and efficiently solved by novel methods either employing depth information only or improving the segmentation performance of earlier methods based on color data with depth information. The interested reader is referred to [40, 61] for a review of the current gesture recognition advancements based on depth data.

This chapter is organized in three sections. Section 9.1 introduces the most relevant approaches for *static gesture recognition* and Sect. 9.2 presents *dynamic gesture recognition*, providing an overview of both the methods based on color data and on the novel approaches designed to leverage the peculiar structure of depth data. For brevity of treatment we will assume body and hands correctly detected and segmented by proper detection and segmentation techniques.

9.1 Static Gesture Recognition

Gesture recognition, either static or dynamic, can be framed as a family of pattern recognition tasks including the extraction from the object of interest of one or more feature sets describing relevant pattern properties, and the comparison of features' values with a classification model previously trained. The goal is detecting the most likely entry from a given "gesture dictionary" that generated the actual gesture.

While in this chapter these concepts will be examined in connection with static gesture recognition, the pipeline of Fig. 9.2 shares several steps also with the hand or body pose estimation task of Chap. 8. Pose estimation can be configured as a pattern recognition problem with each pattern describing a different pose from the relative positions and orientations of the skeletal joints. This information can also be used as a valuable clue for recognizing static gestures described by their skeletal's configuration.

Static gesture recognition, as previously stated, strongly relies on detecting relevant feature sets having high discrimination capabilities and leveraging the underling structure of the acquired data. While the recognition approaches based on color information have been deeply analyzed in the literature [52, 70], this chapter focuses on feature extraction from depth data.

Static gesture descriptors based on depth data characterize the performed gestures mainly from the shape of the body or hand. This shape characterization can be performed either in the image plane by measuring properties of the body or hand contour, or in the Euclidean space by measuring properties of the body or hand surface or the occupied volume.

The remainder of the current section provides an overview of the most common descriptors adopted in the literature for static gesture characterization. Note how certain approaches analyze the segmented depth map Z, while other methods rely either on the point cloud $\mathcal{P}$ generated by back-projecting (inverting (1.4)) the body or hand's samples in Z or on the surface mesh $\mathcal{M}$.

9.1.1 Pose-Based Descriptors

Pose-based descriptors assume that the body or hand surface is represented by a tridimensional mesh $\mathcal{M}$ undergoing deformations induced by the pose of the associated kinematic chain according to a skinning algorithm, e.g., linear blend skinning. The solution of the inverse kinematic problem with the algorithms presented in Chap. 8 leads to reveling the current body or hand pose, identified by the roto-translations $(\mathbf{R}_j, \mathbf{t}_j)$ of each skeletal joint j, where $\mathbf{R}_j$ and $\mathbf{t}_j$ denote respectively the relative rotation matrix and the offset vector with respect to the previous joint in the kinematic chain. We should recall from Chap. 8 that $\mathbf{R}_j$ is often represented by the yaw α, pitch β and roll γ angles of the $\mathbf{R}_j$ decomposition into elementary rotations around x,y, and z axes, namely $\mathbf{R}_j = \mathbf{R}_{j,z}(\alpha)\mathbf{R}_{j,y}(\beta)\mathbf{R}_{j,x}(\gamma)$, or by the axis-angle representation $\boldsymbol{\theta} = \theta\mathbf{e}$ with θ the rotation value around unit vector $\mathbf{e}$, or by means of quaternions.

A pose-based descriptor consists in the simple concatenation of the relative orientations $\mathbf{R}_j$ and the offsets $\mathbf{t}_j$ of the skeletal joints [20, 33, 62, 65, 72, 78, 80]. Note that $\mathbf{R}_1 = \mathbf{I}$ is assigned by definition to the root joint $j = 1$ and the joints relative offsets $\mathbf{t}_j, j = 2, \ldots, N$ are often not considered in the global pose descriptor. In the case of static hand gesture recognition, the same approach is directly applicable to the skeletal hand joints. An alternative descriptor may account directly for the positions of the fingertips in the Euclidean space with respect to a local coordinate system on the hand palm. Such coordinate system, shown in Fig. 9.3, usually has origin in the hand center P_C and is defined by the principal hand direction $\mathbf{h}$, often obtained by performing principal component analysis (PCA) on the hand point cloud $\mathcal{H}$, the normal $\hat{\mathbf{n}}$ to the palm region points $\mathcal{P} \subset \mathcal{H}$, and the cross-product of $\hat{\mathbf{n}}$ and $\mathbf{h}$. The palm region is often described with a plane π fitted to the palm region points $\mathcal{P}$ by means of SVD in a RANSAC framework. Let us denote with $[x_\pi, y_\pi, z_\pi]^T$ the coordinates in this reference system, where the first coordinate corresponds to the principal hand direction $\mathbf{h}$, the second to the cross-product $\hat{\mathbf{n}} \times \mathbf{h}$ and the last to $\hat{\mathbf{n}}$.

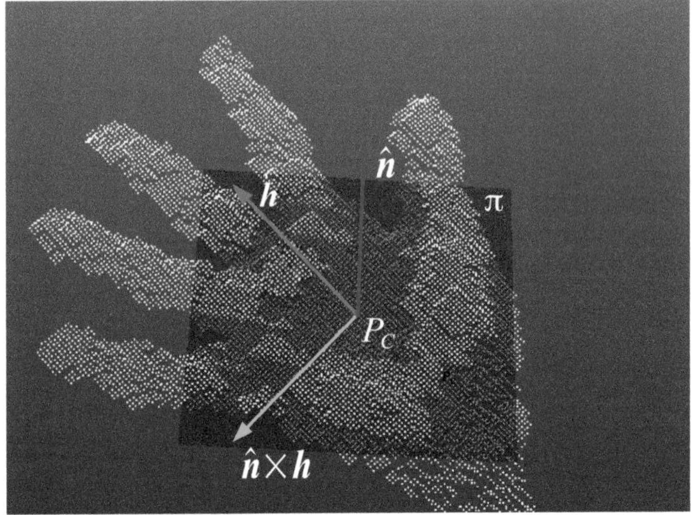

Fig. 9.3 Local 3D coordinate system of the hand with P_C denoting the palm center, $\mathbf{h}$ the main hand direction, $\hat{\mathbf{n}}$ the normal direction to the palm region and $\hat{\mathbf{n}} \times \mathbf{h}$ the remaining axis orthogonal to $\hat{\mathbf{n}}$ and $\mathbf{h}$

9.1.2 Contour Shape-Based Descriptors

Contour shape-based descriptors extract relevant properties of the body or hand silhouette, which is often peculiar to each different gesture. Hand silhouettes, for example, are strongly characterized by the concavities and convexities around the fingertips and between each pair of adjacent fingers. For instance, the contour of the hand clenched in the fist is overall convex, while the contour of the hand with all the fingers extended presents a similar number of convexities (around fingertips) and concavities (between adjacent fingers). This section analyzes only the case of hands, although the presented concepts can be easily extended to bodies.

Given the hand contour pixel p^i in the image plane, sorted according to adjacency criteria (e.g., p^{i+1} is the adjacent contour pixel to p^i in a clock-wise or counter-clockwise fashion), a possible contour representation consists in the Euclidean distance between each hand contour pixel p^i with coordinates $\mathbf{p}^i = [u^i, v^i]^T$ and the estimated hand center p_C with coordinates $\mathbf{p}_C = [u_C, v_C]^T$ [56]. Namely, a plot $\mathscr{C}(p_i)$ is built converting the p^i Cartesian coordinates to their associated polar coordinates $[r(p^i), \theta(p^i)]^T$ with respect to p_C, defined as

$$\begin{aligned} r(p^i) &= \|\mathbf{p}^i - \mathbf{p}_C\|_2 \\ \theta(p^i) &= \text{atan2}\left(v^i - v_C, u^i - u_C\right). \end{aligned} \tag{9.1}$$

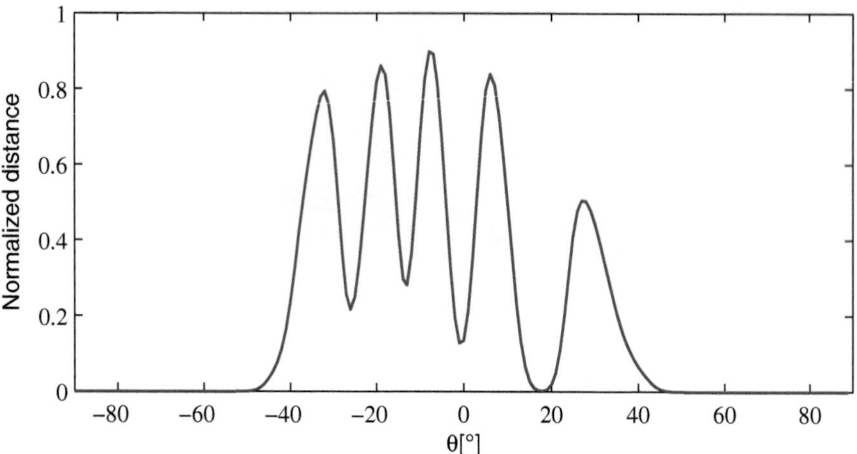

Fig. 9.4 Example of distance plot representing the open hand contour in the image plane

Angle $\theta(p^i)$ is defined by the segment $\overline{p^i p_C}$ connecting p^i and p_C, and the image plane v-axis. Figure 9.4 shows an example of distance plot generated from the open hand.

Hand contour representation of (9.1), however, is affected by the palm detection problems: the higher the hand inclination with respect to the image plane, the higher the warping of the hand shape projection on it, and the shorter the fingers appear. In the worst case, even an hand featuring all the fingers extended may appear as it was a fist on the image plane. In order to make the distance plot invariant to the hand inclination with respect to the camera image plane, (9.1) can be reformulated into (9.2) by replacing the hand center p_C and the hand contour pixels p^i with their back-projected points P_C and P^i and computing the Euclidean distance in the 3D space

$$d(p^i) = \|\mathbf{P}^i - \mathbf{P}_C\|_2. \tag{9.2}$$

Both hand contour representations of (9.1) and (9.2) are affected by two problems: the support of the function strongly depends on the hand's contour length and the complete fingers configuration is lost whenever one or more fingers are partially folded over the palm, since the hand contour on the image plane is not able to capture this information. In order to overcome such limitations, a fixed length distance plot can be directly generated from the 3D hand point cloud without first extracting the hand's contour in the image plane.

Let $\mathscr{F} = \{P^i \in \mathscr{H} \setminus \mathscr{P}\}$ denote the 3D point cloud of the finger samples. For each 3D point $P^i \in \mathscr{F}$, (9.3) computes its normalized Euclidean distance $d(P^i)$ from the palm center P_C, and the angle $\theta(P^i)$ between the projection $\mathbf{P}^i_\pi = [x^i_\pi, y^i_\pi, z^i_\pi]^T$

on the palm plane π, expressed in the palm local coordinate system, and the main hand direction $\mathbf{h}$ by

$$\begin{aligned} d(P^i) &= \|\mathbf{P}^i - \mathbf{P}_C\|_2 \\ \theta(P^i) &= \operatorname{atan2}(y^i_\pi, x^i_\pi) \end{aligned} \tag{9.3}$$

The range $[0, 2\pi)$ of the possible values of $\theta(P^i)$ is then partitioned into Q uniform angular bins of width $\Delta = \lfloor 2\pi/Q \rfloor$ and the 3D points P^i are binned according to their angle $\theta(P^i)$. The resulting histogram $h(\theta_j), j = 1, \ldots, Q$ made by extracting the maximum length $d(P^i)$ within each angular bin θ_j is

$$h(\theta_j) = \max_{P^i \in \mathscr{P}_j} d(P^i) \tag{9.4}$$

with $\mathscr{P}_j$ denoting the set of points $P^i \in \mathscr{F}$ belonging to the jth angular bin θ_j of the desired fixed-length 3D distance plot, analogous to the one showed in Fig. 9.5.

Finally, the distance plots of (9.1), (9.2) or (9.3) are often smoothed by the convolution with a gaussian kernel of short support in order to avoid spurious peaks due to the measurement noise. Although such plots are highly discriminative, and intuitively could be directly used as mono-dimensional feature vectors, there are a few issues to be solved before:

- the plot's amplitudes are different even for instances of the same gesture;
- the plot's support is different for each framed gesture due to the varying hand contour lengths;
- plots representing instances of the same gesture often present a non-negligible phase offset.

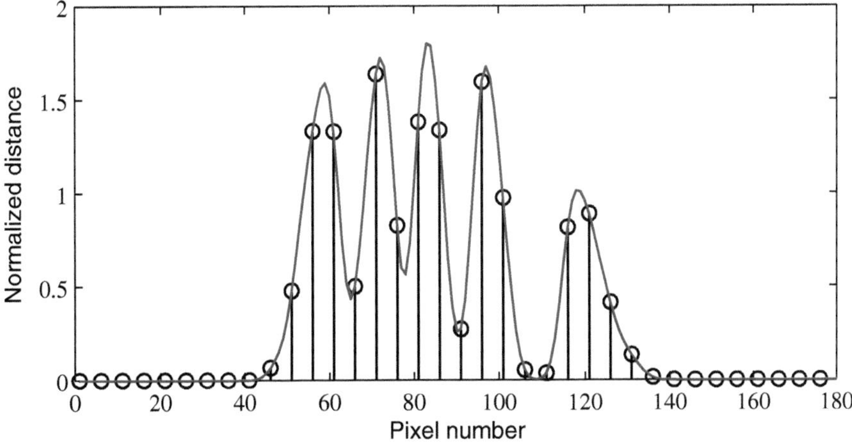

Fig. 9.5 Example of hand contour distances from the palm center on the image plane. The sampled contour distances are represented with *circles*

While the first issue can be easily solved by normalizing the plot's amplitude, e.g., scaling the distances from the palm center by the maximum measured distance or the palm radius [56], the second and third problems require more processing. For instance, the phase offset can be removed by leveraging the estimated hand orientation to rotate the segmented hand depth map and align the main hand direction in the image plane with the v-axis. In this way the first hand contour pixel always corresponds to the one intersecting the v-axis near the wrist center. The second issue can be solved by down-sampling the hand contour by a uniform quantization step with size depending on the hand contour length and the desired feature vector length [38]. The final feature vector is made by the juxtaposition of the sampled distances from the aligned plot of (9.2).

Another possibility to jointly solve the first two issues consists in generating the 3D fixed-support normalized distance plots of (9.3) and find its best alignment with each gesture template, namely the alignment maximizing the value of an appropriate distance metric (such as the zero-mean cross correlation) [19]. In order to overcome the misalignment due to partial matching, the final descriptor could be the concatenation of the histogram peaks extracted within selected regions of each gesture template. The rationale behind this idea is that the peaks in the selected regions are most likely associated with the fingertips, and in the case of misalignment the peak values differ sensibly. Figure 9.6 shows an example of peak extraction from two regions R_1 and R_2 of a gesture template. A variation of the previous approach uses a concatenation of the maximum similarity values of the hand's contour with respect to each gesture template in place of the extracted peaks [38].

Similar to the idea just described, hand gestures may be also characterized by the distances of the finger samples $P^i \in \mathscr{F}$ from the palm plane π [17]. The rationale is

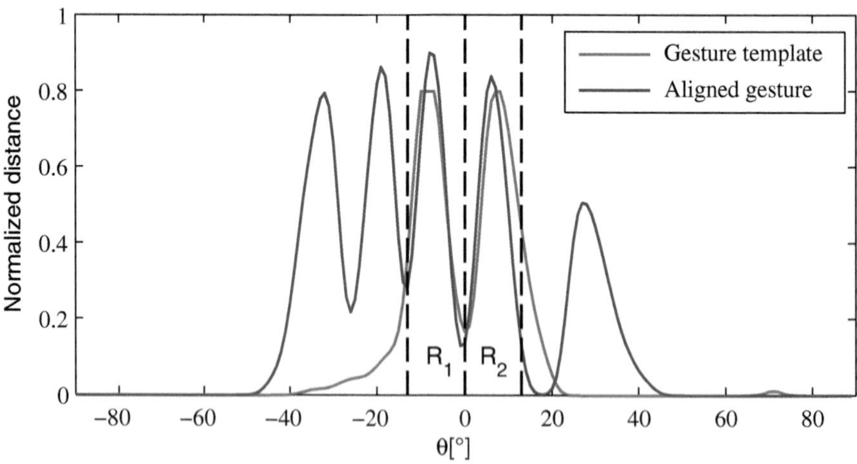

Fig. 9.6 Example of hand contour distances from the palm center in the Euclidean space. The regions of interest for the peak extraction are R_1 and R_2

that the considered gesture dictionary may contain gestures sharing similar distances from the palm center P_C but different fingertip positions in the Euclidean space, that is a property not captured by the descriptor just described.

Let $e(P^i)$ denote the signed distance of the 3D point P^i from the palm plane π, computed as

$$e(P^i) = \text{sgn}\left([\mathbf{P}^i - \mathbf{P}^i_\pi] \cdot \mathbf{h}\right) \|\mathbf{P}^i - \mathbf{P}^i_\pi\|_2 \tag{9.5}$$

with P^i_π denoting the projection of P^i on the palm plane π. The sign of $e(P^i)$ accounts for the fact that P^i may belong to any of the two semi-spaces defined by π, that is, P^i may either be in front or behind π.

Feature extraction can be obtained adapting the previous idea. First build a plot $E(\theta_j)$ representing the signed distance of each sample P^i from the palm plane π, analogously to the distance plot of (9.4), then align $E(\theta_j)$ to each gesture template and extract a set of distances between the fingertips and the palm plane from the selected regions within the aligned plots, as follows

$$E(\theta_j) = \begin{cases} \max\limits_{P^i \in \mathcal{P}_j} e(P^i) & \text{if } \left|\max\limits_{P^i \in \mathcal{P}_j} e(P^i)\right| > \left|\min\limits_{P^i \in \mathcal{P}_j} e(P^i)\right| \\ \min\limits_{P^i \in \mathcal{P}_j} e(P^i) & \text{otherwise} \end{cases} \tag{9.6}$$

with θ_j denoting the jth angular interval with the same width of the one used for the previous distance descriptor. Figure 9.7 shows an example of elevation plot generated by (9.6). While the distance plots generated by (9.3) are rather robust with respect to the same gesture instances, the plots generated by (9.6) are strongly affected by the reliability of the plane π fitting and may thus sensibly differ also in the case of repetitions of the same gesture. For this reason, the descriptor should avoid the alignment of $E(\theta_j)$ with each gesture elevation template and rely, instead, on the angular shifts computed for the alignment of $h(\theta_j)$ with the respective gesture templates. The final vector is composed in the same way of the previous descriptor.

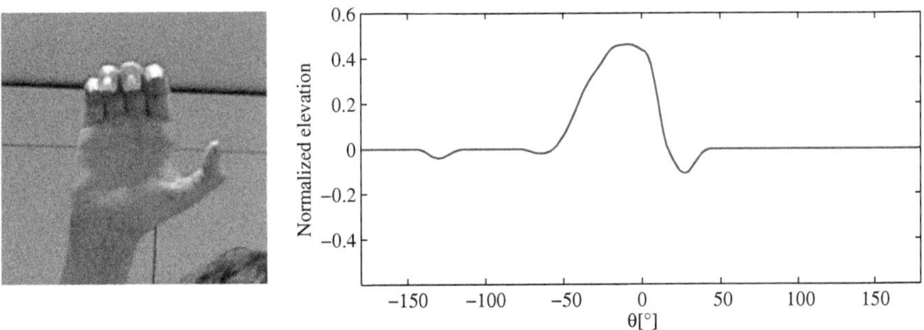

Fig. 9.7 Example of hand contour elevation: input gesture (*left*) and generated elevation plot (*right*)

Hand contour curvature information is another shape property having a dramatic discrimination capability, since the hand contour is strongly characterized by the concavities and the convexities around the fingers, as already stated in the beginning of this section.

For clarity's sake, let $\Gamma = \{(x(u), y(u))|u \in [0, 1]\}$ denote a planar curve representing the hand contour in the image plane, where u is the normalized arc length. The curvature function $\kappa(u)$ of Γ can be expressed as follows

$$\kappa(u) = \frac{\dot{x}(u)\ddot{y}(u) - \ddot{x}(u)\dot{y}(u)}{[\dot{x}(u)^2 + \dot{y}(u)^2]^{3/2}} \tag{9.7}$$

with $\dot{x}(u), \dot{y}(u)$ and $\ddot{x}(u), \ddot{y}(u)$ denoting respectively the first and second order derivatives of the curve components.

Equation (9.7) suggested two approaches employing the hand contour curvature as a gesture recognition clue: the first method represents the hand contour by mean of a *Curvature Scale Space Image* (CSS) [11] and the second one computes the distribution of local curvature descriptors extracted from patches of varying size centered at the hand contour points p^i [18].

The CSS image, an example of which is shown in Fig. 9.8, is a parameterization of the curvature (9.7) of curve Γ in terms of the normalized arc length u and the variance of a Gaussian kernel $g(u, \sigma)$ with variance σ convolved to Γ

$$\kappa(u, \sigma) = \frac{X_u(u, \sigma)Y_{uu}(u, \sigma) - X_{uu}(u, \sigma)Y_u(u, \sigma)}{[X_u(u, \sigma)^2 + Y_u(u, \sigma)^2]^{3/2}} \tag{9.8}$$

with $X_u(u, \sigma)$ and $Y_u(u, \sigma)$ denoting the convolution of $x(u)$ and $y(u)$ with the first-order derivative of $g(u, \sigma)$ with respect to u, and $X_{uu}(u, \sigma)$, $Y_{uu}(u, \sigma)$ the convolution of $x(u)$ and $y(u)$ with the second-order derivative of $g(u, \sigma)$. The zeros of $\kappa(u, \sigma)$ disclose peaks associated with the hand fingers disregarding the hand orientation, and the peak number, height and arrangement is characteristic of each different gesture.

The idea of the local curvature descriptors of the second approach [18], instead, is motivated by the fact that the hand contour curvature analysis by differential operators, such as $\kappa(u)$, is impaired by the noise characterizing depth data and their planar projections describing any non-parametric curve made by the coordinates $\mathbf{p}^i = [u^i, v^i]^T$ of the hand contour pixels. Integral invariants [28, 35] are instead rather robust with respect to noise and then can be usefully employed. Consider a set of S circular masks $M_s(p^i)$, $s = 1, \ldots, S$ of radius r_s centered at each edge pixel p^i of the hand contour. Let $C(p^i, s)$ denote the curvature in p^i at scale level s, expressed as the ratio of the number of hand samples within the mask $M_s(p^i)$ over $M_s(p^i)$ size, computed for each hand contour pixel p^i and mask scale s as exemplified in Fig. 9.9.

The values of $C(p^i, s)$ strongly depend on the hand size. A first possible solution for compensating the different hand sizes would be scaling the hand region to a

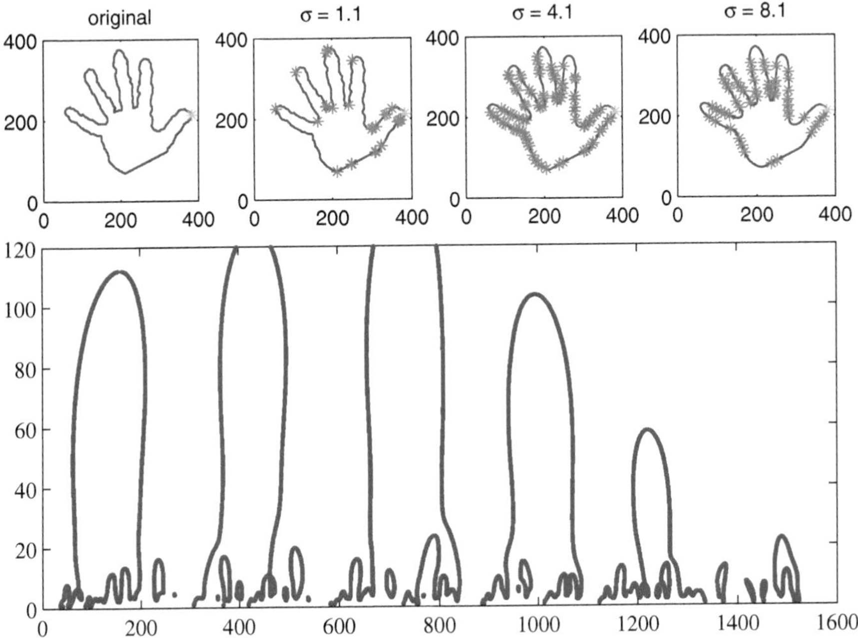

Fig. 9.8 Example of CSS image generated from the hand contour

predefined size prior extracting the local features. An alternative solution consists in setting the adopted mask sizes in metrical units (e.g., by setting each mask radius in millimeters) and exploiting the hand depth information to map the fixed mask radiuses in metrical units to their respective lengths in pixels, according to the hand distance from the acquisition setup. This provision ensures the invariance of the extracted descriptors from the different hand sizes on the image plane due to projective geometry.

The values of $C(p^i, s)$ range from 0 (extremely convex shape) to 1 (extremely concave shape), with $C(p^i, s) = 0.5$ corresponding to the curvature of a straight edge. Each curvature value $C(p^i, s)$ is then binned into a bi-dimensional histogram made of N curvature bins and S mask size bins. The columns of the histogram represent the distribution of the curvature values estimated for each histogram row, namely for each considered mask size. The rows of the histogram can then be concatenated within a unique feature vector, or alternatively a set of image features can be extracted from the 2D histogram.

Although circular masks ensure the descriptor's rotation invariance, the computational complexity of the curvature descriptor's extraction increases dramatically with the number and size of the employed masks. For this reason the feature extraction performance can improve if one replaces the circular masks with rectangular patches and computes the density of the hand pixels in constant time, within each patch disregarding its size, using integral images [14].

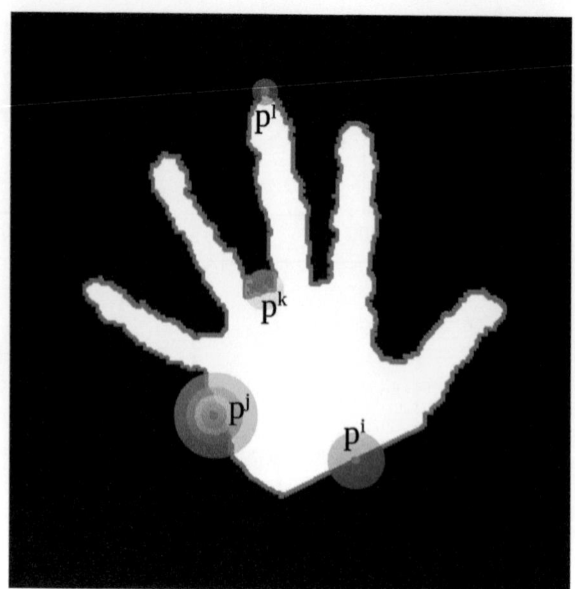

Fig. 9.9 Example of curvature extraction with masks of different size

9.1.3 Surface Shape Descriptors

Surface shape features describe relevant geometric properties of the body or hand surface, that the contour shape descriptors of Sect. 9.1.2 are not able to capture, e.g., the surface morphology. Such features are extracted either directly from the depth values of the 3D point cloud, or indirectly from a parametric surface model, e.g., 3D mesh, fitted to the point cloud.

Spin-images is a well-known local 3D shape descriptor, originally proposed for object recognition [26]. Spin-images represent the local 3D surface $\mathcal{S}_P$ with respect to a 3D point P in terms of the distribution of the 3D points $P^i \in \mathcal{S}_P$, and consist in 2D histograms of $N \times M$ cells accumulating the contributions of the points $P^i \in \mathcal{S}_P$ falling in each cell. The 2D histograms, bin the neighboring point positions expressed in the cylindrical coordinate system (α, β) centered at P and oriented like the normal $\hat{\mathbf{n}}_P$ to the local surface (as shown in Fig. 9.10), according to

$$S_P(P^i) \to [\alpha, \beta] = \left[\sqrt{\|\mathbf{P}^i - \mathbf{P}\|_2^2 - \hat{\mathbf{n}}_P \cdot (\mathbf{P}^i - \mathbf{P})^2}, \hat{\mathbf{n}}_P \cdot (\mathbf{P}^i - \mathbf{P}) \right]. \tag{9.9}$$

Since the coordinates $[\alpha, \beta]$ of the mapped point P^i are not integer, the contribution of P^i to the accumulation histogram is split by bilinear interpolation among the four neighboring cells to $[\alpha, \beta]$ according to its distance from each bin center.

Spin-images in gesture recognition [2] are generally computed for a variable subset of key points extracted from a sampled point cloud, as the size of the original point cloud generated from the segmented body or hand samples is usually

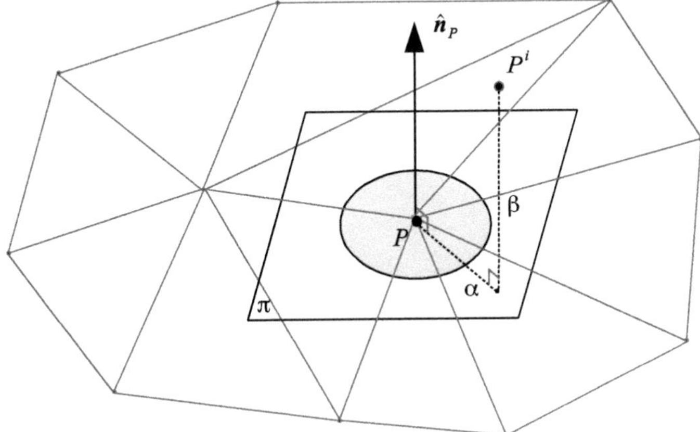

Fig. 9.10 Spin-images: local coordinate system

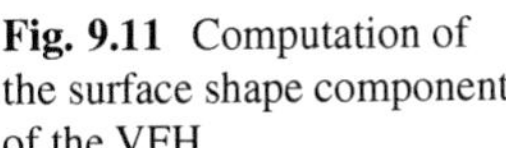
Fig. 9.11 Computation of the surface shape component of the VFH

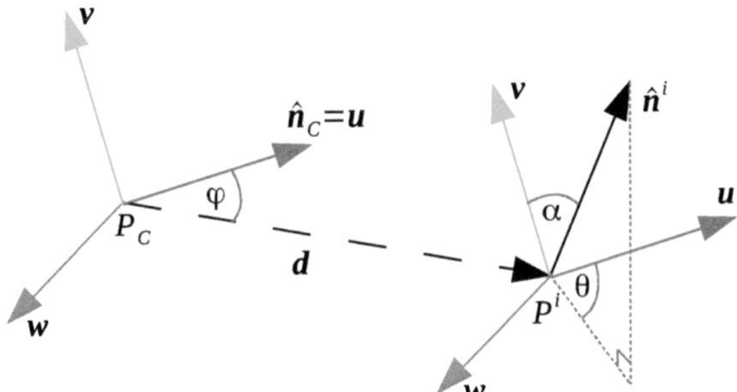

rather high. The number of extracted features depends on the size N of the sampled point cloud, and the final surface descriptor can either represent the spin-images distribution or be a single feature vector made by the concatenation of the first K principal components of the spin-images space $S = \{S_1, \ldots, S_N\}$ after dimensionality reduction by PCA or its variations.

Viewpoint Feature Histogram (VFH) is another 3D surface global descriptor [27, 59], used for object recognition and pose estimation, that aims to represent the body or hand surface in terms of its 3D point normals distribution. Its concept is recalled next with the aid of Fig. 9.11.

Let P_C denote the center of mass of the body or hand point cloud $\mathscr{P}$, $\hat{\mathbf{n}}_C$ the normal to P_C obtained by averaging all the normals $\hat{\mathbf{n}}^i$ of the 3D points $P^i \in \mathscr{P}$. Let also $d = \|\mathbf{P}^i - \mathbf{P}_C\|_2$ denote the Euclidean distance between a point P^i and the center of mass P_C, $\mathbf{v} = (\mathbf{d}/|\mathbf{d}|) \times \hat{\mathbf{n}}_C$, $\mathbf{w} = \hat{\mathbf{n}}_C \times \mathbf{v}$, $\cos\alpha = \mathbf{v} \cdot \hat{\mathbf{n}}^i$, $\cos\varphi = \hat{\mathbf{n}}_C \cdot \mathbf{v}$, $\theta = \arctan\left(\mathbf{w} \cdot \hat{\mathbf{n}}^i / \hat{\mathbf{n}}_C \cdot \hat{\mathbf{n}}^i\right)$. The final feature vector is made by the juxtaposition of two components: the first component collects an histogram of the values θ, $\cos\alpha$, $\cos\varphi$, $|\mathbf{d}|$ computed for each point P^i, while the second component is an histogram of the angles that each viewpoint direction makes with each normal. In [27],

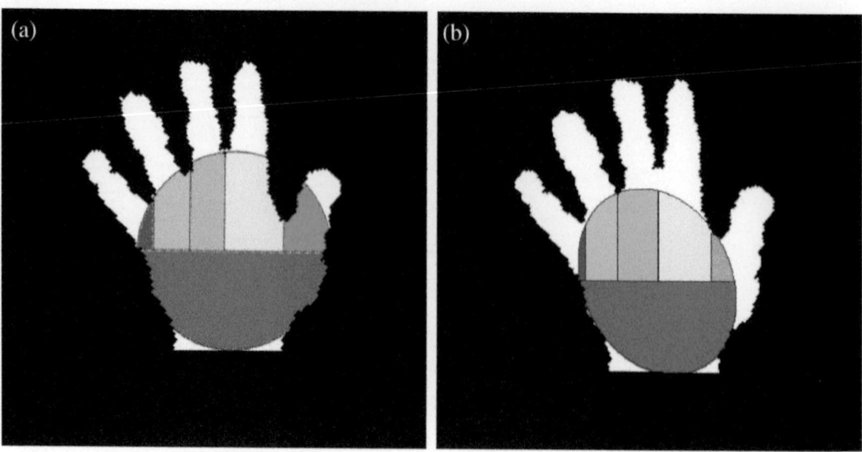

Fig. 9.12 Example of palm region partitionings: (**a**) partitioning by a circle: (**b**) partitioning by an ellipse

for example, the first component has 128 bins and the second 45. A variation of the global descriptor consists in partitioning the volume in a regular tridimensional grid, extracting the descriptor for each cell and concatenating all the local descriptors in a single feature vector.

Palm morphology features, devised for hand gesture recognition only, are extracted from the point cloud of the palm samples $P^i \in \mathcal{P}$, in order to detect which fingers are likely to be folded over the palm on the basis of the deformation the palm surface undergoes when a finger is folded or extended [18]. The samples corresponding to the fingers folded over the palm belong to $\mathcal{P}$ and thus are not considered by feature sets describing the hand contour, but provide relevant information on the fingers opening status. Starting from a 3D plane π fitted to $\mathcal{P}$, under the assumption that the distances of the actual palm points from π tend to 0, the palm surface where a finger is folded is locally deformed and the average Euclidean distance from the finger 3D points to π is not negligible (e.g., 10 [mm] or more). A possible way of exploiting such information consists, then, in partitioning $\mathcal{P}$ into six disjoint sets $\mathcal{P}_L$, $\mathcal{P}_j, j = 1, 2, 3, 4, 5$, where $\mathcal{P}_L$ denotes the lower half of the palm and $\mathcal{P}_j$ the subset of the upper-half of the palm where the jth finger may fold. In this case, after assigning each point $P^i \in \mathcal{P} \setminus \mathcal{P}_L$ to the respective subset $\mathcal{P}_j$, the final descriptor is made by simply juxtaposing the average point distances within each single subset $\mathcal{P}_j$ from a new plane π_L fitted on $\mathcal{P}_L$, as the palm deformation induced by the folded fingers is likely to corrupt the estimate of π. Figure 9.12 shows two examples of palm partitioning, depending on whether the palm region is defined by a circle or by an ellipse [36].

9.1.4 Area and Volume Occupancy Descriptors

Area occupancy descriptors aim at representing the body or hand shape in the image plane from their sample distribution, under the rationale that such a distribution highly characterizes the performed gesture. The algorithms extracting area occupancy features begin either by partitioning the area enclosing the body or hand samples into a regular grid of $W \times H$ cells (Fig. 9.13a), or by other partitioning schemes like a pie chart [29] (Fig. 9.13b) or a radial histogram [64] (Fig. 9.13c). Each cell acts as a bin of an histogram accumulating the samples belonging to it, and the final descriptor consists in the concatenation of the results of a predefined smoothing function applied to each cell value.

Volume occupancy descriptors extend the rationale of area-based descriptors to three dimensions by representing the body or hand shape in the Euclidean space in terms of the point distribution of the segmented point cloud. Again, this family of descriptors adopts a uniform partitioning of the volume enclosing the body or hand point cloud into adjacent cells and uses a final feature vector made by the juxtaposition of the features extracted from each cell.

Usually, each cell c contains the ratio between the accumulated points and the overall number of points in the point cloud

$$N(c) = \frac{\sum_{P \in \mathscr{P}} I(P,c)}{|\mathscr{P}|} \tag{9.10}$$

where $I(P,c)$ is an indicator function assuming value 1 if point $P \in \mathscr{P}$ belongs to cell c and 0 otherwise. The point cloud size $|\mathscr{P}|$ normalizes the feature value $N(c)$ for cell c in order to compare the features extracted from point clouds of different size.

Shape context descriptors, as first step, tessellate the Euclidean space by a grid of $W \times H \times D$ cubes (Fig. 9.14a), or by a sphere centered on the body center of

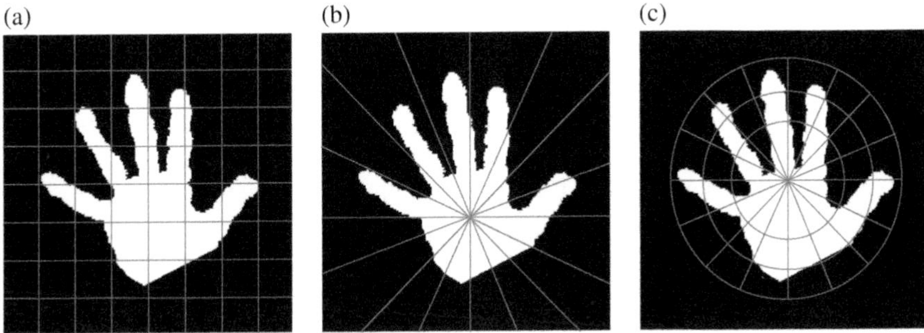

Fig. 9.13 Example of different partitioning schemes for the hand region: (**a**) grid tessellation; (**b**) pie chart partitioning and (**c**) radial histogram

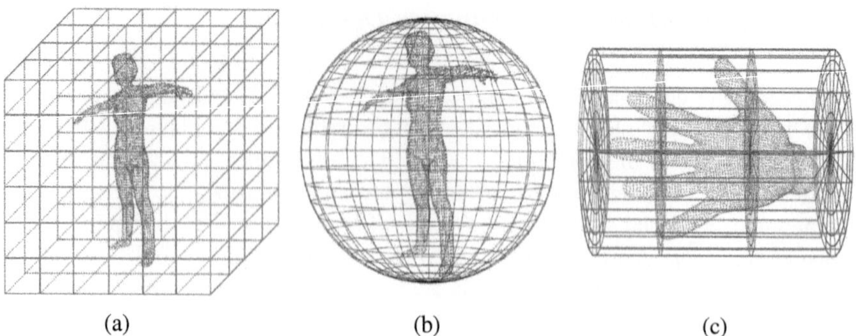

Fig. 9.14 Example of different partitioning schemes for the body or hand region: (a) cube tessellation; (b) sphere partitioning; (c) cylinder partitioning

mass and made by $S \times T \times U$ cells with S, U, and T denoting respectively the sphere latitude, longitude and radius direction (Fig. 9.14b) [24]. The sphere partitioning, in particular, has the advantage of making the descriptor invariant to body or hand rotations, since the sphere is oriented as the body or hand local coordinate system. In the case of static hand gesture recognition, [63] proposed instead to partition the hand volume with a 3D cylinder oriented along the hand main axis estimated by PCA (Fig. 9.14c). By accumulating the descriptor values for each cell along the main axis, namely for each horizontal cross-section, the slip-streamed 3D descriptor is analogous to the area occupancy descriptor extracted in Sect. 9.1.3 using a radial histogram.

The 3D points are then binned in the volume partition cells and the final descriptor is built by concatenating the features extracted with (9.10) from each cell. A variation of the basic approach, named *six point distribution histograms* [39], improves the descriptor robustness by repeating the feature extraction for different tessellations of the body or hand volume using an increasing number of cells thus by reducing their size, and then by concatenating the results.

Local Occupancy Patterns (LOP) [72, 73] belong to the family of volumetric features describing the local body or hand appearance around a given key point in 3D space, where the key points in this case correspond to the positions $P_j, j = 1, \ldots, N$ of the N joints in the body or hand skeleton. For each joint j, the algorithm partitions its local volume (namely, the limited 3D space centered at the joint position P_j) in a grid of $N_x \times N_y \times N_z$ cells (or bins) of size S_x, S_y, S_z as in Fig. 9.15, and then bins the 3D points P^i in the jth joint neighborhood $N(j)$ into the just defined cells. The final descriptor for a single joint j consists in the concatenation of the local feature $\sigma_{j,k}$ extracted from each cell $c_{j,k}$ of the partitioned local joint volume

$$\sigma_{j,k} = \delta\left(\sum_{P \in c_{j,k}} I(P, j, k)\right) \tag{9.11}$$

Fig. 9.15 Example of local volume partitioning for a single joint

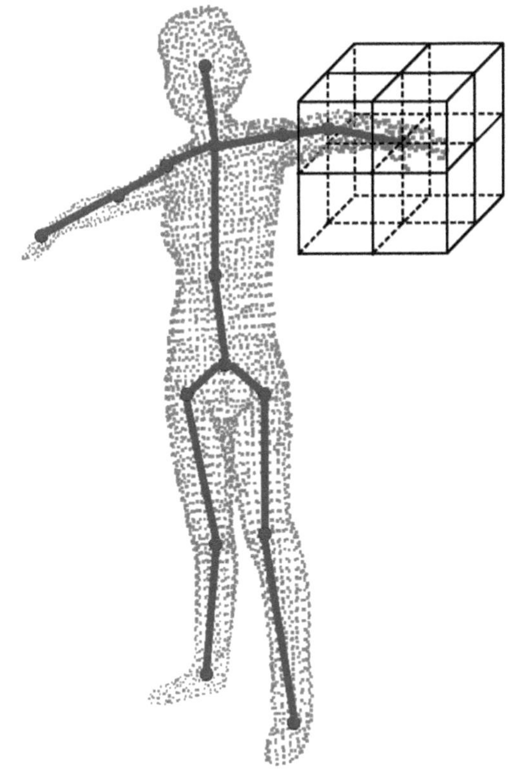

where $I(P, j, k) = 1$ if point P within the point cloud in joint j local volume is contained in cell $c_{j,k}$, and $I(P, j, k) = 0$ otherwise. Function $\delta(x) = 1/(1 + e^{-\beta x})$ is a sigmoid normalization function weighting the belonging of each point P to a given cell, since the point coordinates are not discrete.

Histogram of 3D Joint Locations (HOJ3D) [76] is another volumetric descriptor representing the body or hand shape in the 3D space by means of the distribution of the skeletal joint positions according to the selected volume partition. Following the same rationale of the Local Occupancy Patterns but within a global viewpoint, the body volume is partitioned by a sphere centered at the body center of mass made by $S \times T$ radial cells, with S and T respectively denoting the sphere latitude and longitude directions (Fig. 9.16). Namely, the body volume is partitioned by a set of S vertical angularly equispaced planes passing through the body center of mass and T equispaced parallel horizontal planes with the reference plane passing through the body center of mass.

Different from Local Occupancy Patterns, in this case the point distribution only accounts for the contributions of the 3D joint positions. As the number of joints is much lower than the number of points of the body point cloud and the joint position estimates are likely to be corrupted by noise, the discrete voting on a single bin is replaced by a probabilistic voting by a Gaussian weighting function. Namely,

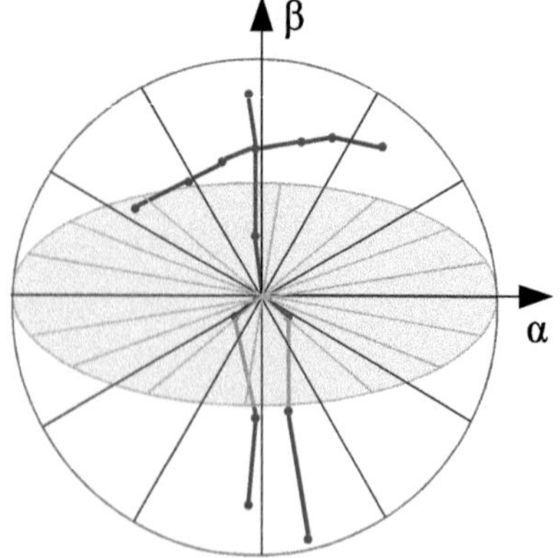

Fig. 9.16 Example of spherical partitioning of the body volume

the unitary vote of each joint location (s, t) is split among the cells neighboring the (s, t) cell by assigning each cell (α, β) a vote fraction inversely proportional to the distance of (α, β) from (s, t). The joint vote corresponds, then, to the probability that the joint belongs to a given cell. The final descriptor is the concatenation of the accumulated votes of the joints for each spherical histogram bin.

9.1.5 Depth Image-Based Descriptors

Image-based features describe important textural characteristics of the segmented body or hand and are a classical tool of automatic gesture recognition. Recall that image-based features are often extracted either from the whole input image represented in gray levels, or from local image patches, and then aggregated within a single final descriptor. The same feature extraction approaches devised for color images can be directly applied to depth maps treated as gray-scale images [43], namely, by mapping the depth values within range $[z_{min}, z_{max}]$ to gray level values within range $[0, 255]$ or $[0, 1]$. Different from color images, depth pixels assume similar gray values for samples having similar depth values disregarding the scene lighting conditions. In this case, depth data make classical color-based descriptors more resilient to the effects of lighting conditions with improved segmentation quality.

Image-based feature extraction algorithms usually follow the pipeline of Fig. 9.17. Image cropping uses the bounding box enclosing the segmented body or hand region to limit the feature extraction area, while image masking prevents the image descriptor from being biased by the information coming from the background pixels of the cropped image. Since most feature extraction techniques aggregate the descriptors computed for each pixel, e.g., by creating a histogram of the descriptors distribution in the whole image or within limited regions, the masking avoids to account for the contributions of the erroneously retained background pixels. The

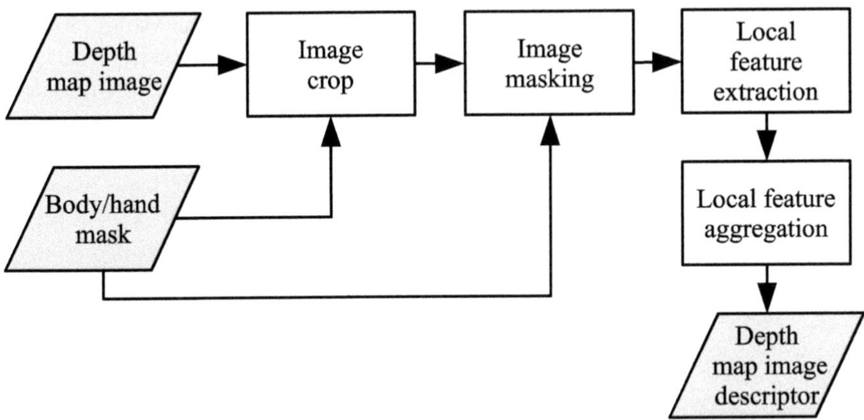

Fig. 9.17 Architecture of a generic feature extraction algorithm from depth maps treated as gray-scale images

final steps of the pipeline is the computation of a color descriptor for each pixel, which is usually a local descriptor encoding the texture information within a patch centered at the pixel. Pixel descriptors are either aggregated in a unique feature vector characterizing the image or collected in normalized histograms representing the descriptor distribution within the image.

The remainder of this section describes how depth information can be used in gesture recognition methods improving the discrimination capabilities of earlier color image feature extraction approaches with some examples.

Recall from Chap. 1 that depth cameras are affected by several image degradation problems due to low quality camera sensors or optics, e.g., lens distortion or misalignment, scene objects motion with respect to the camera and unfavorable lighting conditions. Image degradation, beside making the acquired images visually unpleasant, often leads most computer vision algorithms to fail. The blurring effect is one of the most commonly encountered degradation, which may arise from:

- framed scene out of focus
- motion of the objects in the scene with respect to the camera
- particular atmospheric conditions

Local Phase Quantization (LPQ) [55] is an image descriptor based on the rationale that the local texture phase information is not corrupted by the previously described image blurring effect, hence the local phase spectrum well describes the depth map, treated as a gray-scale image, in presence of blurring.

Feature extraction begins by computing the Short-Time Fourier Transform (STFT) within a patch of size $m \times m$ centered at each pixel p^i of the segmented body or hand depth map. The value of the patch size m is a trade-off between the image blurring sensibility and the descriptor discrimination capability: when using small values of m the lower frequencies capture more details from the depth patch but reduce the insensitivity of the method to blurring, while higher values of m lead to a better blur insensitivity but reduce the descriptor discrimination capability.

Let $F(p^i, \mathbf{f})$, with $\mathbf{f} = [f_u, f_v]^T$ denoting the horizontal and vertical frequency vector, be the STFT of the patch centered at pixel p^i with coordinates $\mathbf{p}^i = [u^i, v^i]^T$. Note how $\mathbf{f} \in \{\mathbf{f_1}, \mathbf{f_2}, \ldots, \mathbf{f_L}\}$, with L the support length of $F(p^i, \mathbf{f})$ and $\mathbf{f}$ is a frequency of F for which the blur Fourier transform is positive. While the concatenation of all the $F(p^i, \mathbf{f})$ phase vectors could be directly used as a texture descriptor, its length and the space required to encode the real phase values would make it impractical. For this reason, the LPQ algorithm first performs a quantization of the SFTF phase according to

$$Q(F(p^i, \mathbf{f})) = \text{sgn}(\Re(F(p^i, \mathbf{f}))) + 2\,\text{sgn}(\Im(F(p^i, \mathbf{f}))) \tag{9.12}$$

where $Q \in \{0, 1, 2, 3\}$, $\Re(\cdot)$ and $\Im(\cdot)$ respectively denote the real and imaginary part of the frequency response and the sign function assigns 1 to positive values and 0 to the negative ones. The quantized phase values for the L frequencies are then represented by 2-bit code words aggregated in a unique code word $C_W(p^i)$ of $2L$ bits defined as

$$C_W(p^i) = \sum_{j=1}^{L} Q(F(p^i, \mathbf{f}_j))2^{2(j-1)} \tag{9.13}$$

which ranges from 0 to $2^{2L} - 1$. After computing the code words encoding all the local patches, the method builds an histogram with $2^{2L} - 1$ bins (one per code word) representing the distribution of the code words for the depth map. As stated above, the depth mask computed in this step can prevent the descriptor from being biased by the background pixels. The histogram is then normalized in order to make unitary the sum of each scaled code word frequency value. The final descriptor is the normalized histogram of the code word frequencies.

Local Binary Patterns (LBP) [46] and *Local Ternary Patterns* (LTP) [23] are other computationally efficient non-parametric local image descriptors widely used in computer vision applications like face detection. They encode the information of a patch of size $m \times m$ centered at pixel p^i with a code word that will be either used as a single feature or aggregated in local histograms representing the code word distribution in a wider region.

Both LBP and LTP extraction algorithms, formalized by the following equations

$$LBP(p^i) = \sum_{j \in N(i)} 2^j s_b(p^i, p^j) \tag{9.14}$$

$$LTP(p^i, p^j, t) = \sum_{j \in N(i)} 3^j s_t(p^i, p^j, t) \tag{9.15}$$

with p^i and p^j denoting respectively the intensities of the central pixel of the image patch and of the jth pixel in its neighborhood $N(i)$. Intensities p^i and p^j can be replaced without loss of generality by the depth values z^i and z^j of the extracted depth patch. $s_b(\cdot)$ and $s_t(\cdot)$ respectively denote a binary and a ternary thresholding functions

$$s_b(p^i, p^j) = \begin{cases} 1 & \text{if } p^j \geq p^i \\ 0 & \text{otherwise} \end{cases} \tag{9.16}$$

$$s_t(p^i, p^j, t) = \begin{cases} 1 & \text{if } p^j \geq p^i + t \\ 0 & \text{if } |p^j - p^i| < t \\ -1 & \text{if } p^j \leq p^i - t. \end{cases} \tag{9.17}$$

The first step of both LBP and LTP is a binary thresholding of the pixel intensities in the neighborhood (usually a 8-neighbors) on the basis of the intensity of the patch central pixel p^i. The second step is the encoding in a binary or ternary code word composed by concatenating the binary or ternary digits associated with the neighboring pixels $p^j \in N(i)$. Because of $s_b(\cdot)$, LBP is highly sensitive to the image noise since very high or very low pixel intensities due to noise may corrupt the extracted code word. LTP, instead, thanks to the global threshold t, is less sensitive to code word corruption due to local fluctuations of the pixel intensities caused by noise. Also note how the LTP descriptor for computational convenience is usually decomposed in two LBP descriptors, as in Fig. 9.18, where the second is obtained by inverting the thresholding function $s_b(\cdot)$. The two sub-descriptors are aggregated only in the final phase.

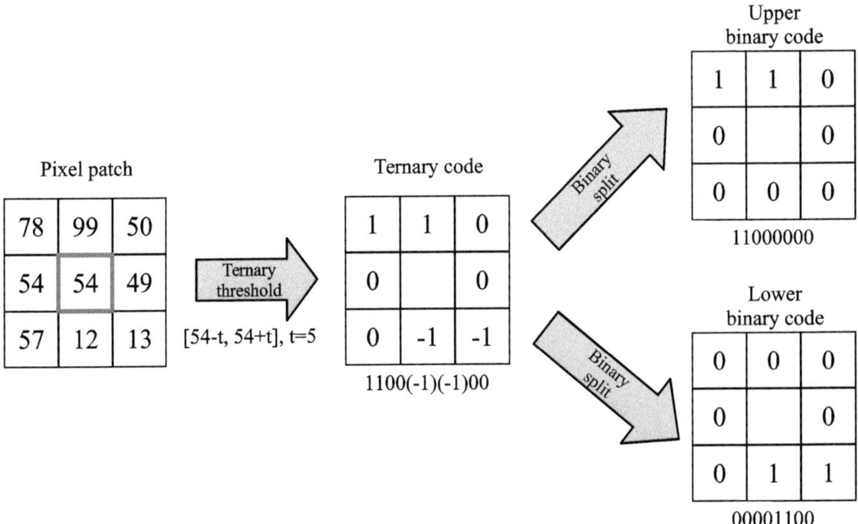

Fig. 9.18 LTP descriptor split in two LBP code words

The feature extraction algorithm first partitions the segmented body or depth image in a grid of $N \times M$ cells and encodes with a LBP or LTP code word each pixel, eventually discarding the code words generated by the background pixels in the associated depth mask. Then a code word histogram is built for each image patch and the final descriptor is made by the concatenation of the normalized histograms of each patch.

Histogram of Oriented Gradients (HOG) [15] is a textural descriptor for images widely used on color images or videos for people and object detection purposes, e.g., pedestrians or vehicles, and recently also on depth maps [60]. It is based on the idea of dividing the image into small connected regions called cells, and building a histogram of gradient directions for the pixels of each cell. The concatenation of the histograms computed for each cell gives the final descriptor. In order to improve resilience with respect to changes in illumination or shadowing, the local histograms can be normalized by a measure of the intensity computed across large image regions called blocks, and by using this value to normalize all the cells within the block. This normalization, however, is not necessary with depth maps as they are not affected by illumination changes and shadowing. The rationale behind HOG is that the object shapes are well characterized by the distribution of local intensity gradients, namely by the occurrences of the gradient orientations in localized portions of an image. An example of HOG descriptor extraction from an hand depth map is shown in Fig. 9.19.

Feature extraction begins by partitioning the segmented body or hand depth map into a regular grid of $N \times M$ cells and computing the horizontal and vertical image gradients G_x and G_y, e.g., by Sobel operator. The algorithm then computes the pixel gradient direction distribution within each cell by building an histogram with bins denoting a limited number of gradient direction ranges (e.g., by partitioning the unitary circle into nine uniform angular regions), optionally discarding the contributions of the background pixels indicated in the depth mask. The final descriptor consists in the concatenation of the normalized histograms extracted from each cell.

Gabor filters [21] are often used in several image processing algorithms and recently also for gesture recognition [51, 53] due to their capability of revealing local patterns in images, e.g., edges. They are known for their capability of accurately modeling the response of the perceptive fields of the orientation-selective simple cells of the human visual cortex. A Gabor filter for 2D image processing consists in a Gaussian kernel function $g(u, v)$ modulated by a sinusoidal plane wave, as

$$g(u, v, \lambda, \theta, \varphi) = \exp\left(-\frac{(u\cos\theta + v\sin\theta)^2 + (-u\sin\theta + v\cos\theta)^2}{2\sigma^2}\right) \cos\left(\frac{2\pi(u\cos\theta + v\sin\theta)^2}{\lambda} + \varphi\right) \tag{9.18}$$

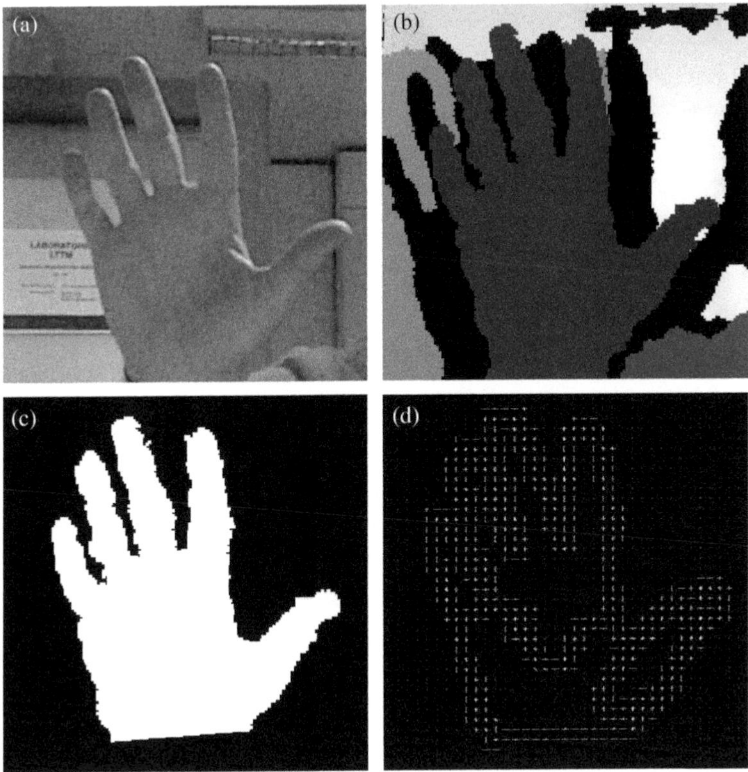

Fig. 9.19 Example of HOG descriptor extraction from the hand region: (**a**) color image; (**b**) depth map; (**c**) hand region mask; (**d**) extracted HOG features

with λ denoting the wavelength of the sinusoidal wave, φ its phase offset and θ the orientation of the normal to the parallel stripes of a Gabor function. The parameter σ^2 is the variance of the Gaussian envelope. In (9.18) only the real part of the filter is considered.

Since a single Gabor filter is only able to effectively detect local structures having a given orientation, in order to extract all the local patterns in the segmented body or hand depth map $Z(u, v)$, the depth map is firstly scaled to a fixed size (e.g., 128×128 pixels) and then provided as input to a filter bank made by Gabor filters with different scales and rotations. $Z(u, v)$ is thus convolved with each kth filter of the bank for $k = 1, \ldots, N$ and the resulting images $J_k(u, v)$, having the same size of $Z(u, v)$, are averaged across overlapping Gaussian basis functions positioned on a regular grid

$$C(i, j, k) = \sum_{u,v} J_k(u, v) \exp\left(-\frac{(u - u^i)^2 + (v - v^j)^2}{2s^2}\right) \tag{9.19}$$

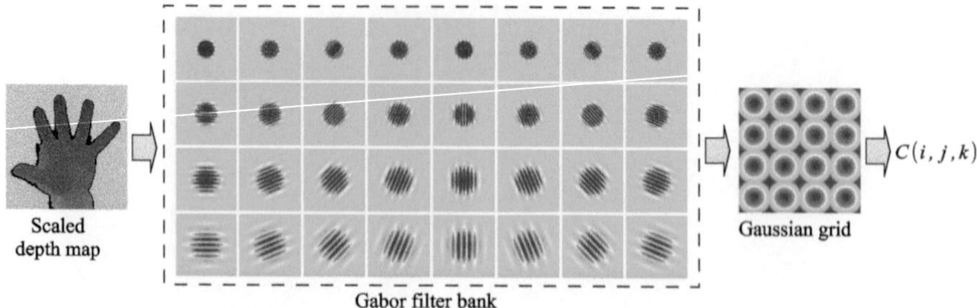

Fig. 9.20 Gabor feature extraction pipeline

where i and j indicate the horizontal and vertical index of the Gaussian basis function, $u^i = 16i - 8$ and $v^j = 16j - 8$ for a grid made of 8×8 basis functions, k indicates the Gabor filter index and $s = 8$ is the Gaussian filter standard deviation controlling the filter shape. Since the 2D Gaussian functions assume non-zero values only within a finite support window (e.g., 16×16 pixels for a grid of 8×8 functions and 128×128 input images), $C(i, j, k)$ simply computes the weighted average of the pixel values of a 16×16 pixels patch extracted from the partitioning of $J_k(u, v)$ into a grid of 8×8 patches.

The final descriptor consists in the concatenation of the $C(i, j, k)$ values computed for the whole grid and all the bank filters. A pictorial representation of the descriptor generation process is shown in Fig. 9.20.

Wavelet decomposition is a well-known tool widely used in several applications, like image and video compression. In particular, Haar wavelets or "Haarlets" have been successfully employed in earlier object recognition approaches, like Viola-Jones face detector [69]. Recall that, analogously to Fourier transform expressing signals as linear combinations of a set of basis functions weighted by Fourier coefficients, wavelets adopt the same rationale, leveraging different families of basis functions effectively describing abrupt signal variations (e.g., edges in images). Haar functions form the simplest wavelet basis and allow the recursive decomposition of an input image into a low resolution component and a set of high resolution components describing the details not captured by the low resolution term. While in data compression applications the details are usually discarded for the sake of a more compact description of the input image, in object recognition approaches the details are preserved for disclosing properties of the scene object to detect.

Haar-like feature extraction algorithms begin by recursively decomposing the segmented body or hand depth map into four sub-maps (Fig. 9.21) having each one half of the size of the original depth map, where the top-left sub-map corresponds to the low resolution coefficients of the wavelet transform and the other sub-maps contain the differences along the sampled rows, columns or diagonal direction and correspond to the detail coefficients of the transform. The low resolution image is then usually decomposed in the same way and the detail coefficients collected until the desired decomposition level is reached. The final descriptor consists in the concatenation of the Haar wavelet transform [67] and can be given to a standard machine-learning approach like SVM or Random Forests.

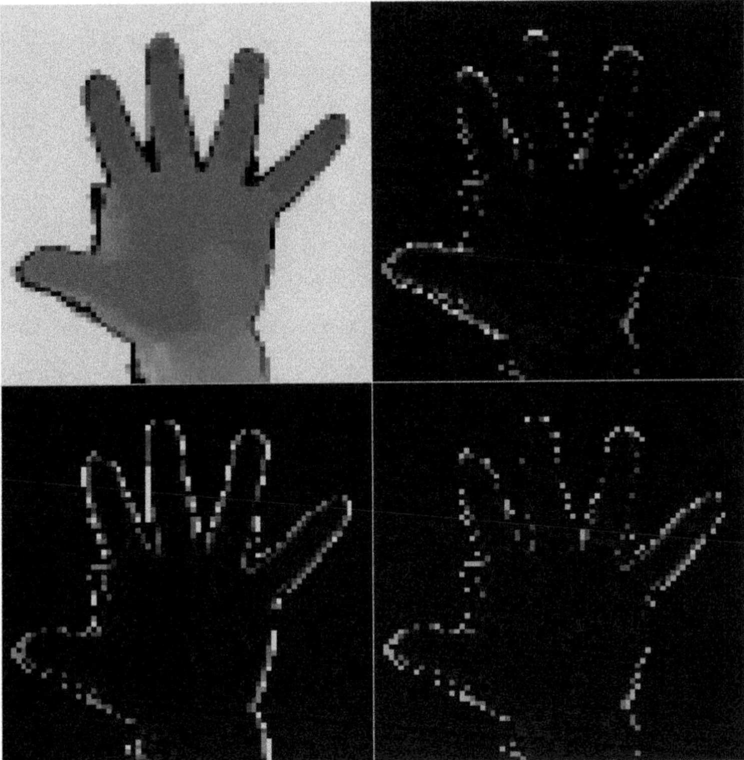

Fig. 9.21 Example of 1-level image decomposition with Haar wavelets: (*top-left*) low resolution depth map, (*top-right*) horizontal, (*bottom-left*) vertical and (*bottom-right*) diagonal detail submaps

9.1.6 Convex Hull-Based Descriptors

The convex hull computed on the body or hand shape in the segmented depth map is another useful clue adopted in various gesture recognition schemes like [19, 38, 50, 51]. The convexity defects, namely the differences between the convex hull outline and the concave regions of the body or hand contour (e.g., between consecutive fingers), indeed, strongly characterize each gesture. Figure 9.22, for example, shows that the fist outline is nearly convex on the overall, while the outline of the open hand gesture presents several concave and convex regions. It follows that the fist is hardly misrecognized as another sign with one or more raised fingers by an analysis based on the convex hull outline.

Several features can be extracted from the information encoded by the convex hull, e.g.:

- number of convex hull vertices
- hand contour perimeter ratio

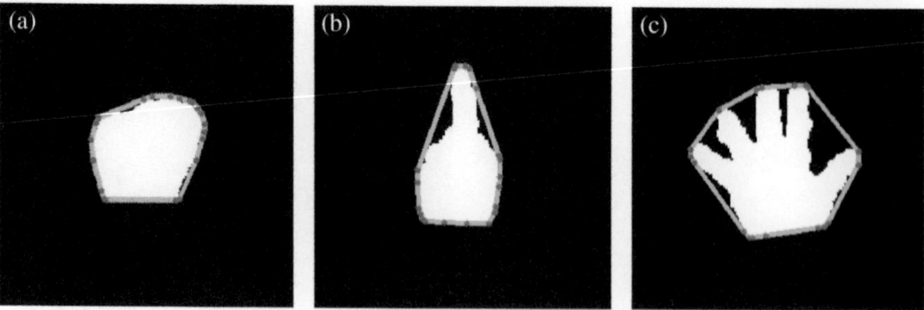

Fig. 9.22 Example of extracted convex hulls from a few hand shapes: (**a**) fist, (**b**) pointing and (**c**) open hand gestures. The convex hull vertices are represented by *red dots*

- hand region area ratio
- empty regions between fingers areas ratio.

The extracted convex hulls may present several spurious vertices due to measurements noise that could compromise the robustness of the extracted descriptors. In particular, as the fingertip positions in the hand silhouette ideally correspond to convex hull vertices (although not all the convex hull vertices necessarily correspond to fingertips), the presence of spurious vertices is likely to make failing any gesture recognition approach based on this clue only. In order to make the convex hull-based descriptors more resilient to depth noise, proper simplification techniques can be applied to the original convex hull data to filter the spurious vertices.

9.1.7 *Feature Classification*

Feature classification, the last step of the gesture recognition pipeline of Fig. 9.2, is commonly adopted in object and pattern recognition tasks and consists in predicting the performed gesture from the extracted feature vector. Classification approaches can be divided in two main families, namely *non-parametric* and *parametric*.

Non-parametric classification approaches predict the class c of a test feature vector $\mathbf{f}$ without relying on any gesture recognition model. Among them, *Nearest Neighbor* (NN) classifier is the most simple and intuitive method. Given a training set $\mathscr{F} = \{\mathbf{f}_1, \ldots, \mathbf{f}_N\}$ of N samples, any feature vector $\mathbf{x} \notin \mathscr{F}$ is assigned the class c of the "nearest" training vector $\mathbf{f}_i \in \mathscr{F}$ according to a selected metric assessing the similarity degree between $\mathbf{x}$ and $\mathbf{f}_i$, as

$$c(\mathbf{x}) \leftarrow \underset{c}{\operatorname{argmin}}\{d(\mathbf{x}, \mathbf{f}_1), \ldots, d(\mathbf{x}, \mathbf{f}_N)\} \tag{9.20}$$

with $c \in \{c_1, \ldots, c_G\}$ the gesture class within the gesture dictionary of G classes and $d(\mathbf{x}, \mathbf{f}_i)$ the similarity value between $\mathbf{x}$ and $\mathbf{f}_i$. The most adopted metrics with generic feature vectors are the Euclidean distance, e.g., $d(\mathbf{x}, \mathbf{f}_i) = \|\mathbf{x} - \mathbf{f}_i\|_2$ or the correlation [24, 48], although several other metrics are preferable, instead, with specific descriptors. For example, cosine similarity $d(\mathbf{x}, \mathbf{f}_i) = (\mathbf{x} \cdot \mathbf{f}_i)/(\|\mathbf{x}\|_2 \|\mathbf{f}_i\|_2)$ is common in information retrieval and text mining, the Hamming distance when the feature vectors $\mathbf{x}$ and $\mathbf{f}_i$ are sequences of symbols or strings ($d(\mathbf{x}, \mathbf{f}_i)$ corresponds to the number of positions where $\mathbf{x}$ and $\mathbf{f}_i$ differ) and the finger earth-mover distance [81] when the feature vector is the hand contour plot of (9.1). Finally, in the case of image patches comparison, e.g., in [16, 74, 75], since the pixel-wise differences (usually in the least-square sense) are highly prone to errors due to the depth measurement noise, methods like [45] propose to evaluate the proximity of the two patches instead of computing the pixel-wise distances from the patches superposition. Chamfer and Haussdorff distances [9, 16] are two proximity metrics often used for this purpose. Chamfer distance is evaluated from the edge images extracted from the two depth patches mapped to gray-scale images and it is defined as the average Euclidean distance between each template pixel p and the nearest edge pixel q in the query edge map, with the idea that the less is the distance the more similar are the two patches. Haussdorff distance, instead, is evaluated as well on the two patches edge maps and it is defined as the maximum among all the minimum distances from each edge pixel p in the template patch to each edge pixel q in the query patch.

Alternative extensions of the original nearest neighbor approach partition the training set into G clusters, each one containing training vectors of the same class, and assign $\mathbf{x}$ to the class label c of the cluster minimizing the distance between $\mathbf{x}$ and the cluster center $\mathbf{f}_g$ [39] or between $\mathbf{x}$ and the cluster vector distribution (Mahalanobis distance). Another common extension of nearest neighbor, named *K-nearest neighbor*, assigns $\mathbf{x}$ the class c obtaining the highest number of votes about $\mathbf{x}$ membership to c from only the first K "nearest" vectors in $\mathcal{F}$ according to the selected distance metric. Such variations are said to compute the distance-to-class of vector $\mathbf{x}$ to class c rather than its distance-to-vector.

Parametric classification approaches, instead, make use of a gesture recognition model previously learned on the training set by proper machine learning techniques. Approaches based on single classifiers like Support Vector Machines (SVM) [7, 18, 49, 51, 63, 64], Decision Trees, Random Forests [53], Neural Networks, genetic algorithms and many others have been proposed in the literature to tackle automatic gesture recognition and several other computer vision problems. Among them, SVM is the most adopted classifier, and its kernel parameters are often optimized by *Grid Search* or its variations. In particular, [19] proposed a Grid Search variation devised to leverage the peculiar composition of the datasets for gesture recognition and to improve the generalization capability of models trained on gesture datasets with a low cardinality. More advanced methods like [66] and [44] employ, instead, ensembles of classifiers designed to leverage the different structure of heterogeneous features and to assign each feature vector the gesture class that gained the highest consent among all the single classifiers in the ensemble.

A complete taxonomy of the various machine learning methods in the literature is beyond the scopes of this chapter, focused only on the main feature families employed in gesture recognition. An extensive treatment of classification can be found in [6].

9.1.8 Feature Selection

Classic machine learning based on single classifiers relies on the intuitive idea that the higher the number of extracted features of different types, the higher the recognition model accuracy due to a larger amount of information available for the class assignment decision. The high dimensionality of the extracted feature vectors, however, implies heavy computational loads for feature extraction and a higher sensitivity of the classifier with respect to small variations in the feature values. The ensembles of classifiers of Sect. 9.1.7 aim at reducing the computational load by splitting the original classification problem within a divide-and-conquer framework in a sequential or parallel training of single learners on subsets of the initial dataset or subsets of the original feature space, while improving the generalization of the estimated classification model.

Both single classifiers or ensembles, often gain only a modest improvement in the overall gesture recognition accuracy, since they are unaware of the structure of the training data, at the expenses of a significant increment of the feature vector length and a consequent higher computational load. The reason of this unfavorable behavior is due to the possible correlations among the extracted features. Highly correlated features, indeed, do not carry any additional information when jointly used, since in front of a variation of some of them the others change accordingly. Conversely, highly uncorrelated features may carry a considerable amount of information when used together, as the changes of some of them are independent from the changes of the remaining ones. Moreover, the simple juxtaposition of uncorrelated yet significant features does not necessarily boost the classification model accuracy, as a limited set of them may already contain all the information needed to determine the class of the performed gesture with the maximum possible accuracy.

For these reasons, feature selection methods are often used to disclose the most relevant features for a given dataset in order to train a robust classifier with the minimum amount of features leading to the maximum classification accuracy [38], e.g.:

- feature selection based on PCA
- feature selection based on F-Score
- feature selection based on Random Forests
- sequential feature selection.

The first three methods firstly assign each feature $g_j, j = 1, \dots, M$ of the feature set $\mathcal{G} = \{g_1, \dots, g_M\}$ a score in direct proportion with its discrimination power.

Feature selection based on PCA ranks the features according to non increasing variance, under the rationale that highly varying features encode more information with respect to slowly changing features, and selects the first L features. Feature selection based on F-Score [12] is analogous to PCA but ranks the features according to their F-Score, based on the harmonic mean of recall and precision. Feature selection based on Random Forests leverages the Out-of-Bag (OOB) error [8] of the model trained with a Random Forest to measure the feature importance: once the OOB error has been estimated, the values of one of the features in the dataset are permuted and the OOB error is estimated again. The procedure is repeated for each feature and the importance of each feature is given by the normalized average increase of the OOB error after the permutation. Finally, sequential feature selection [58] is an effective scheme based on iteratively constructing the minimal set of most relevant features maximizing the accuracy of the classification model. Starting from the empty set, the greedy algorithm first evaluates all the possible singletons and selects the one that maximizes the recognition accuracy of the classifier trained on it. In the further steps, the algorithm extends incrementally the feature set with one feature following the same rationale, until adding new features does not lead to any significant performance improvement.

9.1.9 Static Gesture Recognition with Deep Learning

The static gesture recognition methods described above share the same pipeline of Fig. 9.2 and all have in common that the extraction of a relevant set of features to be classified by proper machine learning techniques requires a considerable amount of pre-processing in order to segment the body or the hand from the background. Segmentation is fundamental to prevent the extracted features from being biased by the background, but is a time consuming task and still an open problem. Assuming the body or hand is correctly separated from the background, the gesture recognition accuracy mostly depends on the quality of the trained recognition model and of the extracted features. The latter are often tailored to the specific application and tuned on the employed training set, thus becoming ineffective when changing application or dataset.

Nowadays, favored by the rapid evolution of CPUs and better of GPUs, Deep Learning techniques like, Convolutional Deep Neural Networks (CNN), Deep Belief Networks (DBN) and Recurrent Neural Networks (RNN) [30] are raising a high interest in research and industry for their wide application potential ranging from big data analysis to object recognition. Different from classic gesture recognition techniques, Deep Learning methods do not require an accurate pre-processing of the acquired data and feature engineering, since such systems are designed to learn appropriate features directly on the unsegmented body or hand data. On the other hand, Deep Learning models are highly subject to model over-fitting and require a much higher amount of training data with respect to the other methods.

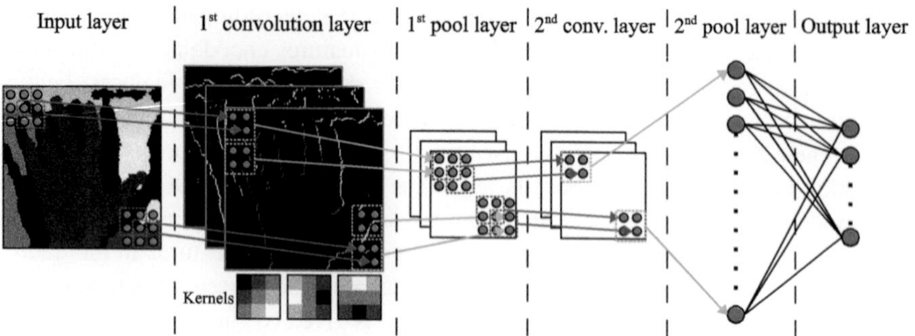

Fig. 9.23 Example of convolutional neural network with two convolution and pooling layers

Deep Learning approaches are all extensions of classic Neural Networks that aim at reducing the high computational training load by relieving the hidden neurons from being fully connected to the adjacent inner layer, relying instead on sparse connection designs to leverage the peculiar structure of the processed data.

CNNs (Fig. 9.23) are the most adopted method for static gesture recognition, and mimic the visual cortex neurons behavior in object recognition. They are made by a set of sparsely connected *convolution* and *pooling* layers, enabling the neural network to extract and enhance implicit features of an image, followed by a classic fully connected multi-layer perceptron network (MLP) receiving as input the high level features extracted by the convolution and pooling layers. The input layer neurons are directly associated with the color image or depth map pixels and each $n \times m$ (usually with $n = m$) neuron window centered at each input neuron is mapped to a single hidden neuron in the first convolution layer. Note how the input image or depth map is often previously scaled to a small predefined size to make the problem tractable. Each neuron in the first convolution layer is associated a value computed by evaluating its activation function on the weighted contributions from the mapped $n \times m$ neurons of the input layer. Since the weights associated with each input neuron, learned from the training set with back-propagation strategies are constrained to be the same for each mapped $n \times m$ window, the values of the hidden neurons in the first convolution layers can be thought as being computed by applying a 2D linear filter to the color image or depth map whose kernel function is defined by the mapping weights. The kernel describes a single filter that, in the first level, analogously to Gabor filters activates the input neurons along a predefined edge direction. Multiple kernels are usually learned to extract different features on the same layer.

The adjacent pooling layer, instead, performs a subsampling to reduce the dimensionality of intermediate representations. For this purpose, the convolution layer is often partitioned by a regular grid of $k \times l$ neurons and the maximum or the average value of the activation function within each cell is selected and sent to a unique neuron in the subsequent convolution layer. Because of the pooling layers, the inner hidden neurons are no longer associated with the image or depth map pixels but are arranged in a 2D feature map. The process is repeated by interleaving several convolution and pooling layers until a desired network depth has been reached. Clearly the deeper the convolution layer, the more global the extracted features.

Variations of the classical CNN approach deal with input data of different types, e.g., a depth map and a color image (Multi-channel Convolutional Neural Network (MCNN) [3]), or adopt data augmentation techniques to reduce the potential overfitting and improve the generalization of the gesture classifier [41]. In the first case each data type, or channel, is processed by an independent CNN, and the final high level features are aggregated with the ones coming from the other channels and sent as input to a single MLP.

9.2 Dynamic Gesture Recognition

Static gesture recognition of Sect. 9.1 is mainly adopted in applications driven by hand gestures where each gesture is highly characterized by the hand shape or pose, e.g., the automatic decoding of a static sign language alphabet [47]. Static body gesture recognition applications, instead, are less common and are rather associated with the analogue body pose estimation problem of Chap. 8.

Dynamic gesture recognition is recently receiving great attention both in research and in industrial applications oriented to video surveillance (e.g., crowd analysis) and to natural touch-less interfaces driven by hand gestures. Different from static gestures, dynamic gestures are mostly characterized by the shape of the trajectory followed by the hand center or the joints of a body or hand skeleton model, or their speed variations, throughout the whole input sequence.

Recall that human motion is a continuous process and, in general, people seamlessly commute between several activities without any clear context change. The first important problem of dynamic gesture recognition is *temporal segmentation*, namely the task of partitioning the streamed input sequences framing the user motion into subsequences associated with single dynamic gestures. While a manual separation [77] is the most intuitive and easiest way of solving the gesture temporal segmentation, this solution cannot be adopted for most applications, which must rather rely on automatic approaches detecting the dynamic gesture beginning and end in the input sequence. The literature offers a few approaches for this task:

- Use a single pose [51] (Fig. 9.24a) or a pair of reserved body or hand poses [50] (Fig. 9.24b) as start and stop markers in the input sequence. While in the first case only one static gesture is used as marker and must be hold for the whole dynamic gesture duration, in the latter case the dynamic gesture beginning and end are determined by the recognition of two different static gestures, that can no longer be included in the considered static gesture dictionary for the application.
- Pause detection: assume all the dynamic gestures in the input sequences are interleaved by periods of lack of movements.

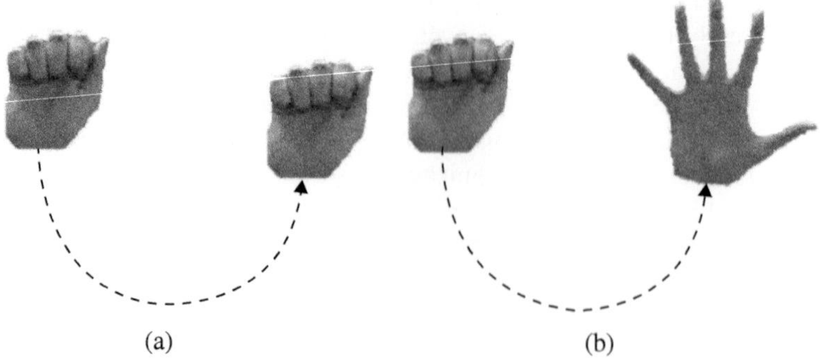

Fig. 9.24 Example of temporal segmentation schemes: (**a**) "hold" and (**b**) "start/stop"

Applications like natural interfaces driven by hand gestures, often adopt hybrid gesture recognition approaches employing both static and dynamic gestures to impart commands to the system interface. Static gestures in this case can be either used to give single commands or as start and stop markers for more complex dynamic gestures.

Dynamic gesture recognition approaches can be classified in two main families: *deterministic recognition* and *stochastic recognition*. While the methods of the first family directly recognize the performed gestures from a set of features extracted from the input sequence frames, the approaches of the second family estimate the most likely dynamic gesture generated by the noisy observations by solving MAP optimization problems.

9.2.1 Deterministic Recognition Approaches

Deterministic dynamic gesture recognition is the natural extension of the static recognition of Sect. 9.1 on multiple frames, and can be classified again into two families of approaches. The first family treats the acquired action sequence frames individually and independently from the others with respect to feature extraction. The separate descriptors extracted from each frame are then juxtaposed in a single feature vector to be classified with one of the methods presented in Sect. 9.1.7. The second family, instead, extracts a single descriptor for the whole acquired sequence, generally by aggregating the descriptors extracted from each frame, and classify it, again, with one of the methods of Sect. 9.1.7.

The first family of approaches suffers from several problems related to the gesture execution speed, even when the gestures are performed by the same user, providing sequences of different length, making impracticable the simple juxtaposition of the extracted feature vectors from each frame.

A commonly adopted solution is a prior *alignment* of the acquired gesture sequence with gesture sequence templates, representing a dynamic gesture class, in order to adapt the length of the acquired sequence to the one of the appropriate gesture template. The sequence alignment allows to compare frame by frame the descriptors extracted from both the input and template sequences and to recognize the performed gesture by standard classification approaches (e.g., SVM). Given two sequences of different lengths to compare, a first intuitive solution would be either compressing by decimation or expanding by interpolation the input sequence in order to match the template support. In the case of dynamic gestures described by the hand trajectory in the image plane or in the Euclidean space, the alignment can be either performed independently on the single trajectory components or only on the dominating one [75]. While a decimation is likely to lose relevant information on the sampled sequence as it is not aware of the peculiar dynamic of the performed gesture (e.g., rapid movements may not be detected), an interpolation is likely to introduce spurious information that may lead to wrong assumptions on the gesture dynamic.

A better solution commonly adopted in the literature consists in a smart alignment of the input and template gesture sequences by mean of Dynamic Time Warping (DTW) algorithm [5], which makes the input sequence similar to the reference sequence by locally warping the input sequence support in order to minimize the dissimilarity of the warped sequence with respect to the template. DTW algorithm has been often used for several tasks, like speech recognition [42] with the goal of determining the best alignment of two analogue speech signals following the rationale that two instances of the same speech will have a similar profile with possibly different realizations in time. Namely, two different instances of the same speech ideally differ only by the speaker speed in pronouncing the single words. Therefore they can be simply matched by locally compressing or expanding (warping) the time axis in the test sequence to match its crests with the crests of the template. The alignment is often associated with a score directly proportional with the similarity degree of the test sequence with respect to the template. Dynamic hand gesture recognition can similarly use DTW to align the hand center trajectory with the reference trajectories of the dynamic gestures to recognize [57, 77]. Figure 9.25 shows an example of two instances of the same dynamic gesture that can be successfully aligned with DTW.

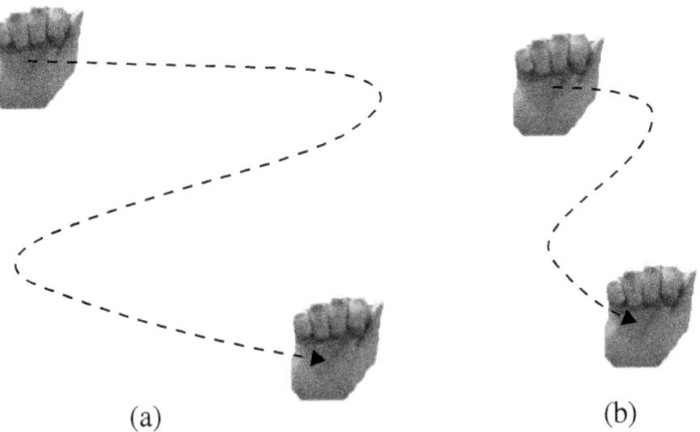

Fig. 9.25 Example of two instances of the same dynamic gesture

Let $S = \{S_1, \ldots, S_N\}$ denote a generic input sequence (e.g., the hand trajectory defined by the positions of the hand center in each acquired frame) and $R = \{R_1, \ldots, R_M\}$ denote a generic finite template sequence, with S_i and R_j generally multidimensional entities. DTW begins by computing a matrix $D \in \mathbb{R}^{N \times M}$ with each entry d_{ij} representing the distance between S_i and R_j. The distance function is usually the Euclidean distance between S_i and R_j, although other metrics may be preferable in case S_i and R_j do not represent signal amplitudes or vectors in a metric space. The warping path W between sequences S and R is defined as a contiguous sequence of elements $W_k = (i,j), k = 1, \ldots, T$

$$W = \{W_1, \ldots, W_T\}, \max(m,n) \leq T \leq m + n + 1 \tag{9.21}$$

subject to the following constraints:

- **Boundary conditions:** $W_1 = (1,1)$ and $W_T = (m,n)$, namely, the path starts from the first element of both sequences and ends at the last element.
- **Contiguity:** given $W_{n-1} = (S_i, R_j)$ and $W_n = (S_k, R_l)$, then $|i - k| \leq 1$ and $|j - l| \leq 1$, namely the next step in path W must happen within the 8-neighborhood of the current step position in D.
- **Monotonicity:** given $W_{n-1} = (S_i, R_j)$ and $W_n = (S_k, R_l)$, then $k - i \leq 1$ and $l - j \leq 1$, namely the next step in path W cannot happen "back in time".

The optimal warping path $\widehat{W}$ is the one minimizing the following equation subject to the previously described constraints

$$\widehat{W} = \underset{W}{\operatorname{argmin}} \left\{ DTW(S,R) = \underset{W}{\operatorname{argmin}} \frac{1}{T} \sqrt{\sum_{n=1}^{T} W_n} \right\} \tag{9.22}$$

with $DTW(S,R)$ denoting the score of the warping path W aligning S with R and T compensating for the different lengths of the possible warping paths. Equation (9.22) can be efficiently solved with dynamic programming by the following recursion on the cumulative distance $\gamma(i,j)$

$$\gamma(i,j) = d_{ij} + \min\{\gamma(i-1, j-1), \gamma(i-1, j), \gamma(i, j-1)\}. \tag{9.23}$$

The best warping path $\widehat{W}$ score computed by (9.22) can be directly used to discriminate the dynamic gesture from the aligned trajectory, e.g., by using the DTW score as similarity metric within a nearest neighbor framework [77]. Alternatively, the DTW algorithm can be employed as an intermediate step for aligning feature sequences of different number of frames before feeding them into standard classifiers requiring fixed-length feature vectors.

Since DTW can also align generic multivariate sequences, namely sequences of frames each one described by a multidimensional feature vector (e.g., the relative joint positions and orientations in the case of body action decoding), it may become computationally prohibitive for highly dimensional feature vectors. Moreover, since

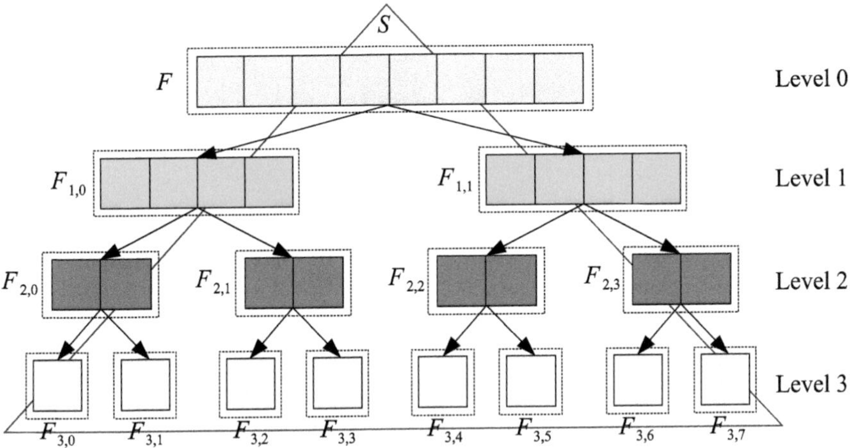

Fig. 9.26 Example of Fourier temporal pyramid

not all the features are generally highly discriminative, the warping cost at each step W_n may sensibly vary for local fluctuation of the feature values thus leading to unreliable sequence alignments. For this reason, a proper weight set can be learned from the dataset in order to modulate the contribution (and the importance) of each feature in the computation of the warping cost at each step [57].

While DTW is mostly often adopted for dynamic hand gesture recognition, an alternative approach well suited to body gesture sequences alignment, with the sequences representing the dynamic evolution of the tracked skeletal joint relative orientations and offsets, is the Fourier Temporal Pyramid (FTP) [72, 73] (Fig. 9.26). The method consists in hierarchically partitioning the input sequence S in $N_l, l = 1, \ldots, L$ segments for each pyramid level l, with N_l usually set as $N_l \triangleq 2^{l-1}$. Short-time Fourier Transform (STFT) is then applied to each temporal subsequence corresponding to a segment and the first F Fourier coefficients corresponding to the lower frequencies of the transform are collected for each segment and juxtaposed in a single feature vector $\mathbf{F}_l$. The final descriptor is made by the concatenation of all the feature vectors $\mathbf{F}_l$. Note how the final feature vector length is ensured to be fixed as the overall number of retained Fourier coefficients does not depend on the input subsequence length but only on the predefined temporal pyramid depth.

9.2.2 Stochastic Recognition Approaches

A human action can be represented with a higher level perspective by the sequence of poses assumed by the person performing the action. Each pose can be defined by a multidimensional feature vector belonging to an infinite continuous space, since the possible body configurations are countless. However, by considering each body pose

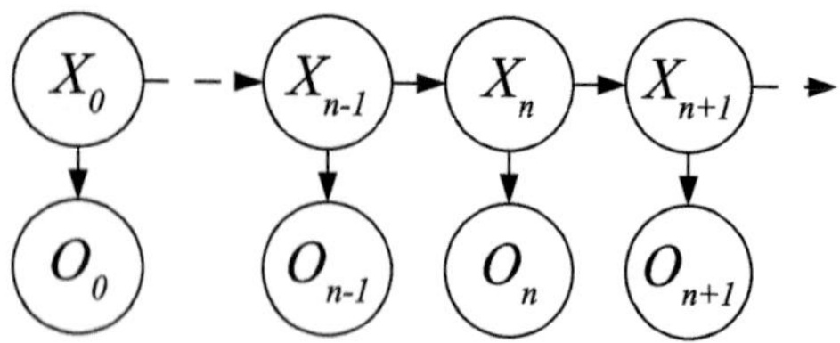

Fig. 9.27 Bayesian network representing a hidden Markov model for dynamic gesture recognition

as an interpolation of a subset of canonical poses drawn from a finite discrete space, a single action can be modeled as a dynamic system with states corresponding to the canonical poses and with evolution depending both on the system dynamic and on the observations from each frame. As each recognizable body action of a predefined action dictionary can be univocally represented by a finite state sequence, the goal of stochastic action recognition approaches is disclosing the most likely state sequence the system underwent during the action execution and which generated the observations on the input frames. The rationale is that the predicted action is the one whose state-sequence template is the most similar to the most likely state sequence generating the input sequence.

9.2.2.1 Dynamic Gesture Recognition with Hidden Markov Models

Hidden Markov Models (HMM) have been widely used in dynamic gesture recognition approaches based on color data [52] and, recently, also in approaches based on depth data [20, 22, 50, 51, 64, 76, 80]. They model each action $\psi_g \in \Psi = \{\psi_1, \ldots, \psi_G\}$ to be decoded by a separate Markov Chain (Fig. 9.27) with random variables $X_n, n = 1, \ldots, T$ associated with the canonical body poses $s_k \in \mathscr{S} = \{s_1, \ldots, s_K\}$ involved in the action.

Each action ψ_g is then considered as a dynamic system whose states represent the poses s_k, and whose evolution from instant $n = 1$ to instant $n = T$ is described by the sequence of random variables $X = \{X_1, \ldots, X_T\}$ drawing values from $\mathscr{S}$. Note how the random variables X_n are not univocally mapped to the poses s_k, hence the system evolution X may contain repetitions of the values (states) in $\mathscr{S}$. By definition the random variables X_n are hidden and their values can only be inferred from their observations $O_1, \ldots, O_T$ drawing values from the set of measurements (symbols) $\mathscr{V} = \{v_1, \ldots, v_M\}$. The observation sequence $O = \{O_1, \ldots, O_T\}$ is said to be generated by the random variables X_n. In the case of dynamic hand gesture recognition, when only the hand trajectory is relevant to discriminate the different gestures in the selected dictionary, the system states often represent the basic possible trajectory direction changes (e.g., upward, downward, diagonal) and the observations correspond to the measured trajectory angles, positions, or speeds [50, 51, 77].

Given the previously defined notation, the action decoding problem can be formulated as

$$\hat{\psi} = \underset{\psi_g \in \Psi}{\operatorname{argmax}}\, P(\{O_1, \ldots, O_T\} \mid \psi_g). \tag{9.24}$$

Namely, the goal of action decoding is disclosing the most likely action model $\hat{\psi}$ generating the observations $O_n, n = 1, \ldots, T$. The Bayesian network of Fig. 9.27 describes the joint distribution probability

$$P(X_n, O_n) = P(X_n|X_{n-1})P(O_n|X_n) \tag{9.25}$$

with $P(O_n|X_n)$ denoting the likelihood of observation O_n on the hidden state X_n, and $P(X_n|X_{n-1})$ denoting the conditional state transition probability depending only on the previous state X_{n-1}. According to the HMM theory [4], each action model ψ_g is defined by specifying: $\boldsymbol{\pi} = [\pi_1, \ldots, \pi_K]$ the stochastic vector with entries $\pi_k, k = 1, \ldots, K$ denoting the probability that the initial system state X_0 is $s_k \in \mathcal{S}$. Let $\mathbf{A} \in \mathbb{R}^{K \times K} = \{a_{ij}\}$ represent the state transition matrix, with entries a_{ij} denoting the probability of commuting from state s_i to state s_j in the next instant $n+1$ disregarding n, and $\mathbf{B} \in \mathbb{R}^{K \times M} = \{b_{ij}\} = P(O_n = v_j | X_n = s_i)$ represent the observation model. Note how $P(O_n|X_n)$ is analogous to the observation models described in Chap. 8 for body or hand tracking, e.g., it may evaluate the similarity between real and synthetic depth patches, and in the case of pixel-wise features, under the assumption of a statistically conditional independence among the pixels, it can be rewritten by Bayes' rule as $P(O_n|X_n) = \prod_{j=1}^{J} P(O_n^j|X_n^j)$ with J denoting the number of depth map pixels and X_n^j, O_n^j respectively denoting the random variable and the observation associated with the jth pixel at instant n.

The action decoding problem of (9.24) can be reformulated in terms of (9.25) as the optimization on ψ_g of the observation likelihood

$$\begin{aligned} &P(\{O_1, \ldots, O_T\}|\psi_g) = \\ &\quad = \sum_{X \in \mathcal{S}^T} P(\{O_1, \ldots, O_T\}|\{X_1, \ldots, X_T\}, \psi_g) P(\{X_1, \ldots, X_T\}|\psi_g) \end{aligned} \tag{9.26}$$

with $X = \{X_1, \ldots, X_T\}$ denoting one of the possible state evolutions after T instants drawn from $\mathcal{S}^T = \mathcal{S} \times \mathcal{S} \times \cdots \times \mathcal{S}$. The $P(\{O_1, \ldots, O_T\}|\{X_1, \ldots, X_T\}, \psi_g)$ term is the observation likelihood of (9.25), and can be reformulated in terms of HMM action model $\psi_g(\mathbf{A}, \mathbf{B}, \boldsymbol{\pi})$ as

$$P(\{O_1, \ldots, O_T\}|\{X_1, \ldots, X_T\}, \psi_g) = b(X_1, O_1) b(X_2, O_2) \cdots b(X_T, O_T) \tag{9.27}$$

and $P(\{X_1, \ldots, X_T\}|\psi_g)$ can be reformulated as well in terms of ψ_g as

$$P(\{X_1, \ldots, X_T\}|\psi_g) = \pi(X_1) a_{X_1 X_2} \cdots a_{X_{T-1} X_T}. \tag{9.28}$$

Finally, (9.24) from (9.27) and (9.28) can be rewritten as

$$\begin{aligned} &P(\{O_1, \ldots, O_T\}|\psi_g) = \\ &\quad = \sum_{X \in \mathcal{S}^T} \pi(X_1) b(X_1, O_1) a_{X_1 X_2} b(X_2, O_2) \cdots a_{X_{T-1} X_T} b(X_T, O_T). \end{aligned} \tag{9.29}$$

Equation (9.29) can be interpreted as follows: at instant $n = 1$ the system starts in state X_1 with probability $\pi(X_1)$ and generates the symbol O_1 with probability $b(X_1, O_1)$. Then in the next instant $n = 2$ the system makes a transition to state X_2 from state X_1 with probability $a_{X_1X_2}$ and generates the symbol O_2 with probability $b(X_2, O_2)$, and so on until instant $n = T$. Although (9.29) is formally correct, its naive solution is not practicable even for a limited number of system states as it involves for each action model ψ_g the evaluation of an exponential number ($O(2TK^T)$ with K the number of system states) of possible system state sequences of length T in order to find the one that most likely generates the observation sequence O [54]. More efficient techniques have been developed in order to make tractable action decoding, among which the forward-backward procedure is still the most adopted algorithm. This method, instead of facing the optimization of $P(\{O_1, \ldots, O_T\}|\psi_g)$ on the full observation sequence, solves it inductively on the observation subsequence length by introducing the concept of forward variable

$$\alpha_n(k) = P(\{O_1, \ldots, O_n\}, X_n = s_k|\psi_g). \tag{9.30}$$

The forward-variable $\alpha_n(k)$ considers the probability that the partial observation sequence from instant 1 to instant n was generated by model ψ_g and lead to state s_k at instant n. With the aid of forward-variables, the decoding problem can be inductively solved as

$$\begin{aligned} \alpha_1(k) =& \pi(X_1 = s_k) b(X_1 = s_k, O_1), 1 \leq k \leq K \\ \alpha_{n+1}(j) =& \left[\sum_{k=1}^{K} \alpha_n(k) a_{s_k s_j}\right] b(X_{n+1} = s_j, O_{n+1}), 1 \leq n \leq T-1, 1 \leq j \leq K. \end{aligned} \tag{9.31}$$

Eventually, for $n = T$ the last step returns the desired value of $P(\{O_1, \ldots, O_T\}|\psi_g)$, which is

$$P(\{O_1, \ldots, O_T\}|\psi_g) = \sum_{k=1}^{K} \alpha_T(k). \tag{9.32}$$

The forward-backward procedure computational load is dramatically lower (in the order of $O(TK^2)$ calculations) with respect to that of the naive solution method. A complete treatment of this topic is beyond the scope of this chapter and the interested reader is referred to [54].

Assuming the state evolution X of a HMM parameterized by $\psi_g = (\mathbf{A}, \mathbf{B}, \boldsymbol{\pi})$ has already been disclosed in the previous step, the action model ψ_g definition requires the specification of matrices $\mathbf{A}$, $\mathbf{B}$ and of the initial stochastic state vector $\boldsymbol{\pi}$. Since the specification of $\mathbf{A}, \mathbf{B}, \boldsymbol{\pi}$ requires a thorough knowledge of the action dynamic, which not only remains undisclosed but it is also likely to lead to wrong assumptions, the action model ψ_g is usually automatically learned from an action dataset by finding the parameters maximizing its likelihood $\mathscr{L}(\psi_g|\{O_1, \ldots, O_T\}) = P(\{O_1, \ldots, O_T\}|\psi_g)$ with respect to the observation sequences in the dataset.

Unfortunately, there is no close-form solution for the maximization of the model likelihood, hence the action recognition methods based on HMM rather rely on global or local optimization techniques based on gradient descent or on iterative approaches improving the model estimate $\hat{\psi}_g$ at each iteration until convergence. Among the iterative approaches, the most used one is the Baum-Welch method [54], based on the expectation-maximization framework. The initial model parameter values can be estimated from the dataset as follows:

- $\hat{\pi}_k$: number of train sequences starting with state s_k over the training set cardinality
- $\hat{a}_{ij}$: number of transitions from state s_i to state s_j over the number of transitions from state s_i
- $\hat{b}_{ij}$: number of times the train sequences are in state s_i and observe symbol v_j over the number of times the train sequences are in state s_i.

9.2.2.2 Dynamic Gesture Recognition with Hierarchical Markov Models

The dynamic gesture recognition approaches described in the previous sections do not account for the hierarchical structure of human activities, namely the fact that each activity can be partitioned in a sequence of atomic sub-activities executed in chronological order. For example, the simple action of drinking from a glass may be indeed split into three basic sub-actions: first the person brings the glass to its mouth, usually with a smooth arm movement, then the glass and the head are tilted for drinking and eventually the glass is put back in its original position, again with an arm gesture similar to the one performed to bring the glass to the mouth. A few approaches like [62] aim to capture the hierarchical nature of human gestures using a two-layered Markov Model (MEMM), as exemplified in Fig. 9.28.

The model consists in an extended HMM where the hidden nodes X_n are associated with the possible body states S_k defined above, observed by the features extracted from the input sequence frames (nodes O_n), and in turn acts as

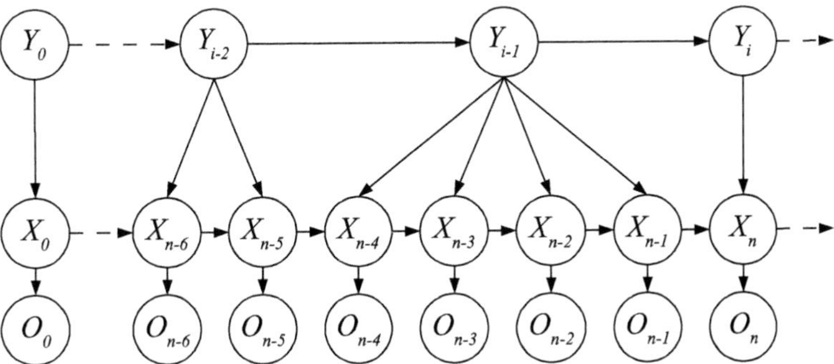

Fig. 9.28 Example of a hierarchical hidden Markov model for activity recognition

observations for a second layer of hidden nodes Y_i denoting the possible sub-activities the observed states X_n refer to. The complete activity model is determined by the following quantities:

- $P(O_n|X_n)$: corresponds to the standard observation model of HMM.
- $P(X_{n_i-m}|X_{n_i-m-1}, Y_i), m = 0, \ldots, n_i - n_{i-1} - 1$: replaces the standard state transition probability $P(X_n|X_{n-1})$ of HMM modeling the fact that the system state transition depends not only from the currently observed pose but also from the sub-activity Y_i the pose is referred to.
- $P(Y_i|Y_{i-1})$: denotes the fact that activities evolve over time and that certain activities are more likely to happen when a particular activity has been previously executed. It corresponds to the state transition probability of standard HMM.

While the system operation is analogous to HMM, an additional problem to solve is the disclosure of the sub-activities Y_i to recognize, which can be learned from the training set for each action either by manually labeling the frames or by using an automatic clustering approach associating each cluster a different sub-activity, as done in HMM.

Let now $O_i = \{O_{n_{i-1}+1}, \ldots, O_{n_i}\}$ denote an observation sequence and Y_{i-1} the previous activity node. The goal can be formalized as the computation of

$$\begin{aligned} P(Y_i, \{X_{n_{i-1}+1}, \ldots, X_{n_i}\}|O_i, Y_{i-1}) = \\ = P(Y_i|Y_{i-1}) \sum_{X_{n_i-1}} \frac{P(X_{n_{i-1}+1}|X_{n_{i-1}}, Y_i)P(X_{n_{i-1}+1}|O_{n_{i-1}+1})}{P(X_{n_{i-1}+1})} \\ P(X_{n_{i-1}}) \prod_{n=n_{i-1}+2} \frac{P(X_n|X_{n-1}, Y_i)P(X_n|O_n)}{P(X_n)}. \end{aligned} \tag{9.33}$$

The reader interested in the details of the derivation of (9.33) and its inference is referred to [62]. Note only how the main problem in the maximization of the likelihood of (9.33) for action decoding is determining to which high level activity Y_i each state X_n should be connected, namely, it is a problem of graph structure selection for the MEMM model, which cannot be solved in practice by evaluating all the $O(2^n)$ possible substructures but can be solved efficiently by dynamic programming.

9.2.3 *Dynamic Gesture Recognition with Action Graphs*

Hidden Markov models have been widely used for human action decoding since the early approaches based on video analysis, due to their high capability of modeling human actions. One major drawback of HMMs for this task is the need of defining a separate HMM for each single action to decode, and the need of acquiring the

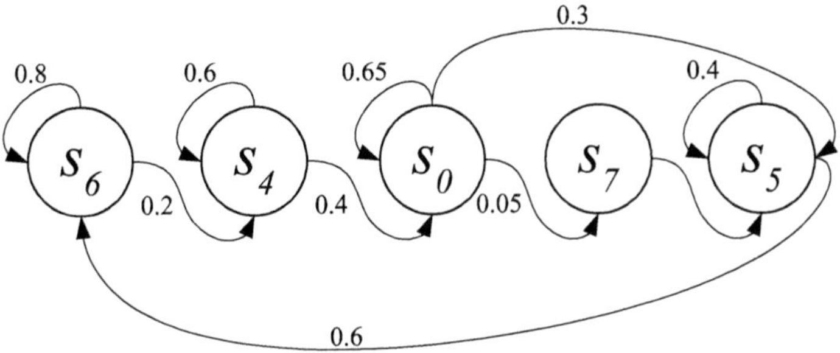

Fig. 9.29 Example of action graph: the involved poses are a subset of $\mathscr{S} = \{s_0, \ldots, s_7\}$ and the numbers above the arcs denote the probability $P(s_j|s_i)$ of moving to pose s_j from pose s_i

full action sequence, namely HMMs are not able to decode an action from the partially framed sequence. An alternative graphical model to HMM, increasingly adopted in novel action decoding schemes, are the Action Graphs [29, 31, 32, 68], which operate similar to HMM but are able to decode actions from partially framed sequences and allow pose sharing among different actions. Figure 9.29 shows an example of Action Graph.

Let $O = \{O_1, \ldots, O_T\}$ denote again the sequence of observations up to instant $n = T$ drawing values from the symbol set $\mathscr{V} = \{v_1, \ldots, v_M\}$, $X = \{X_1, \ldots, X_T\}$ the sequence of the poses generated by O drawing values from the pose set $\mathscr{S} = \{s_1, \ldots, s_K\}$, and $\Psi = \{\psi_1, \ldots, \psi_G\}$ the set of actions to decode. Action decoding with Action Graphs can be formulated as the following optimization problem

$$\hat{\psi} = \underset{\psi_g \in \Psi, X \in \mathscr{S}^T}{\operatorname{argmax}} \; P(\{O_1, \ldots, O_T\}, \{X_1, \ldots, \mathscr{S}^T\}, \psi_g) \tag{9.34}$$

with $P(\{O_1, \ldots, O_T\}, \{X_1, \ldots, \mathscr{S}^T\}, \psi_g)$ denoting the probability that action ψ_g generated the observation sequence O by visiting the pose sequence X. According to Bayes' rule, (9.34) can be reformulated as

$$\hat{\psi} \propto \underset{\psi_g \in \Psi, X \in \mathscr{S}^T}{\operatorname{argmax}} \; P(\psi_g) P(\{X_1, \ldots, \mathscr{S}^T\}|\psi_g) P(\{O_1, \ldots, O_T\}|\{X_1, \ldots, X_T\}, \psi_g) \tag{9.35}$$

with $P(\psi_g)$ the prior probability of action ψ_g (learnable from the training set action distribution), $P(\{X_1, \ldots, X_T\}|\psi_g)$ the likelihood of pose sequence X on action ψ_g and $P(\{O_1, \ldots, O_T\}|\{X_1, \ldots, X_T\}, \psi_g)$ the posterior that the observation sequence O was generated by action ψ_g visiting the pose sequence X. For action recognition purposes, a few assumptions are typical, namely:

- observation O_n is statistically independent from action ψ_g given X;
- observation O_n only depends on X_n;
- pose X_n only depends on X_{n-1}.

The first two assumptions claim that each observation O_n statistically depends only on the pose X_n that generated it disregarding the action, while the third assumption claims that each pose X_n is only statistically dependent from the previous pose X_{n-1} in X, as also assumed in Markov chains. Given the previous assumptions and notation, an action Graph Γ is defined as the trained system $\Gamma = (\mathscr{S}, \Psi, \Lambda, \mathscr{G}, \mathbf{A})$ with $\Lambda = \{\lambda_{ij}\}$ denoting the probability $P(O_n = v_j|X_n = s_i)$ that pose s_i generates symbol v_j at instant n (measurement model) disregarding the action, and $\mathscr{G} = \{(\mathscr{S}, \mathscr{A}_g)\}_{g=1}^{G}$ denoting the set of graphs $(\mathscr{S}, \mathscr{A}_g)$, one for each action $\psi_g \in \Psi$, built on the pose space $\mathscr{S}$ as vertex set and the set of arcs $\mathscr{A}_g$ between the poses in $\mathscr{S}$ involved in action ψ_g, and $\mathbf{A} = \{a_{ij}\}$ the global transition matrix describing the probability $P(X_{n+1} = s_j|X_n = s_i)$ of commuting from pose s_i to pose s_j at instant n disregarding the action. Note how $\mathscr{A}_g$ is associated with a transitions probability matrix $\mathbf{A}_g$ denoting the probability $P(X_{n+1} = s_j|X_n = s_i, |\psi_g)$ of commuting from pose s_i at instant n to pose s_j at instant $n+1$ depending on the considered action ψ_g. Equation (9.35) can be rewritten as

$$\hat{\psi} = \underset{\psi_g \in \Psi, X \in \mathscr{S}^T}{\operatorname{argmax}} P(\psi_g) P(\{X_1, \ldots, \mathscr{S}^T\}|\psi_g) \prod_{n=1}^{T} P(O_n|X_n). \tag{9.36}$$

Note how the first term of (9.36) corresponds to a Markov chain and models the specific knowledge, while the second term models the shared knowledge.

Given the observation sequence $\{O_1, \ldots, O_T\}$ generated by the considered action ψ_g, action decoding undergoes the following steps:

1. find in each graph $(\mathscr{S}, \mathscr{A}_g)$ the most likely path (pose sequence) generating O;
2. compute the likelihood for each action $\psi_g \in \Psi$;
3. assign to $\hat{\psi}$ the class g of the action ψ_g maximizing the likelihood of (9.36) and accept the result if the likelihood is higher than a given threshold T_ψ.

There are mainly three action decoding schemes for Action Graphs [31], listed next:

- **Action-specific Viterbi decoding:** is an optimal decoding scheme evaluating the likelihood $\mathscr{L}(O; \psi_g)$ of each action $\psi_g \in \Psi$ for the observation sequence O given by

$$\mathscr{L}(O; \psi_g) = \max_{X \in \mathscr{S}^T} P(\psi_g) \prod_{n=1}^{T} P(X_n|X_{n-1}, \psi_g) \prod_{n=1}^{T} P(O_n|X_n). \tag{9.37}$$

  Let $\mathscr{L}(O; \hat{\psi})$ denote the maximum likelihood among the actions ψ_g. $\hat{\psi}$ is accepted as the actual action generating O only if

$$\frac{\mathscr{L}(O;\hat{\psi})}{\sum_{g=1}^{G}\mathscr{L}(O;\psi_g)} \geq T_{\psi}. \tag{9.38}$$

Note how this method, although being optimal, is both highly time and memory consuming since it searches for the optimal path with respect to each possible action.

- **Global Viterbi decoding:** is a sub-optimal decoding scheme looking for the most probable sequence of poses $\hat{X}$ generating O, disregarding the action, instead of evaluating all the possible paths and defined as

$$\hat{X} = \underset{X\in\mathscr{S}^T}{\operatorname{argmax}} \prod_{n=1}^{T} P(X_n|X_{n-1})P(O_n|X_n). \tag{9.39}$$

The decoded action is the one maximizing the likelihood of (9.39) according to the *unigram* or *bigram* schemes defined by the next two equations respectively

$$\mathscr{L}(O;\hat{\psi}) = \underset{\psi_g\in\Psi}{\operatorname{argmax}}\, P(\psi_g) \prod_{n=1}^{T} P(\hat{X}_n|\psi_g) \tag{9.40}$$

$$\mathscr{L}(O;\hat{\psi}) = \underset{\psi_g\in\Psi}{\operatorname{argmax}}\, P(\psi_g) \prod_{n=1}^{T} P(\hat{X}_n|\hat{X}_{n-1},\psi_g). \tag{9.41}$$

Note how this method requires only about $1/G$ of the computational resources of the action-specific Viterbi decoding and considers only the global transition probability matrix $\mathbf{A}$.

- **Maximum-likelihood decoding:** is another sub-optimal decoding scheme searching the sequence of the most likely poses

$$\hat{X} = \underset{X\in\mathscr{S}^T}{\operatorname{argmax}} \prod_{n=1}^{T} P(O_n|X_n). \tag{9.42}$$

Note that (9.42) is different from the previous scheme which, instead, searches the most likely sequence of poses. The decoded action is, again, the one maximizing the likelihood of (9.40) or (9.41) according to the unigram or bigram scheme.

Analogously to the HMM models for action decoding of Sect. 9.2.2.1 the most significant poses assumed by the body in the action set Ψ, associated with the system states, can be learned from the training set, assuming the feature vectors are uniformly distributed among the different actions recognizable by the trained system. Different techniques have been proposed in the literature for this purpose, dominated by the ones clustering the pose measurements with K-means [34, 68, 76] or mean-shift [13] using the Euclidean distance or other metrics tailored to the type of data to cluster. The pose distribution $P(v_i|s_k)$ in the dataset within each cluster k is usually represented in a more compact form by the parameters of a Gaussian Mixture Model (GMM) fitted by the EM algorithm to the frame observation values v_i of pose s_k in the dataset [31, 32, 68] by

$$P(v_i|s_k) = \sum_{j=1}^{J} \pi_{j,k} \mathcal{N}(v_i; \mu_{j,k}, \Sigma_{j,k}) \tag{9.43}$$

with s_k denoting the pose described by cluster k, J the number of the mixture components, $\pi_{j,k}$ the weight associated with the jth mixture component of cluster k, and $\mathcal{N}(v_i; \mu_{j,k}, \Sigma_{j,k})$ the normal distribution described by mean $\mu_{j,k}$ and covariance matrix $\Sigma_{j,k}$. An alternative to (9.43) are the so called *bags of 3D points* [32].

The last components of action graph Γ to be determined are the global transition matrix $\mathbf{A}$ and the transition matrices $\mathbf{A}_g, g = 1, \dots, G$ for each action $\psi_g \in \Psi$, which can be learned as well from the training data. Let $s_i, s_j \in \mathcal{S}$. The global transition probability from pose s_i to pose s_j and the transition probability with respect to action $\psi_g \in \Psi$ can be computed as

$$P(s_j|s_i) = \frac{\sum_{m=1}^{M} P(s_j|v_m)P(s_i|v_{m-1})}{\sum_{m=1}^{M} P(s_j|v_m)} \tag{9.44}$$

$$P(s_j|s_i, \psi_g) = \frac{\sum_{m=1}^{M_g} P(s_j|v_m, \psi_g)P(s_i|v_{m-1}, \psi_g)}{\sum_{m=1}^{M_g} P(s_j|v_m, \psi_g)} \tag{9.45}$$

with M denoting the pose training set size, and M_g the number of training vectors for action ψ_g. Quantities $P(s_j|v_m)$ and $P(s_j|v_m, \psi_g)$ respectively denote the likelihood of pose s_j in the whole training set and in the training set referred to action ψ_g only, often modeled as uniform distributions as

$$P(s_j|v_m) = \frac{M_j}{M} \tag{9.46}$$

$$P(s_j|v_m, \psi_g) = \frac{M_{j,g}}{M_g} \tag{9.47}$$

with N_j denoting the number of observations associated with pose s_j in its training set, disregarding the action ψ_g, and $N_{j,g}$ the number of pose s_j observations referred to action ψ_g. Finally, with respect to the hidden Markov models adopted in other approaches, the action decoding can be performed "online" by evaluating the maximum likelihood for each action by incremental extensions of the observation sequence O (namely by only considering, for each step, an input sequence O only incremented by an observation O_T) until the computed maximum likelihood exceeds a minimum acceptance likelihood threshold [31].

9.2.4 Descriptors for Dynamic Gesture Recognition

Body and hand motion information plays an important role in dynamic gesture recognition, even because often the actions are better characterized by the space-time evolution of the body parts of interest, rather than by the body or hand shape which, instead, are fundamental for the recognition of static gestures of Sect. 9.1. As previously stated, describing human motion generally requires the extraction of appropriate descriptors from each acquired frame and their aggregation in a unique feature vector to represent the whole input sequence gesture class to decode. This section describes the most common feature extraction methods adopted in dynamic gesture recognition, grouped in two families: features describing a single frame and features describing the whole input sequence. Different from the descriptors of Sect. 9.1.7 for static gesture recognition that mostly leverage on body and hand shape information, the features adopted for dynamic gesture recognition mostly rely on the information about the volume occupied by the body or hand samples within the input sequence. Moreover, several approaches based on the dynamic volume occupancy as recognition clue are direct extensions to four dimensions (three dimensions for the coordinates in Euclidean 3D space and one for the time axis) of the volume occupancy methods of Sect. 9.1.7.

Frame-wise descriptors for dynamic gesture recognition extract a separate descriptor for each frame and concatenate all the descriptors in a single feature vector to be fed to one of the classifiers of Sect. 9.1.7. Note how, for clarity's sake, the input sequences are assumed to be aligned with each gesture template by DTW or equivalent methods. Even though any of the feature families of Sect. 9.1 can potentially be used also for the dynamic gestures case, most methods in the literature prefer frame descriptors with low dimensions to minimize the overall length of the final feature vector of the input sequence.

As already stated in Sect. 9.1.1, body or hand skeleton pose can be effectively described by the relative orientations $\mathbf{R}_j, j = 1, \ldots, N$ of the model joints and the absolute position $\mathbf{t}_1$ of the root joint in the Euclidean space. An intuitive temporal extension is a simple concatenation of the joint orientations in each frame within a unique feature vector. Alternatively, [78] extracts features measuring the offset of each joint j with respect to the other joints $k \neq j$ within the same frame, within different frames or with respect to the initial frame F_0. In order to reduce the feature vector space and the measurement noise effects, PCA can be applied to each joint feature evolution. The compressed descriptor for each joint is called *Eigen-Joint* [78] recalling that only the most significant eigenvalues of PCA are retained.

Sequence-wise feature extraction algorithms, instead, operate on a sequence basis, namely by obtaining a single descriptor for the whole input sequence by aggregating the descriptors extracted from each single frame in place of their juxtaposition.

Gesture recognition methods based on color images and videos often resorted on *motion history images* [1], a view-based temporal template method for representing an arbitrarily long motion sequence with a unique gray-scale image I_M whose pixel intensities are associated with a given function of the motion density. Image I_M is often associated with a binary *motion energy image* I_E describing the motion shape and its spatial distribution. Without any loss of generality, the same techniques designed for color images can be directly applied to the acquired sequence of depth maps $\{Z_1, \ldots, Z_T\}$, as shown in Fig. 9.30.

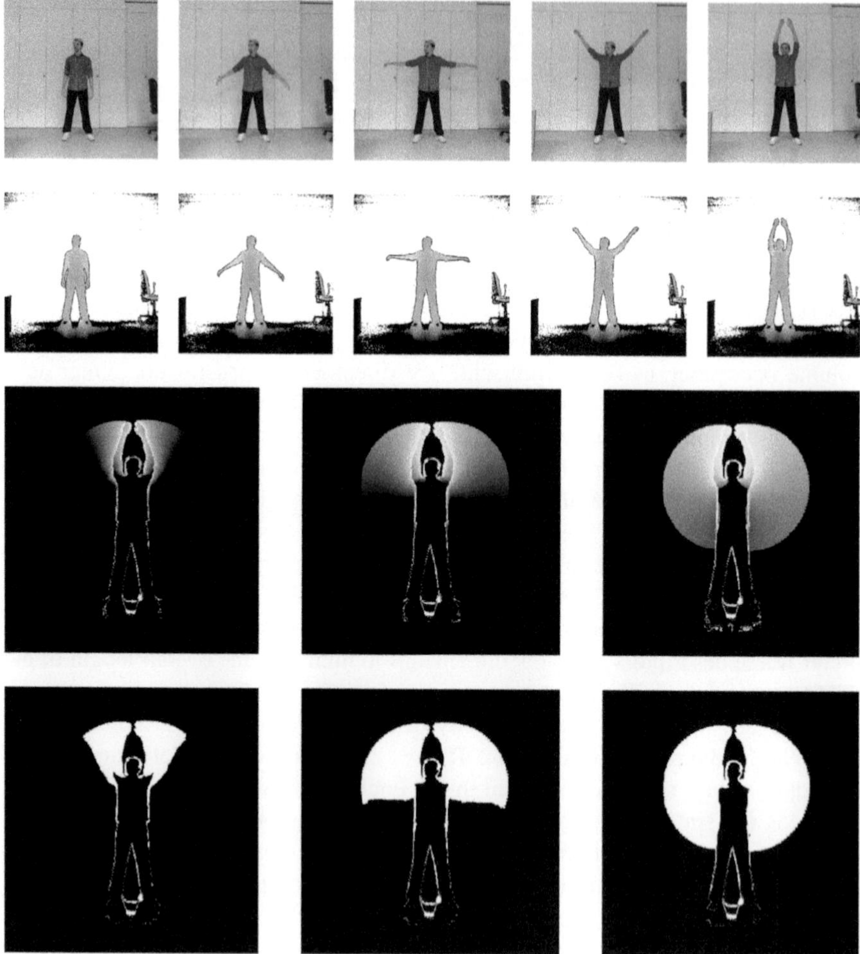

Fig. 9.30 Example of motion template generated from a depth sequence: (*first row*) key color frames of the acquired sequence; (*second row*) associated key depth maps; (*third and fourth row*) generated MHIs and MEIs

Let $\psi(p^i, n)$ denote a dynamic indicator function assuming value 1 if pixel p^i observed any motion at instant n or 0 otherwise, with $1 \leq n \leq T$, and τ denoting the motion history length (in number of frames). The motion history value $H_\tau(p^i, n)$ at pixel p^i at instant n is defined as

$$H_\tau(p^i, n) = \begin{cases} \tau & \text{if } \psi(p^i, n) = 1 \\ \max(0, H_\tau(p^i, n-1) - \delta) & \text{otherwise} \end{cases} \tag{9.48}$$

with $\delta > 0$ denoting the decay rate (usually $\delta = 1$) of the motion density in time. In particular, $H_\tau(p^i, T)$ is the final motion history image pixel p^i value for the full sequence. Function $\psi(p^i, n)$ can leverage background subtraction as

$$\psi_B(p^i, n) = \begin{cases} 1 & \text{if } |z_n^i - z_B^i| \geq \zeta \\ 0 & \text{otherwise} \end{cases} \tag{9.49}$$

or frame differencing as

$$\psi_D(p^i, n) = \begin{cases} 1 & \text{if } |z_n^i - z_{n-1}^i| \geq \zeta \\ 0 & \text{otherwise} \end{cases} \tag{9.50}$$

with B denoting the reference background depth map, z_n^i the depth value of pixel p^i at instant n and ζ a preset depth difference threshold. Function $\psi(p^i, n)$ can also be defined leveraging motion flow or more complex techniques to detect motion on each pixel p^i.

The parameters τ and δ of (9.48), usually determined empirically, modulate the length of the trail left by the moving body in H_τ and the speed of its fading: if the history length τ is lower than the sequence length T and $\delta = 1$, H_τ starts losing the initial motion information right after τ frames. Conversely, a history length τ much higher than the sequence length T allows to retain the full motion information and, if the decaying rate δ is negligible, it leads to a slowly changing motion gradient. A high decaying value δ, instead, favors the contribution of the most recent motion by discarding most of the motion information related to the initial trait of the motion sequence and leads to a discrete quantization of the motion within the input sequence.

Equation (9.48) is also called update function, since for each instant n it updates the motion status of pixel p^i. The cumulative binary motion image I_E, instead, is defined as

$$E_\tau(p^i, n) = \bigcup_{j=1}^{\tau-1} \psi(p^i, n-j) = \begin{cases} 1 & \text{if } H_\tau(p^i, n) \geq 1 \\ 0 & \text{otherwise.} \end{cases} \tag{9.51}$$

Namely, $E_\tau(p^i, n)$ assumes value 1 if pixel p^i observed any motion in the recent motion history before instant n. It is straightforward to see that E_τ can be operatively obtained by a simple binary thresholding of the motion history image H_τ.

H_τ and E_τ together form a temporal template and carry complementary information, since H_τ only describes how the motion happened while E_τ only describes where the motion happened. An appropriate descriptor is then extracted from the two images (e.g., Hu moments [25] or HOG) and fed into a classification scheme of Sect. 9.1.7 for static gesture recognition (e.g., SVM or K-NN).

Finally, the standard MHI descriptor described above, although being more robust with respect to its original version computed on color videos thanks to the independence of depth information from lighting conditions, is not able to effectively capture the human motion in the 3D space due to perspective projection. A possible solution to overcome this limit consists in generating a separate MHI of each view of interest of the body 3D point cloud (front, side and top) and concatenating the set of image descriptors extracted from each MHI in a single feature vector to be fed to SVM or other classifiers [79]. The MHI for each view, or in this case the *Depth Motion Maps* (DMM), are generated from the orthogonal projection of the body 3D point cloud to each virtual image plane $\pi_{front}, \pi_{side}, \pi_{top}$ identified by the faces of the minimum 3D bounding box enclosing the body point cloud.

Random Occupancy Patterns (ROP) [71] are volumetric descriptors that, similar to Shape Context [24], represent the framed action sequence by the time varying occupation of the cells of a random partition of the 4D volume (Fig. 9.31). The acquired depth maps are turned into 3D point clouds by back-projection and treated as a 4D volume with the first three dimensions associated with the coordinates of the points in the Euclidean space and the fourth associated with the time axis. Analogously to Shape Context, the ROP descriptor of a single cell c simply consists in a soft-thresholded count of the 4D points occupying a given cell defined as

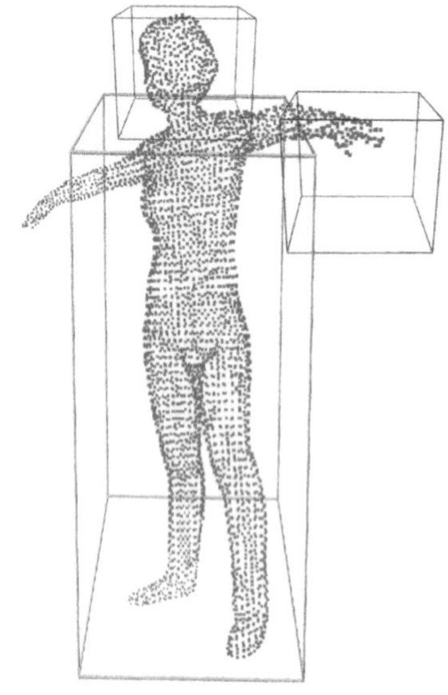

Fig. 9.31 Example of volume sampling in ROP

$$N(c) = \delta\left(\sum_{n=1}^{T}\sum_{P\in\mathscr{P}_n} I(P,c)\right) \tag{9.52}$$

with $\mathscr{P}_n$ denoting the 3D point cloud generated from the segmented body or hand depth map at instant n, $I(P,c)$ an indicator function assuming value 1 if the 3D point $P \in \mathscr{P}_n$ belongs to cell c at instant n and 0 otherwise, and $\delta(\cdot)$ a sigmoid smoothing function.

Equation (9.52) is evaluated at several cell locations and cell sizes and the final descriptor is a simple juxtaposition of the features extracted from each cell into a single feature vector to be fed to SVM or other classifiers. Alternatively, the final feature vector can be expressed by sparse coding schemes in terms of the basis elements of a dictionary from a lossy compression problem [71].

Space-Time Occupancy patterns (STOP) [68] are volumetric descriptors that, analogously to Random Occupancy Patterns, begin by partitioning the 4D volume in cells and compute the cell occupancy by the 4D points of the time-varying body or hand point cloud. The feature extraction algorithm evaluates

$$N(c) = \begin{cases} 1 & \text{if } |A_c| \geq T_P \\ \dfrac{|A_c|}{T} & \text{otherwise} \end{cases} \tag{9.53}$$

where T_P is a predefined saturation value, and $|A_c|$ denotes the number of 4D points falling into cell c within the acquired depth sequence. Count $N(c)$ is evaluated for each cell c of the volume partitioned by a regular grid, and the cell descriptors are concatenated into a single feature vector. Due to the high number of cells, the final feature vector is generally too long to be efficiently treated and feature selection strategies like PCA are often employed for dimensionality reduction. Figure 9.32 depicts the occupancy of the cells associated with the body point cloud during the execution of an action.

Finally, the *Histogram of Oriented 4D Normal Orientations* (HON4D) [49] characterizes the human motion by the distribution of the directions of the normals to the vertices of a 3D mesh $\mathscr{M}$ modeling the deformable body surface. Analogously to ROP and STOP descriptors, the normals and the vertices are denoted, again, by 4D vectors with the first three dimensions associated with the normal direction and the last one with the time axis.

The normal orientation distribution is estimated by building a 4D orientation histogram, with each bin associated with a region of the 4D space defined by a uniform quantization that, different from the 2D and 3D spaces, cannot be performed by bidimensional or tridimensional tessellations (e.g., by square or rectangle grids for HOG descriptor or by a cubic honeycomb for STOP). The 4D space is indeed uniformly partitioned by mean of 4D regular geometric objects named *polychorons* (or 4-polytopes), which are 4D extensions of the 2D polygons. The method of [49], for example, suggests to use the 120 4D vertices of a tetraplex to uniformly quantize the 4D space.

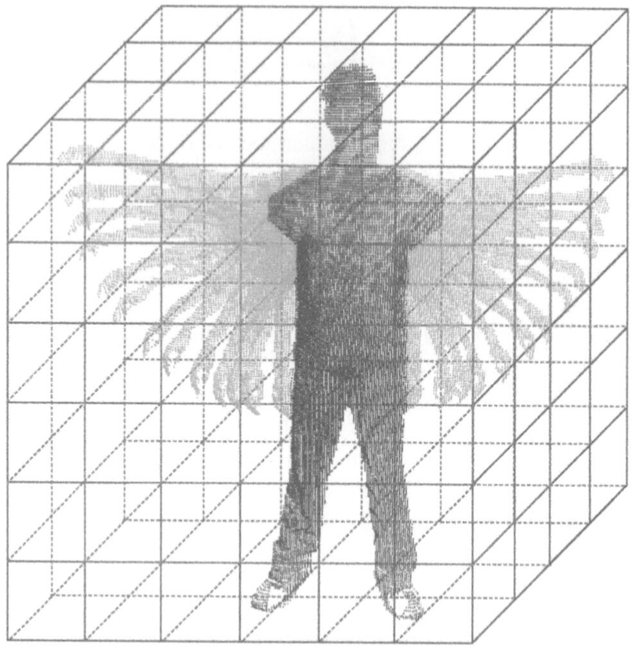

Fig. 9.32 Example of STOP features: the *red points* are contained in cells with more than T_P points

Each tetraplex vertex V^j is called projector, as the 4D vectors joining the space origin 0 with each vertex V^j are used as reference directions to compute the 120 components of each 4D normal $\hat{\mathbf{n}}_i$ at point P^i with respect to the "basis" made by the vectors $\mathbf{V}^j$ as

$$c(\hat{\mathbf{n}}_i, \mathbf{V}^j) = \max(0, \hat{\mathbf{n}}_i^T \mathbf{V}^j) \tag{9.54}$$

where in (9.54) only the non negative components (namely, the ones for which the internal product is not negative) are considered. Finally, the normal direction distribution $P(\mathbf{V}^j|\mathcal{N})$ is modeled as

$$P(\mathbf{V}^j|\mathcal{N}) = \frac{\sum\limits_{\hat{\mathbf{n}}_i \in \mathcal{N}} c(\hat{\mathbf{n}}_i, \mathbf{V}^j)}{\sum\limits_{\mathbf{V}^p \in \mathcal{V}} \sum\limits_{\hat{\mathbf{n}}_i \in \mathcal{N}} c(\hat{\mathbf{n}}_i, \mathbf{V}^p)} \tag{9.55}$$

with $\mathcal{N}$ denoting the set of the 4D surface normals and $\mathcal{V}$ denoting the projector set. In order to introduce clues from the spatio-temporal context, the acquired depth map sequence can be partitioned in spatio-temporal cells and the final feature vector can be made by concatenating the HON4D descriptors extracted from each cell.

9.3 Conclusions and Further Readings

Gesture recognition raised a great interest for its wide set of applications in both research and industry. In the research field this is demonstrated by the vast literature about gesture recognition methods, in the past based on images or videos and more recently exploiting also depth data. The large industrial interest is demonstrated by the introduction of ad hoc cameras and software developments kits for this task, like the Intel RealSense^{TM} family of cameras and software or the SoftKinetic DepthSense products. Another interesting device is the Leap Motion sensor that, different from the other sensors considered in this book, does not capture a depth map but only a few relevant key points such as the 3D positions of the fingertips. The Leap Motion has been used alone or together with other depth cameras for hand gesture recognition applications [37].

This chapter showed how depth data allow to improve gesture recognition methods based on visual information, overcoming the limits of color-based approaches like the dependence on environment lighting conditions and the robustness to self-occlusions. Furthermore the peculiar characteristics of geometry information led to the disclosure of novel and robust descriptors for the body and hand improving the overall recognition accuracy. Although the results obtained by recent gesture recognition approaches based on depth data are rather promising, gesture recognition remains an open problem with challenging issues. There is a very active research on both the academic and industrial side in order to develop novel sensors and recognition algorithms able to provide a reliable solution to the problem in all situations exploiting the geometry information, but also different and new types of information channels. There is also a rapidly growing interest for novel applications of depth based gesture recognition in surveillance applications and in natural touchless interfaces but both these fields are still in the embryonic stage and novel sensors and algorithms are required for these applications.

References

1. M.A.R. Ahad, J.K. Tan, H. Kim, S. Ishikawa, Motion history image: its variants and applications. Mach. Vis. Appl. **23**(2), 255–281 (2012)
2. B. Apostol, C.R. Mihalache, V. Manta, Using spin images for hand gesture recognition in 3D point clouds, in *Proceedings of IEEE International Conference on System Theory, Control and Computing* (2014), pp. 544–549
3. P. Barros, S. Magg, C. Weber, S. Wermter, A multichannel convolutional neural network for hand posture recognition, in *Proceedings of International Conference on Artificial Neural Networks* (Springer, Heidelberg, 2014), pp. 403–410
4. L.E. Baum, T. Petrie, Statistical inference for probabilistic functions of finite state Markov chains. Ann. Math. Stat. **37**(6), 1554–1563 (1966)
5. R. Bellman, R. Kalaba, On adaptive control processes. IRE Trans. Automat. Control **4**(2), 1–9 (1959)
6. C.M. Bishop, *Pattern Recognition and Machine Learning. Information Science and Statistics* (Springer, Heidelberg, 2007)

7. K.K. Biswas, S.K. Basu, Gesture recognition using microsoft kinect, in *Proceedings of International Conference on Automation, Robotics and Applications* (2011), pp. 100–103
8. L. Breiman, Random forests. Mach. Learn. **45**(1), 5–32 (2001)
9. P. Breuer, C. Eckes, S. Muller, Hand gesture recognition with a novel IR time-of-flight range camera: a pilot study, in *Proceedings of International Conference on Computer Vision/Computer Graphics Collaboration Techniques* (Springer, Berlin/Heidelberg, 2007), pp. 247–260
10. T.I. Cerlinca, S.G. Pentiuc, Robust 3D hand detection for gestures recognition, in *Intelligent Distributed Computing V*, ed. by F.M.T. Brazier, K. Nieuwenhuis, G. Pavlin, M. Warnier, C. Badica. Studies in Computational Intelligence, vol. 382 (Springer, Berlin/Heidelberg, 2012), pp. 259–264.
11. C.C. Chang, I.Y. Chen, Y.S. Huang, Hand pose recognition using curvature scale space, in *Proceedings of IEEE International Conference on Pattern Recognition* (2002), pp. 386–389
12. Y.W. Chen, C.J. Lin, Combining svms with various feature selection strategies, in *Feature Extraction*, ed. by I. Guyon, M. Nikravesh, S. Gunn, L.A. Zadeh. Studies in Fuzziness and Soft Computing, vol. 207 (Springer, Berlin/Heidelberg, 2006), pp. 315–324
13. D. Comaniciu, P. Meer, Mean shift: a robust approach toward feature space analysis. IEEE Trans. Pattern Anal. Mach. Intell. **24**, 603–619 (2002)
14. F.C. Crow, Summed-area tables for texture mapping, in *Proceedings of ACM SIGGRAPH* (New York, 1984), pp. 207–212
15. N. Dalal, B. Triggs, Histograms of oriented gradients for human detection, in *Proceedings of IEEE Conference on Computer Vision and Pattern Recognition* (2005), pp. 886–893
16. P. Doliotis, V. Athitsos, D. Kosmopoulos, S. Perantonis, Hand shape and 3D pose estimation using depth data from a single cluttered frame, in *Advances in Visual Computing*, ed. by G. Bebis, R. Boyle, B. Parvin, D. Koracin, C. Fowlkes, S. Wang, M.-H. Choi, S. Mantler, J. Schulze, D. Acevedo, K. Mueller, M. Papka. Lecture Notes in Computer Science, vol. 7431 (Springer, Berlin/Heidelberg, 2012), pp. 148–158
17. F. Dominio, M. Donadeo, G. Marin, P. Zanuttigh, G.M. Cortelazzo, Hand gesture recognition with depth data, in *Proceedings of ACM/IEEE International Workshop on Analysis and Retrieval of Tracked Events and Motion in Imagery Stream* (New York, 2013), pp. 9–16
18. F. Dominio, M. Donadeo, P. Zanuttigh, Combining multiple depth-based descriptors for hand gesture recognition. Pattern Recogn. Lett. **50**, 101–111 (2014). Depth Image Analysis
19. F. Dominio, G. Marin, M. Piazza, P. Zanuttigh, Feature descriptors for depth-based hand gesture recognition, in *Computer Vision and Machine Learning with RGB-D Sensors*, ed. by L. Shao, J. Han, P. Kohli, Z. Zhang. Advances in Computer Vision and Pattern Recognition (Springer, Cham, 2014), pp. 215–237
20. D. Droeschel, J. Stuckler, S. Behnke, Learning to interpret pointing gestures with a time-of-flight camera, in *Proceedings of ACM/IEEE International Conference on Human-Robot Interaction* (2011), pp. 481–488
21. D. Gabor, Theory of communication. Part 1: the analysis of information. J. Inst. Electr. Eng. Part III Radio Commun. Eng. **93**(26), 429–441 (1946)
22. V. Ganapathi, C. Plagemann, D. Koller, S. Thrun, Real time motion capture using a single time-of-flight camera, in *Proceedings of IEEE Conference on Computer Vision and Pattern Recognition* (2010), pp. 755–762
23. X.H. Han, G. Xu, Y.W. Chen, Robust local ternary patterns for texture categorization, in *Proceedings of IEEE International Conference on Biomedical Engineering and Informatics* (2013), pp. 846–850
24. M.B. Holte, T.B. Moeslund, P. Fihl, Fusion of range and intensity information for view invariant gesture recognition, in *Proceedings of IEEE Conference on Computer Vision and Pattern Recognition Workshops* (2008), pp. 1–7
25. M.K. Hu, Visual pattern recognition by moment invariants. IRE Trans. Inf. Theory **8**(2), 179–187 (1962)
26. A.E. Johnson, M. Hebert, Using spin images for efficient object recognition in cluttered 3D scenes. IEEE Trans. Pattern Anal. Mach. Intell. **21**(5), 433–449 (1999)

27. T. Kapuscinski, M. Oszust, M. Wysocki, D. Warchol, Recognition of hand gestures observed by depth cameras. Int. J. Adv. Robot. Syst. 12, 12–36 (2015)
28. N. Kumar, P.N. Belhumeur, A. Biswas, D.W. Jacobs, W.J. Kress, I.C. Lopez, J.V.B. Soares, Leafsnap: a computer vision system for automatic plant species identification, in *Proceedings of IEEE European Conference on Computer Vision* (Springer, Berlin/Heidelberg, 2012), pp. 502–516
29. A. Kurakin, Z. Zhang, Z. Liu, A real time system for dynamic hand gesture recognition with a depth sensor, in *Proceedings of European Signal Processing Conference* (2012), pp. 1975–1979
30. D. Li, Y. Dong, Deep learning: methods and applications. Found. Trends Signal Process. **7**(3–4), 197–387 (2014)
31. W. Li, Z. Zhang, Z. Liu, Expandable data-driven graphical modeling of human actions based on salient postures. IEEE Trans. Circuits Syst. Video Technol. **18**(11), 1499–1510 (2008)
32. W. Li, Z. Zhang, Z. Liu, Action recognition based on a bag of 3D points, in *Proceedings of IEEE Conference on Computer Vision and Pattern Recognition Workshops* (2010), pp. 9–14
33. F. Lv, R. Nevatia, Recognition and segmentation of 3-D human action using hmm and multi-class adaboost, in *Proceedings of IEEE European Conference on Computer Vision* (Springer, Berlin/Heidelberg, 2006), pp. 359–372
34. J. Macqueen, Some methods for classification and analysis of multivariate observations, in *Proceedings of Berkeley Symposium on Mathematical Statistics and Probability* (1967), pp. 281–297
35. S. Manay, D. Cremers, B.W. Hong, A.J. Yezzi, S. Soatto, Integral invariants for shape matching. IEEE Trans. Pattern Anal. Mach. Intell. **28**(10), 1602–1618 (2006)
36. G. Marin, M. Fraccaro, M. Donadeo, F. Dominio, P. Zanuttigh, Palm area detection for reliable hand gesture recognition, in *Proceedings of IEEE International Workshop on Multimedia Signal Processing* (2013)
37. G. Marin, F. Dominio, P. Zanuttigh, Hand gesture recognition with leap motion and kinect devices, in *Proceedings of IEEE International Conference on Image Processing* (2014), pp. 1565–1569
38. G. Marin, F. Dominio, P. Zanuttigh, Hand gesture recognition with jointly calibrated leap motion and depth sensor. Multimedia Tools Appl. 75, 1–25 (2015)
39. R.P. Mihail, N. Jacobs, J. Goldsmith, Static hand gesture recognition with 2 Kinect sensors, in *Proceedings of International Conference on Image Processing, Computer Vision, and Pattern Recognition* (2012), p. 1
40. T.B. Moeslund, A. Hilton, V. Krüger, A survey of advances in vision-based human motion capture and analysis. Comput. Vis. Image Underst. **104**(2), 90–126 (2006)
41. P. Molchanov, S. Gupta, K. Kim, J. Kautz, Hand gesture recognition with 3D convolutional neural networks, in *Proceedings of IEEE Conference on Computer Vision and Pattern Recognition Workshops* (2015)
42. C. Myers, L. Rabiner, A.E. Rosenberg, Performance tradeoffs in dynamic time warping algorithms for isolated word recognition. IEEE Trans. Acoust. Speech Signal Process. **28**(6), 623–635 (1980)
43. L. Nanni, A. Lumini, F. Dominio, M. Donadeo, P. Zanuttigh, Combination of depth and texture descriptors for gesture recognition, in *Advances in Machine Learning Research*, ed. by S. Shandilya. Engineering Tools, Techniques and Tables (Nova Science, Commack, 2014)
44. L. Nanni, A. Lumini, F. Dominio, M. Donadeo, P. Zanuttigh, Ensemble to improve gesture recognition. Int. J. Autom. Identif. Technol. 5, 47–56 (2014)
45. I. Oikonomidis, N. Kyriazis, A. Argyros, Efficient model-based 3D tracking of hand articulations using kinect, in *Proceedings of British Machine Vision Conference* (BMVA, Dundee, 2011), pp. 101.1–101.11
46. T. Ojala, M. Pietikäinen, D. Harwood, A comparative study of texture measures with classification based on featured distributions. Pattern Recognit. **29**(1), 51–59 (1996)

47. S.C.W. Ong, S. Ranganath, Automatic sign language analysis: a survey and the future beyond lexical meaning. IEEE Trans. Pattern Anal. Mach. Intell. **27**(6), 873–891 (2005)
48. S. Oprisescu, M. Ciuc, I. Vasile, Hand posture recognition using the intrinsic dimension, in *Proceedings of IEEE International Conference on Optimization of Electrical and Electronic Equipment* (2014), pp. 974–979
49. O. Oreifej, Z. Liu, Hon4d: histogram of oriented 4D normals for activity recognition from depth sequences, in *Proceedings of IEEE Conference on Computer Vision and Pattern Recognition* (2013), pp. 716–723
50. F. Pedersoli, N. Adami, S. Benini, R. Leonardi, Xkin: extendable hand pose and gesture recognition library for kinect, in *Proceedings of ACM International Conference on Multimedia* (New York, 2012), pp. 1465–1468
51. F. Pedersoli, S. Benini, N. Adami, R. Leonardi, Xkin: an open source framework for hand pose and gesture recognition using kinect. Vis. Comput. **30**(10), 1107–1122 (2014)
52. R. Poppe, A survey on vision-based human action recognition. Image Vis. Comput. **28**(6), 976–990 (2010)
53. N. Pugeault, R. Bowden, Spelling it out: real-time asl fingerspelling recognition, in *Proceedings of IEEE International Conference on Computer Vision Workshops* (2011), pp. 1114–1119
54. L. Rabiner, A tutorial on hidden markov models and selected applications in speech recognition. Proc. IEEE **77**(2), 257–286 (1989)
55. E Rahtu, J. Heikkilä V. Ojansivu, T. Ahonen, Local phase quantization for blur-insensitive image analysis. Image Vis. Comput. **30**(8), 501–512 (2012)
56. Z. Ren, J. Yuan, Z. Zhang, Robust hand gesture recognition based on finger-earth mover's distance with a commodity depth camera, in *Proceedings of ACM International Conference on Multimedia* (New York, 2011), pp. 1093–1096
57. M. Reyes, G. Dominguez, S. Escalera, Feature weighting in dynamic timewarping for gesture recognition in depth data, in *Proceedings of IEEE International Conference on Computer Vision Workshops* (2011), pp. 1182–1188
58. T. Rückstieß, C. Osendorfer, P. Van der Smagt, Sequential feature selection for classification, in *Proceedings of International Conference on Advances in Artificial Intelligence* (2011), pp. 132–141
59. R.B. Rusu, G. Bradski, R. Thibaux, J. Hsu, Fast 3D recognition and pose using the viewpoint feature histogram, in *Proceedings of IEEE/RSJ International Conference on Intelligent Robots and Systems* (2010), pp. 2155–2162
60. L. Spinello, K.O. Arras, People detection in RGB-D data, in *Proceedings of IEEE/RSJ International Conference on Intelligent Robots and Systems* (2011), pp. 3838–3843
61. J. Suarez, R.R. Murphy, Hand gesture recognition with depth images: a review, in *Proceedings of IEEE International Symposium on Robot and Human Interactive Communication* (2012), pp. 411–417
62. J. Sung, C. Ponce, B. Selman, A. Saxena, Unstructured human activity detection from RGBD images, in *Proceedings of IEEE International Conference on Robotics and Automation* (2012), pp. 842–849
63. P. Suryanarayan, A. Subramanian, D. Mandalapu, Dynamic hand pose recognition using depth data, in *Proceedings of IEEE International Conference on Computer Vision and Pattern Recognition* (2010), pp. 3105–3108
64. M. Tang, Recognizing hand gestures with microsoft's kinect. Technical report, Department of Electrical Engineering, Stanford University (2011)
65. M.Z. Uddin, D.T. Nguyen, T.S. Kim, Human activity recognition via 3-D joint angle features and hidden markov models, in *Proceedings of IEEE International Conference on Image Processing* (2010), pp. 713–716
66. D. Uebersax, J. Gall, M. Van den Bergh, L. Van Gool, Real-time sign language letter and word recognition from depth data, in *Proceedings of IEEE International Conference on Computer Vision Workshops* (2011), pp. 383–390

67. M. Van den Bergh, L. Van Gool, Combining RGB and ToF cameras for real-time 3D hand gesture interaction, in *Proceedings of IEEE Workshop on Applications of Computer Vision* (2011), pp. 66–72
68. A.W. Vieira, E.R. Nascimento, G.L. Oliveira, Z. Liu, M.F.M. Campos, Stop: space-time occupancy patterns for 3D action recognition from depth map sequences, in *Progress in Pattern Recognition, Image Analysis, Computer Vision, and Applications*, ed. by L. Alvarez, M. Mejail, L. Gomez, J. Jacobo. Lecture Notes in Computer Science, vol. 7441 (Springer, Berlin/Heidelberg, 2012), pp. 252–259
69. P. Viola, M. Jones, Robust real-time object detection, in Int. J. Comput. Vis. (2001)
70. J.P. Wachs, M. Kölsch, H. Stern, Y. Edan, Vision-based hand-gesture applications. Commun. ACM **54**(2), 60–71 (2011)
71. J. Wang, Z. Liu, J. Chorowski, Z. Chen, Y. Wu, Robust 3D action recognition with random occupancy patterns, in *Proceedings of IEEE European Conference on Computer Vision* (2012), pp. 872–885
72. J. Wang, Z. Liu, Y. Wu, J. Yuan, Mining actionlet ensemble for action recognition with depth cameras, in *Proceedings of IEEE Conference on Computer Vision and Pattern Recognition* (2012), pp. 1290–1297
73. J. Wang, Z. Liu, Y. Wu, J. Yuan, Learning actionlet ensemble for 3D human action recognition. IEEE Trans. Pattern Anal. Mach. Intell. **36**(5), 914–927 (2014)
74. R. Wang, S. Paris, J. Popović, 6D hands: markerless hand-tracking for computer aided design, in *Proceedings of Annual ACM Symposium on User Interface Software and Technology* (New York, 2011), pp. 549–558
75. L. Xia, K. Fujimura, Hand gesture recognition using depth data, in *Proceedings of IEEE International Conference on Automatic Face and Gesture Recognition* (2004), pp. 529–534
76. L. Xia, C.C. Chen, J.K. Aggarwal, View invariant human action recognition using histograms of 3D joints, in *Proceedings of IEEE Conference on Computer Vision and Pattern Recognition Workshops* (Providence, 2012), pp. 20–27
77. C. Yang, Y. Jang, J. Beh, D. Han, K. Hanseok, Gesture recognition using depth-based hand tracking for contactless controller application, in *Proceedings of IEEE International Conference on Consumer Electronics* (2012), pp. 297–298
78. X. Yang, Y.L. Tian, Eigenjoints-based action recognition using naïve-bayes-nearest-neighbor, in *Proceedings of IEEE Conference on Computer Vision and Pattern Recognition Workshops* (2012), pp. 14–19
79. X. Yang, C. Zhang, Y. Tian, Recognizing actions using depth motion maps-based histograms of oriented gradients, in *Proceedings of ACM International Conference on Multimedia* (New York, 2012), pp. 1057–1060
80. W. Yong, Y. Tianli, L. Shi, L. Zhu, Using human body gestures as inputs for gaming via depth analysis, in *Proceedings of IEEE International Conference on Multimedia and Expo* (2008), pp. 993–996
81. R. Zhou, Y. Junsong, M. Jingjing, Z. Zhengyou, Robust part-based hand gesture recognition using kinect sensor. IEEE Trans. Multimedia **15**(5), 1110–1120 (2013)

Chapter 10
Conclusions

Depth data acquisition and processing have been confined to research institutes and industrial applications for many years, but the recent introduction of consumer depth cameras has made depth acquisition simpler and less expensive. The widespread availability of depth data paved the way to many new applications besides the ones in the gaming industry for which the first consumer depth camera, i.e., the Kinect^{TM} v1, was originally introduced. Challenging computer vision problems, like human pose estimation or three-dimensional reconstruction, that used to require professional equipment and skilled engineers, can now be tackled by anyone with simple and inexpensive consumer depth cameras. On the other hand, the exploitation of the data from this type of acquisition equipment requires ad hoc algorithms able to extract and process the relevant information for the tasks of interest. The development of these algorithms is a challenging research issue, specially when the target are completely automatic methods to be used in a consumer environment.

This book tries to cover the various different aspects of the technology and applications of consumer depth cameras. It covers both the theoretical principles behind the acquisition devices and the practical implementation aspects of the computer vision algorithms needed for the various applications. Examples with real data are used in order to show the performance of the various algorithms. Different from other books focusing only either on the technological aspects of the sensors or on the applications, this book covers both topics and it shows how the performances and limitations of the depth camera technology affect the design of the algorithms used for the various applications.

The first part of this book addressed the operating principles of structured light and ToF depth cameras entering the products of today's market. The focus of these chapters is on the general working principles rather than on the characteristics of specific depth camera models, providing useful conceptual tools applicable also to future evolutions of these technologies. This is particularly relevant, since the

P. Zanuttigh et al., *Time-of-Flight and Structured Light Depth Cameras*,
DOI 10.1007/978-3-319-30973-6_10

depth camera market is rapidly growing with novel acquisition devices appearing on the market every year, and detailed descriptions of specific products would quickly become outdated.

The second part of this book addressed the methods suited to obtain accurate 3D information from data acquired with depth cameras eventually assisted by standard cameras. This process requires effective calibration procedures and suitable techniques for depth super-resolution and data fusion. Both the photometric and geometric calibration of standard cameras and depth cameras were covered, as well as the joint calibration of multiple cameras. Calibration plays a fundamental role also for super-resolution and data fusion methods that allow one to overcome some of the most critical limitations of current consumer depth cameras by exploiting side information from standard cameras. Both simple deterministic schemes, suited to real-time applications, and more complex approaches based on global optimization techniques were presented for super-resolution and data fusion.

The last part of this book addressed the applications not with the purpose of presenting all the possible scenarios where consumer depth cameras can be used, but with the aim of providing a comprehensive set of tools allowing one to understand the countless arenas where geometric and possibly color information can make a difference. Four topics were treated in detail, namely scene segmentation, 3D scene reconstruction, human pose estimation and tracking and gesture recognition. The solution of matting and scene segmentation is extremely challenging by color data alone, but they can be efficiently solved by combining together color and depth data clues thanks to their complementary characteristics. 3D reconstruction has been studied for several years, and in theory, all the methods and commercial applications developed for professional structured light and laser scanners could be used with consumer depth camera data. However, the peculiar characteristics of the depth data provided by consumer depth cameras require a rethinking of standard approaches and ad hoc techniques as discussed in detail in this book. Body and hand pose estimation and tracking are other important applications where depth camera data can improve the algorithms performance. Indeed, these were the first applications for which consumer depth cameras have been introduced in the gaming market. This book offered a good panorama on current approaches. The recognition of static and dynamic gestures of both the complete body and the hand is a problem strictly related to body and hand pose estimation and tracking, for which this book presented an extensive review of the most effective methods.

The purpose of this book was to offer a bag of tools, useful to students, practitioners and researchers, interested to explore and deepen the potential offered by depth data in light of the fact that depth acquisition devices continue to appear on the market and depth-based computer vision is a rapidly growing field. We hope the content of this book may contribute to stimulate the readers with notions and ideas useful to the development of methodologies and new algorithms bringing to some advancements of this recent but extremely interesting field.

Index

P. Zanuttigh et al., *Time-of-Flight and Structured Light Depth Cameras*,
DOI 10.1007/978-3-319-30973-6